Electromagnetic Modelling and Measurements for Analysis and Synthesis Problems

NATO ASI Series

Advanced Science Institutes Series

A Series presenting the results of activities sponsored by the NATO Science Committee, which aims at the dissemination of advanced scientific and technological knowledge, with a view to strengthening links between scientific communities.

The Series is published by an international board of publishers in conjunction with the NATO Scientific Affairs Division

A Life Sciences	Plenum Publishing Corporation
B Physics	London and New York
C Mathematical	Kluwer Academic Publishers
and Physical Sciences	Dordrecht, Boston and London
D Behavioural and Social Sciences	
E Applied Sciences	
F Computer and Systems Sciences	Springer-Verlag
G Ecological Sciences	Berlin, Heidelberg, New York, London,
H Cell Biology	Paris and Tokyo
I Global Environmental Change	

NATO-PCO-DATA BASE

The electronic index to the NATO ASI Series provides full bibliographical references (with keywords and/or abstracts) to more than 30000 contributions from international scientists published in all sections of the NATO ASI Series.
Access to the NATO-PCO-DATA BASE is possible in two ways:

– via online FILE 128 (NATO-PCO-DATA BASE) hosted by ESRIN,
Via Galileo Galilei, I-00044 Frascati, Italy.

– via CD-ROM "NATO-PCO-DATA BASE" with user-friendly retrieval software in English, French and German (© WTV GmbH and DATAWARE Technologies Inc. 1989).

The CD-ROM can be ordered through any member of the Board of Publishers or through NATO-PCO, Overijse, Belgium.

Series E: Applied Sciences - Vol. 199

Electromagnetic Modelling and Measurements for Analysis and Synthesis Problems

edited by

Bernard de Neumann

Department of Mathematics and CSR,
City University, London, U.K.

Kluwer Academic Publishers

Dordrecht / Boston / London

Published in cooperation with NATO Scientific Affairs Division

Proceedings of the NATO Advanced Study Institute on
Electromagnetic Modelling and Measurements for Analysis and Synthesis Problems
"Il Ciocco", Castelvecchio Pascoli, Lucca, Italy
10–21 August 1987

Library of Congress Cataloging-in-Publication Data

```
NATO Advanced Study Institute on Electromagnetic Modelling and
  Measurements for Analysis and Synthesis Problems (1987 : Lucca,
  Italy)
    Electromagnetic modelling and measurements for analysis and
synthesis problems / edited by Bernard de Neumann.
       p.    cm. -- (NATO ASI series. Series E, Applied sciences ; vol.
  199)
    "Proceedings of the NATO Advanced Study Institute on
  Electromagnetic Modelling and Measurements for Analysis and
  Synthesis Problems, held in 'Il Ciocco', Castelvecchio Pascoli,
  Lucca, Italy, 10-21 August 1987"--T.p. verso.
    "Published in cooperation with NATO Scientific Affairs Division."
    ISBN 0-7923-1265-1 (alk. paper)
    1. Antennas (Electronics)--Data processing--Congresses.
  2. Antennas, Reflector--Data processing--Congresses.
  3. Electromagnetic fields--Data processing--Congresses.    I. De
  Neumann, Bernard.    II. North Atlantic Treaty Organization.
  Scientific Affairs Division.    III. Title.    IV. Series: NATO ASI
  series.    Series E, Applied sciences ; no. 199.
  TK7871.6.N36    1987
  621.382'4--dc20                                              91-13107
```

ISBN 0–7923–1265–1

Published by Kluwer Academic Publishers,
P.O. Box 17, 3300 AA Dordrecht, The Netherlands.

Kluwer Academic Publishers incorporates the publishing programmes of
D. Reidel, Martinus Nijhoff, Dr W. Junk and MTP Press.

Sold and distributed in the U.S.A. and Canada
by Kluwer Academic Publishers,
101 Philip Drive, Norwell, MA 02061, U.S.A.

In all other countries, sold and distributed
by Kluwer Academic Publishers Group,
P.O. Box 322, 3300 AH Dordrecht, The Netherlands.

Printed on acid-free paper

Printed in the Netherlands

Dedicated to the memory of
Jozef K. Skwirzynski
and
Yvonne Skwirzynska

Table of Contents

x

PREFACE

In this volume is presented the proceedings of a NATO
Advanced Study Institute (ASI) on the theme of
Electromagnetic Modelling and Measurements for Analysis and
Synthesis Problems. The ASI was held at Il Ciocco,
Castelvecchio Pascoli, Tuscany, Italy, August 10th - 21st,
1987. It has been my good fortune to act as co-director of
two of Jozef's previous ASIs, and so I am well acquainted
with the JKS format for ASIs. As participants will realise,
I did not attend this ASI, and so I only have a partial
appreciation of the programme. In particular it has not
been possible to include transcripts of any panel
discussions which may have taken place. Readers may recall
that such transcripts have formed a most interesting and
useful part of previous ASI proceedings edited by Jozef
Skwirzynski, and helped to convey the spirit of the
meetings. Unfortunately it has proved impossible to locate
the tapes, despite the best efforts of Jozef's assistant,
Barry Stuart. A further dificulty has arisen through the
untimely death of Jozef's former deputy and colleague at GEC
Research, Ed Pacello, who assisted Jozef with the
organisation of the precursor of this ASI.

The following is taken from original material relating to
the aims of the Advanced Study Institute:

"PURPOSE OF THE INSTITUTE

This Institute is concerned with computer modelling and
with experimental measurements as two complementary tools
for both analysis and synthesis of electromagnetics (EM),
infra-red (IR) and optical problems.

ANTENNA ANALYSIS AND SYNTHESIS
Consideration of techniques for measuring EM fields for

reflector antenna diagnosis with shaped and pencil beams. Holograhic techniques will also be reviewed.

SYNTHESIS PROBLEMS

The problem of finding the source distribution over a given locus of points in space which produces a specified response. A more general synthesis problem is one in which the source geometry is unknown, a situation which leads to the true inverse problem. The difficult problems of transient propagation and scattering, and some aspects of pulse propagation, will be considered.

OBJECT OR TARGET SHAPE RECOGNITION

Several methods for obtaining source shapes from inverse scattering data will be considered which facilitate automated target recognition in electromagnetic, infra-red, and optical fields."

Regrettably Jozef became incapacitated before the editing task began, and died before it could be completed. I am indebted to all the participants for their patience, and support in my completing this task and bringing this account to a close. I know that Jozef would wish me to thank his co-directors, and indeed all those who contributed towards the successful outcome of this ASI. In particular, I am sure that Jozef would wish me to publicly thank his old compatriot, friend and bridge partner, Barry Stuart, for all that he did in assisting the administration of this and previous ASI's. I too add my thanks to these, and also wish to thank in particular Barry Stuart, whose patience knows no bounds, Adrian Wright, and Marjorie Sadler. Many people are deserving of my thanks in this venture, and I apologise to any of them whom I have inadvertently not acknowledged herein.

Bernard de Neumann

Chalkwell,
November 1989.

The following poem, by Leo Felsen, was dedicated to the ASI, and Jozef promised that it would be in the final volume.

Another year has passed us by.
Let's give the new one a fair try.

May peace prevail in every place,
Both here on earth and out in space.

May dollars, DM, francs and yen
Find levels that they can maintain

Between extremes of Boom and Bust,
One a niveau that all will trust.

May politics become a game
Where leaders trade respect not blame,

And where the rules by which they play
Would match their deeds with what they say.

While we must hope, let us be wise;
On earth will be no paradise.

Though cautious, let's not be afraid.
Greetings and HAPPY 88!

Jozef K. Skwirzynski, M.B.E.
1921 - 1989

Jozef Kazimierz Skwirzynski, M.B.E.

Jozef was born in Budzanow, Poland on 21st December 1921. He began his education there, but it was interrupted by the invasions of Poland by both Germany and then the USSR in 1939. He was arrested by the Russians in 1939, and deported to the Ukraine where he spent two years in prison/work camps. He felt that it would be to his advantage to volunteer for skilful tasks whilst being detained, and so, when people with driving skills were called for, he put himself forward. This resulted in his ploughing a lonely furrow in the Ukraine, where he drove a caterpillar tractor towing a plough. Later, in 1941, after Germany invaded the USSR, and after most Russian-held Polish prisoners had been released to form an army in the USSR, which was not armed through fear of insurrection, volunteers were called for who could fly. Naturally Jozef volunteered. This resulted in Jozef being released from the Polish army and his being posted to Britain, via Tehran, Bombay and Cape Town. A somewhat tortuous journey not without incident, which included a night in jail in Cape Town. Upon his arrival in Britain he was sent to Blackpool, Lancashire, where he met his wife-to-be, Yvonne. The cultural shock must have been immense, and the Polish contingent found itself wondering what to do with some of the items which they were presented with upon arrival, such as small jars of Marmite, Cherry Blossom Polish, etc. Jozef had many tales connected with his arrival in the UK, and his friends will, no doubt, recall them with pleasure. It was soon discovered that Jozef had no flying abilities, and so he became, after training, a navigator in the Polish Section of the Royal Air Force. The rapidity of his uptake of the art of aircraft navigation did not go unnoticed, particularly as he had to cope with the disadvantage of learning English simultaneously, and quickly he became a Navigation Instructor to other Poles.

After the war Jozef took a double First in Mathematics and Physics from Imperial College, and afterwards spent two years there as a lecturer. He joined Marconi Research Laboratories, Great Baddow, in 1951. Here he became Chief of the Mathematical Physics and Circuitry Group in 1969, and Manager of Theoretical Support Services in 1977. During his time at Baddow he became an authority on the design of electrical filters, an area which is replete with Poles and poles, and he had a book published on the subject in 1968. He also authored many papers and articles upon theoretical aspects of electronic engineering, many of which were published. At the time of his retirement he was Consultant to Marconi Research Centre.

He represented Marconi and GEC on various national and international bodies, and served on the University Grants Committee mathematics sub-committee for several years. He organised and ran a large number of NATO Advanced Study Institutes - probably more than any other person. In his spare time he, with his wife, until she died, ran a small but widely known antiques business; indeed he was to become an acknowledged expert on oriental antiques.

In the United Kingdom's New Year's Honours List of 1987 he became an additional Member of the Most Excellent Order of the British Empire in recognition of his services. He died 29th October, 1989.

Many people, including myself, have benefited from Jozef's interest, influence and training, and owe much to the opportunities which he gave them. Life was never dull with Jozef around especially with his "Jozefisms" - such as "hunchback" cars! I acknowledge with pride that I worked with him for twenty-three years. He sometimes joked that I would take over his job, and whilst the editing of this volume has been a sad task, I am proud to have been called upon to complete this part of his work.

Bernard de Neumann

Chalkwell,
September 1990.

LIST OF DELEGATES

Dr Bulent Aksoy
Mr Bruno Audone
Ms Meric Bakiler
Prof Bingay Bilgin
Prof M. Tuncay Birand
Prof J.S. Byrnes
Dr Graziano Cerri
Prof Peter Clarricoats
Prof Giuseppe Conciauro
Dr Sidney Cornbleet
Mr Jesus G. Cuevas del Rio
Prof A.L. Cullen
Mr H.Sevki Darendelioglu
Prof Nabil H. Farhat
Prof Leopold B. Felsen
Ms Handan Gurbuz
Miss Sevgi Haman
Dr Dwight L. Jaggard
Dr Jose L. Fernandez Jambrina
Prof Seyfeddin Karagozlu
Prof Mustafa Korkmaz
Prof Stanley J. Kubina
Prof Lev B. Levitin
Mr P.G. Mantica
Mr F. Mercurio
Prof Edmund K. Miller
Mr J.S. Nakhwal
Mr Ercument Ekrem Ozkan
Dr Hugo F. Pues
Dr Yahya Rahmat-Samii
Dr Tapan K. Sarkar
Dr Jozef K. Skwirzynski
Mr Barry Stuart
Prof K. Suchy
Dr Nilgun Tarhan

Dr Andrew J. Terzuoli, Jr
Dr A.G. Tijhuis
Mr J.T. Tokarski
Dr L.A. Trinogga
Mr Miles Upton
Mr K. Van't Klooster
Mr Giuseppe D. Vecchi
Mrs W. Zaworska

Beam-forming applications of polynomials with restricted coefficients†

James S. Byrnes
Prometheus Inc.
21 Arnold Avenue
Newport, RI 02840
USA
jbyrnes@cs.umb.edu
Also at University of Massachusetts at Boston.

1. Introduction

The basic mathematical question to consider in electronic beam steering with a discrete array consisting of omnidirectional elements spaced at equal increments along a line is how coefficients of a polynomial may be chosen so as to arrive at a desired beam pattern. In numerous applications, these coefficients are required to satisfy certain restrictions, such as a bound on their dynamic range. Here, dynamic range refers to the ratio of the largest to the smallest magnitude. Thus, particularly in null steering, it is often advantageous, or even necessary, for the shading coefficients to all have the same magnitude.

Although the mathematical, statistical, and physical problems that arise in the consideration of array shading have been studied for roughly half a century, many interesting questions remain. In the linear array case under consideration, letting n denote the number of elements, the pattern function $G(z)$ is a polynomial of degree $n-1$, z is a point on the unit circle, and the shading coefficients are just the n coefficients of this polynomial. An important reason for performing array shading is to shape the pattern function G so that it has low sidelobes and small beamwidth. As is well known, both of these quantities cannot be minimized simultaneously, and

† Research sponsored by the Air Force Office of Scientific Research (AFSC), under Contracts F49620-88-C-0028 and F49620-90-C-0023.

B. de Neumann (ed.), Electromagnetic Modelling and Measurements for Analysis and Synthesis Problems, 1–15.
© 1991 *Kluwer Academic Publishers. Printed in the Netherlands.*

the choice of shading coefficients results in a tradeoff between these two desirable ends.

Electronic beam steering is another fundamental purpose of array shading, and it is this application that we address here. In addition to permitting the rotation of the main response axis of the pattern function, beam steering also allows the simultaneous formation of a number of beams in different directions. In particular, if sources of interference lie at bearings different from that of the desired source, then the signal-to-noise plus interference ratio (SNIR) may be increased dramatically by directing nulls of the pattern function toward these interfering sources, in spite of the fact that the absolute power of the desired signal is thereby reduced. Adaptive techniques have been developed by which array processing systems can electronically respond to an unknown interference environment. However, although the basic adaptive array principles have been known for some time, their application has been limited by hardware constraints and by the lack of sufficiently robust, real-time algorithms. New approaches to this latter consideration are described herein.

There are many cases when constaints must be placed upon the magnitudes of the coefficients of the pattern function. Thus, as explained by Hudson [Hud81], when coefficients are implemented by attenuation, they must be scaled so that the largest modulus is unity, since the amplitude gain for the desired transmission, and even the overall output signal-to-noise ratio (SNR), can be reduced by large coefficients. In discussing main-lobe constraints on optimal arrays, Hudson observes that when a main-lobe null is created, very large shading coefficients are formed, resulting in enhanced output of uncorrelated noise. Hence, size restrictions on the coefficients are again required.

On the other hand, in a situation such as occurs in an adaptive radar receiver after clutter has decayed due to increasing range, so that there will be few and widely spaced target echoes of minimal power compared to a steady jamming source, it is necessary to constrain the adaptive array so that the shading coefficients are prevented from falling to zero. A similar situation occurs in an adaptive antenna using the least mean square (LMS) algorithm, where the shading coefficients will decay to zero if either the signal level falls to zero, or if the reference signal is absent for some reason. One method of controlling this is to substitute the steered gradient system described by Griffiths for the reference signal LMS antenna, but this has the disadvantage of being very sensitive to errors in the assumed direction of the desired signal.

As mentioned earlier, another approach to these questions is to restrict the dynamic range of the shading coefficients. Although an informal rule of thumb for this range appears to be "2 and everyone is happy, 10 and some are happy, 100 and nobody is happy," a formal mathematical study of the relevant properties of polynomials, whose coefficients are thereby restricted, does not seem to have been previously undertaken. An important thrust of the research effort reported herein has been to initiate such a study and to relate to the above applications the large amount of work that has been accomplished by mathematicians on polynomials with restricted coefficients.

Furthermore, there is an intimate relationship between the engineering questions described above and several areas of classical mathematical analysis. Foremost among the problems of mutual interest is the question of how close to constant the modulus of a polynomial can be along some curve, typically the unit circle. This is of great concern to theoretical mathematical analysts because of the fundamental nature of polynomials and the simplicity and intrinsic beauty of the question. It is equally important to engineers working in such fields as array design, adaptive beamforming and null steering, filter design, peak power limited transmitting, and the design of reflection phase gratings. This paper describes our research into both aspects of this remarkable intertwining of the disciplines of pure mathematics and engineering.

2. Mathematical results

Concerning the purely mathematical aspects of our work, note that properties of polynomials with restricted coefficients have been the subject of much fruitful research in twentieth century mathematical analysis. Of particular interest have been polynomials with coefficients ± 1 or complex of modulus one. The study of such functions was apparently initiated by G.H. Hardy (see Zygmund [Zyg59, p. 199]), and furthered by J.E. Littlewood, P. Erdös and others.

For the purposes of this discussion, it will be convenient to introduce the notation of Littlewood [Lit66]. Thus, let F_n and G_n be, respectively, the class of all polynomials of the form

$$f(z) = \sum_{k=0}^{n} \pm 1 z^k \quad \text{and} \quad g(z) = \sum_{k=0}^{n} \exp(a_k i) z^k,$$

where $|z| = 1$ and the a_k are arbitrary real constants. Clearly, the L^2 norm of g is $\sqrt{n+1}$ for all $g \in G$ (and hence for all $f \in F_n \subset G_n$), and the question "how close can such a g come to satisfying

$$|g| \equiv \sqrt{n+1}\,?"$$

has long been the object of intense study.

The first qualitative result concerning the above question for G_n was obtained by G.H. Hardy [Zyg59, p. 199], who demonstrated the existence of a positive constant C and a sequence $\{g_n\}$, $g_n \in G_n$, satisfying $|g_n(z)| \leq C\sqrt{n}$ for all n and z. The identical result for F_n was obtained by Shapiro [Sha57] and published by Rudin [Rud59]. Littlewood [Lit62] conjectured that there exist positive constants A and B such that, for any n, there is an $f \in F_n$ $(g \in G_n)$ satisfying

$$A\sqrt{n} \leq |f(z)| \leq B\sqrt{n} \qquad (A\sqrt{n} \leq |g(z)| \leq B\sqrt{n})$$

for all z, while Erdös conjectured [Erd57] that there is a positive constant C such that for $n \geq 2$, $\|g\|_\infty \geq (1+C)\sqrt{n}$ for all $g \in G_n$ (and hence for all $f \in F_n$). Analogous conjectures for the L^p norms of $g \in G_n$ were settled in a series of papers by Beller and Newman [Bel71, BN71, BN73]. Beller and Newman [BN74] also proved the Littlewood conjecture for polynomials whose coefficients have moduli bounded by 1, after observing that the proof of this result given by Clunie [Clu59] depended on an erroneous result of Littlewood. In [Kör80], Körner was able to modify the result of Byrnes in [Byr77] to prove the Littlewood conjecture for G_n, and then Kahane [Kah80] showed that the Erdös conjecture is false for G_n. These conjectures for F_n remain unresolved.

One approach to the Erdös conjecture for polynomials in F_n is to consider their L^4 norm. For polynomials in G_n, we have the following result:

Theorem 2.1. For each positive integer n, there is a sequence of coefficients $\{c_k\}_{k=0}^n$ such that all $|c_k| = 1$ and

$$\frac{1}{2\pi} \int_0^{2\pi} \left| \sum_{k=0}^n c_k e^{ik\theta} \right|^4 d\theta < (n+1)^2 + 4(n+1)^{3/2}.$$

Proof. We show that, in fact, the Gauss coefficients,

$$c_k = e^{\pi i k^2/(n+1)},$$

satisfy the required property. Toward that end, note that

$$\sum_{k=0}^{n} c_k e^{ik\theta} \sum_{m=0}^{n} \bar{c}_m e^{-im\theta} = n+1+\sum_{j\neq 0}\left(\sum_{m=0}^{n} c_m + j\bar{c}_m\right) e^{ij\theta}.$$

Therefore, by Parseval's Theorem, assuming for convenience that n is even,

$$\frac{1}{2\pi}\int_0^{2\pi}\left|\sum_{k=0}^{n} c_k e^{ik\theta}\right|^4 d\theta = (n+1)^2 + \sum_{j\neq 0}\left|\sum_{m=0}^{n} c_m + j\bar{c}_m\right|^2$$

$$= (n+1)^2 + 2\sum_{j=1}^{n}\frac{\sin^2(j^2\pi/(n+1))}{\sin^2(j\pi/(n+1))}$$

$$= (n+1)^2 + 4\sum_{j=1}^{n/2}\left(\frac{\sin(j^2\pi/(n+1))}{\sin(j\pi/(n+1))}\right)^2$$

$$\leq (n+1)^2 + 4\sum_{j=1}^{n/2}\min(j^2, (n+1)^2/(2j)^2)$$

$$\leq (n+1)^2 + 4(n+1)^{3/2},$$

where we have used the facts that

$$|\sin jx/\sin x| \leq j, \quad \text{and} \quad |1/\sin x| \leq \pi/2x \quad \text{for } 0 < x < \pi/2.$$

This completes the proof of theorem 2.1.

Another method of constructing polynomials with unimodular coefficients is to form a suitable weighted average of existing ones. For example, we may employ a slight variation of the basic construction in [Byr77] as follows:

For each m, $0 \leq m \leq N^2 - 1$, and for $z = e^{2\pi i\theta}$, let

$$P_m(z) = P_m(\theta) = \sum_{k=0}^{N-1}\sum_{j=0}^{N-1} e^{2\pi i\left(jk/N + (j+kN)m/N^2\right)} z^{j+kN}$$

Clearly, each P_m is a polynomial of degree $N^2 - 1$ with coefficients of modulus one. Furthermore, it follows from [Byr77] that, for a suitable small positive ϵ (i.e., of order N^{-2}), $|P_m(\theta)|$ is essentially flat for

$$\epsilon - m/N^2 \leq \theta \leq 1 - \epsilon - m/N^2$$

Now define $P^*(\theta)$ by

$$P^*(\theta) = \sum_{m=0}^{N^2-1} z^{mN^2} P_m(\theta).$$

P^* is a polynomial of degree $N^4 - 1$ with coefficients of modulus one. Also, by writing

$$P^*(\theta) = \sum_{k=0}^{N-1} \sum_{j=0}^{N-1} e^{2\pi i jk/N} z^{j+kN} \frac{1 - z^{N^4}}{1 - e^{2\pi i (j+kN)/N^2} e^{2\pi i N^2 \theta}}$$

and letting

$$\tilde{\theta} = -N^{-4}(A + BN + CN^2) \qquad \text{for } 0 \le A, B \le N - 1, \quad 0 \le C \le N^2 - 1$$

it is seen that

$$P^*(\tilde{\theta}) = N^2 e^{2\pi i (AB/N - (A+BN)(A+BN+CN^2)/N^4)}$$

so that $|P^*(\tilde{\theta})| = N^2$.

In addition, the essential flatness of $|P^*(\tilde{\theta})|$ in the interval $\epsilon \le \theta \le 1 - \epsilon$, where now ϵ is of order N^{-4}, follows as before. However, numerical evidence suggests that $P^*(N^4/2) = O(1)$, a similar situation to that which occured with the original polynomials [Byr77]. This being the case, P^* is not quite a Kahane-type polynomial, as we had originally hoped.

Note, however, that the above method of constructing $P^*(\theta)$ can also be employed to create new flat spectrum sequences, which are periodic sequences $\{a_k\}_{k=0}^{\infty}$ with the property that their discrete Fourier transform (DFT) has a power spectrum consisting of a very small number (usually one or two) of distinct values. This is because the DFT can be thought of as the values of the polynomial

$$P(z) = \sum_{k=0}^{n-1} a_k z^k,$$

where n is the period, at the n-th roots of unity. Our construction yields polynomials whose spectra are essentially flat at almost all points of the unit circle, not just at the roots of unity. Observe that flat spectrum sequences constructed in this manner satisfy the additional property that all of the terms of the original sequence have the same magnitude. Applications of these concepts to notch filtering and communications are discussed elsewhere in this paper.

Another method of viewing these questions is in the context of interpolation problems. As noted earlier, for any $P \in G_n$, the Parseval Theorem implies that the L^2 norm of P on the unit circle C is $\sqrt{n+1}$. Furthermore, since $|P(z)|^2$ is expressible as $z^{-n} Q(z)$, where Q is of degree $2n$, there can be at most $2n$ distinct points z_k where

$$|P(z_k)| = \sqrt{n+1}$$

Let us call such a set of points an L^2 Interpolating Set for P. A natural question is which, if any, subsets of C consisting of $2n$ points can be an L^2 Interpolating Set for some P of the required form.

In its full generality, this question appears to be quite difficult. For $n = 1$, it is trivial to show that $S = \{a, b\}$ is an L^2 Interpolating Set if and only if $b = -a$. For arbitrary n, observe that for $S = \{z_k\}_{k=1}^{2n}$ to be an L^2 Interpolating Set, the coefficients of P must be chosen so that

$$Q(z) - (n+1)z^n = \alpha \prod_{k=1}^{2n} (z - z_k),$$

where α is a constant of modulus one. Furthermore, the coefficient of z^n on the left side of this equation vanishes, so the same must be true on the right side. Clearly, this will be a very rare occurance, so that most sets will not be L^2 Interpolating Sets. In fact, it is not at all obvious that for $n > 1$, there exist *any* L^2 Interpolating Sets. Thus far, we are only able to show that if S is to be such a set, its elements cannot be too close to each other. More precisely,

Theorem 2.2. For any n, there is an $\epsilon > 0$ such that no S of the form

$$S = \{e^{i\theta_k}\}_{k=1}^{2n}, \quad \text{with } |\theta_k| \leq \epsilon \text{ for } 1 \leq k \leq 2n,$$

is an L^2 Interpolating Set for any $P \in G_n$.

Proof. Assume the contrary. Fix n. Then, for any $\epsilon > 0$, there is a set

$$S = S(\epsilon, n) = \{e^{i\theta_k}\}_{k=1}^{2n}, \quad \text{with } |\theta_k| \leq \epsilon \text{ for } 1 \leq k \leq 2n,$$

such that S is an L^2 Interpolating Set for some P, say

$$P_{\epsilon,n}(z) = \sum_{k=0}^{n} a_{\epsilon,k} z^k, \quad \text{with all } |a_{\epsilon,k}| = 1.$$

Choose a sequence of position ϵ's, say $\{\epsilon_j\}_{j=1}^{\infty}$, approaching 0.

For each k, $0 \le k \le n$, the sequence $\{a_{\epsilon_j,k}\}_{j=1}^{\infty}$ is bounded, and all terms of each of these $n+1$ sequences have modulus one. By the standard method of choosing a convergent subsequence for one k at a time, we can find a strictly increasing sequence of positive integers

$$\{m_j\}_{j=1}^{\infty}, \quad \text{and a set} \quad \{a_k\}_{k=0}^{n}$$

of complex numbers all of modulus one, such that

$$\{a_{\epsilon_{m_j},k}\}_{j=1}^{\infty}$$

converges to a_k for every k, $0 \le k \le n$. Since

$$|P_{\epsilon_{m_j}}(e^{i\theta})| - \sqrt{n+1}$$

can't change sign for

$$|\epsilon_{m_j}| \le \theta \le \pi,$$

we can assume, by taking another subsequence if necessary, that either

$$|P_{\epsilon_{m_j}}(e^{i\theta})| - \sqrt{n+1}$$

is always positive or always negative for

$$|\epsilon_{m_j}| < \theta \le \pi.$$

Suppose the former (the argument being the same in the latter case), and define

$$P_0(z) = \sum_{k=0}^{n} a_k z^k.$$

Clearly,

$$\{P_{\epsilon_{m_j}}(z)\}_{j=1}^{\infty}$$

converges uniformly to $P_0(z)$ on $|z| = 1$, so that

$$|P_0(e^{i\theta})| \ge \sqrt{n+1} \qquad \text{for } 0 < \theta \le 2\pi.$$

Since the L^2 norm of P_0 is $\sqrt{n+1}$, this is impossible, and the proof of theorem 2.2 is complete.

Also of interest is the locations of the zeroes of polynomials with unimodular coefficients. This is directly related to many other problems

discussed herein and has obvious importance in the choice of pattern functions for null steering. To quantify this question, let $r_j e^{i\alpha_j}$, $1 \leq j \leq n$, be the zeros of $P_n \in G_n$, normalize P_n so that the coefficient of z_n is 1, and define

$$\lambda_n = \max \min_j |1 - r_j| \quad \text{and} \quad \lambda_{n,q} = \max \left(\sum_{j=1}^n |1 - r_j|^q \right)^{1/q},$$

where the maximum is taken over all such $P_n(z)$.

Since any $P_1(z) = z - e^{i\alpha_1}$ for some real α_1, it is obvious that

$$\lambda_1 = \lambda_{1,q} = 0.$$

Considering the case $n = 2$,

$$P_2(z) = z^2 - (r_1 e^{i\alpha_1} + r_2 e^{i\alpha_2})z + r_1 r_2 e^{i(\alpha_1 + \alpha_2)}$$

so that

$$r_1 r_2 = |r_1 e^{i\alpha_1} + r_2 e^{i\alpha_2}| = 1.$$

Assume that $r_1 \geq 1$. Since

$$1 = |r_1 e^{i\alpha_1} + r_2 e^{i\alpha_2}| \geq r_1 - r_2 = r_1 - 1/r_1,$$

the maximum value for $r_1 - 1/r_1$ (hence the maximum value for $r_1 - 1$) is achieved when $r_1 - 1/r_1 = 1$, or

$$r_1 = \frac{1 + \sqrt{5}}{2} \quad \text{and} \quad r_2 = \frac{2}{1 + \sqrt{5}}.$$

In this case,

$$r_1 - 1 = \frac{\sqrt{5} - 1}{2} \quad \text{and} \quad 1 - r_2 = \frac{\sqrt{5} - 1}{1 + \sqrt{5}} < \frac{\sqrt{5} - 1}{2},$$

so that

$$\lambda_2 = \frac{\sqrt{5} - 1}{\sqrt{5} + 1} = \frac{3 - \sqrt{5}}{2}.$$

Also,

$$\lambda_{2,2}^2 = \max_r ((r - 1)^2 + (1 - 1/r)^2).$$

By an elementary calculus argument, it is seen that this maximum occurs for $r = (\sqrt{5} + 1)/2$. Thus,

$$\lambda_{2,2} = \sqrt{5 - 2\sqrt{5}}.$$

We leave as an open question the behavior of other values of λ_n and $\lambda_{n,q}$.

3. Applications

As mentioned in section 1, applications of polynomials with restricted coefficients abound in the engineering world. Those which we focus on herein include null steering, adaptive beamforming, notch filtering, peak power limited transmitting, and the synthesis of low peak-factor signals and flat spectrum sequences.

Several new designs of analytic null steering algorithms for linear arrays are described in [BN88]. Two of them, the β-Technique and the Positive Coefficient Model, allow for placing an arbitrary number of nulls in arbitrary directions while maintaining main beam and sidelobe level control. A method of incorporating these deterministic null steering techniques into existing adaptive algorithms is proposed. The resulting Direct Adaptive Nulling System offers the possibility of significant increases in array performance at very little cost.

A major reason for combining deterministic methods with existing techniques is that arrays must ordinarily deal with significant random noise. In these cases, one has no a priori information about the direction or nature of such unwanted signals. Thus, in such applications, as well as in cases where advance knowledge of jammer characteristics is lacking, indirect statistical methods are unavoidable, although their efficiency may be greatly increased by combining them with analytic approaches.

There exist applications, however, where much is known in advance about the characteristics of both the desired signals and the undesired noise. This is especially true where one has control of the generation of these waveforms. Thus, in the case where one system is producing both offensive signals (i.e., searching for and homing in on targets) and defensive signals (i.e., identifying and tracking incoming weapons), so that mutual interference becomes a predominant concern, the problem is almost exclusively deterministic in nature. In such cases, robust and computationally efficient analytic algorithms controlling both the individual performance of the offensive and defensive signals and the interactive jamming between them are crucial to mission success.

A related problem is the determination of optimal shading coefficients for a conformal array. As is well known, using various measures of optimality, this is a computational problem of order n^3, where n is the number of array elements. Thus, the computational load will be reduced by a factor of 8 if the coefficients may be restricted to be real. Circumstances where this occurs are described in [Byr88a]. A different method of improving computational efficiency, namely a convex programming approach, will be an important focus of further research.

Another interesting application of our concepts is to notch filters. A nearly ideal notch filter employing coefficients of equal magnitude is given in [Byr88b]. The construction is based upon earlier work of the author involving polynomials with restricted coefficients [Byr77]. The fundamental idea employed in [Byr88b] to construct a notch filter with a single notch may be combined with the concept of an n-nomial [Byr73] to produce nearly ideal filters with multiple notches. Furthermore, as noted elsewhere, zero coefficients do not affect the dynamic range, so that these multi-notch filters maintain the property of having unit dynamic range.

In addition to their use in the construction of notch filters, Byrnes Polynomials [Byr77, Kah80, Kör80] have potential applications to the design of peak power limited transmitters and the synthesis of low peak-factor signals and flat spectrum sequences. In transmitter design, for example, one is often faced with a peak power constraint. Under various conditions, the transmitter output may be modeled as a polynomial. Here the maximum modulus of the polynomial on the unit circle represents the peak power, while the L^2 norm of the polynomial is the average power. Thus, the classical engineering problem of minimizing the peak-to-average ratio becomes the mathematical question of minimizing the ratio of the sup norm to the L^2 norm of a polynomial on the unit circle.

In the trivial case where one frequency is to be transmitted (i.e., the polynomial can be a monomial), clearly the ideal value 1 for the peak-to-average ratio is achieved, and the polynomial is indeed of constant modulus on the unit circle. For the more interesting and practical case of transmitting many linerarly increasing frequencies, it is usually desired to transmit each frequency at the same power, which should be as large as possible. As the power of each individual frequency is represented by the modulus of the corresponding coefficient, the mathematical question naturally arises of how close to constant the modulus of a polynomial with equimodular coefficients can be on the unit circle.

More precisely, if n pure tones are transmitted with frequencies of

the form $f_0 + k\Delta$, where f_0 is the fundamental frequency and Δ is the increment, then the waveform is

$$x(t) = \sum_{k=0}^{n-1} A_k \cos(2\pi(f_0 + k\Delta)t + \theta_k)$$

$$= S(t)\cos(\arg S(t) + 2\pi f_0 t).$$

Here,

$$S(t) = \sum_{k=0}^{n-1} A_k e^{i\theta_k} e^{i2\pi k\Delta t},$$

$\theta_k = $ phase, and $A_k = $ power in kth tone.

As mentioned, almost always all frequencies are transmitted with equal power, so that $A_k \equiv 1$. To minimize the peak power of $x(t)$, the maximum (over t) of $|x(t)|$ must be minimized (over θ_k). It is relatively straightforward to see that the exact problem is to obtain

$$\min_{\theta_k} \max_{t} \left| \sum_{k=0}^{n-1} e^{i\theta_k} e^{i2\pi k\Delta t} \right|,$$

a job which is performed by the Byrnes polynomials [Byr77] in nearly ideal fashion.

The adaptation of such polynomials to these problems is important, since in applications like the Link 11 Communications System, the average power is usually maintained at one tenth or less of its theoretical ideal to prevent transmitter overload. Employing concepts such as those described above should yield a significant reduction in the peak-to-average ratio, thereby allowing a large increase in average power, hence a more efficient communications system. These considerations also show that the Byrnes construction has direct application to the synthesis of low peak-factor signals.

Now consider the problem of designing a flat spectrum sequence $\{a_k\}_{k=0}^{\infty}$ as defined earlier. These sequences have direct use in such diverse areas as concert hall acoustics, the quieting of an object's response to radar and active sonar, and speech synthesis. Schroeder [Sch85] presents many of the fascinating details of these applications.

As we observed, the DFT can be thought of as the values of the polynomial

$$P(z) = \sum_{k=0}^{n-1} a_k z^k,$$

where n is the period, at the n-th roots of unity. The Byrnes construction [Byr77] yields polynomials whose spectra are essentially flat at almost all points of the unit circle, not just at the roots of unity. Furthermore, they have the additional property that all of the terms of the original sequence, $\{a_k\}$, have the same magnitude. Applications of these concepts to notch filtering and communications are discussed elsewhere.

In our final application, we have begun to exploit the great success of J.P. Kahane [Kah80] in solving the Littlewood conjecture. As we note in section 2, Kahane showed that there indeed exist polynomials with unimodular coefficients whose modulus is essentially constant on the unit circle. It is our opinion that the breakthrough of Kahane was due to his ingenious use of randomness and probability in his construction. Behind his and previous approaches was the idea of Gauss, viz. the "Gauss Sums." To put it quite simply, we feel that Littlewood's problem was vanquished by the "equation"

$$\text{Kahane} = \text{Gauss Sums} + \text{Probabalisitic Choices}$$

Our idea is to exploit the Kahane breakthrough by developing methods to judiciously make the "Probabalistic Choices" referred to above, and thereby convert Kahane's "randomized" proof into a constructive one. This would not only result in exciting new mathematics, but would also be directly applicable to several important engineering problems. In addition to the areas of peak power limited transmitting and flat spectrum sequences discussed earlier, such polynomials would find immediate use in the design of reflection phase gratings, and therefore would be employable in solving concert hall acoustics problems and in quieting the response of an object to sonar or radar. Another potential application of this "educated randomness" construction is in the synthesis of multi-element omnidirectional beam patterns.

In the concert hall acoustics application of reflection phase gratings, it is desired to design the ceiling so that sound is widely scattered *except* in the specular direction. As described earlier, in the context of notch filter design, the Byrnes polynomials [Byr77] place a null in any given direction while the coefficients maintain their other desirable properties of being both flat spectrum and low correlation sequences. Thus, they might even be perferable to the Kahane polynomials in this context. This also appears to apply to monostatic radar, where the null would be placed in the direction of the radar. For bistatic radar, on the other hand, the receiver direction is often unknown. Thus, if a construction based upon the Kahane polynomials could be employed, radar energy would be reflected

equally in all directions, thereby reducing the probability that there would be enough energy reflected in any particular direction to enable detection. A possible undersea application of these ideas occurs in the design of baffles used to quiet machinery noise from submarines, in an attempt to prevent the noise from escaping the hull. Note that our constructions would complement the coatings that are already in use or are being designed to attack these problems, since these coatings provide uniform attenuation. Furthermore, surface structures based upon the Byrnes polynomials would have the highly diffusing property over a large set of frequencies. It is not yet clear whether the Kahane polynomials also yield this important property. The design of two-dimensional arrays so that energy may be scattered with equal intensity over the solid angle is also of considerable interest. It appears that a straightforward product formulation gives the desired results for the Byrnes polynomials, but the situation is not so clear for the Kahane polynomials. We continue to focus our research on the many fascinating questions raised in this final paragraph.

4. Bibliography

[Bel71]   E. Beller. Polynomial extremal problems in L^p. *Proc. Amer. Math. Soc.*, 30:249–259, 1971.

[BN71]   E. Beller and D.J. Newman. An l_1 extremal problem for polynomials. *Proc. Amer. Math. Soc.*, 29:474–481, 1971.

[BN73]   E. Beller and D.J. Newman. An extremal problem for the geometric mean of polynomials. *Proc. Amer. Math. Soc.*, 39:313–317, 1973.

[BN74]   E. Beller and D.J. Newman. The minimum modulus of polynomials. *Proc. Amer. Math. Soc.*, 45:463–465, 1974.

[BN88]   J.S. Byrnes and D.J. Newman. Null steering employing polynomials with restricted coefficients. *IEEE Trans. Antennas and Propagation*, 36(2):301–303, 1988.

[Byr73]   J.S. Byrnes. L^2 approximation with trigonometric n-nomials. *J. of Approx. Th.*, 9:373–379, 1973.

[Byr77]   J.S. Byrnes. On polynomials with coefficients of modulus one. *Bull. London Math. Soc.*, 9:171–176, 1977.

[Byr88a] J.S. Byrnes. The minimax optimization of an antenna array employing restricted coefficients. *Scientia*, 1:25–28, 1988.

[Byr88b] J.S. Byrnes. A notch filter employing coefficients of equal magnitude. *IEEE Trans. Acoustics, Speech, Signal Processing*, 36(11):1783–1784, 1988.

[Clu59] J.C. Clunie. The minimum modulus of a polynomial on the unit circle. *Quarterly J. of Math.*, 10:95–98, 1959.

[Erd57] Paul Erdös. Some unsolved problems. *Mich. Math. J.*, 4:291–300, 1957.

[Hud81] J.E. Hudson. *Adaptive Array Principles*. Peter Peregrinus Ltd., New York, 1981.

[Kah80] Jean-Pierre Kahane. Sur les polynomes a coefficients unimodulaires. *Bull. London Math. Soc.*, 12:321–342, 1980.

[Kör80] T.W. Körner. On a polynomial of Byrnes. *Bull. London Math Soc.*, 12:219–224, 1980.

[Lit62] J.E. Littlewood. On the mean values of certain trigonometric polynomials, II. *Illinois J. Math*, 6:1–39, 1962.

[Lit66] J.E. Littlewood. On polynomials $\sum \pm 1 z^m, \sum e^{\alpha_m i} z^m$, $z = e^{i\theta}$. *J. Lon. Math. Soc.*, 41:367–376, 1966.

[Rud59] W. Rudin. Some theorems on Fourier coefficients. *Proc. Amer. Math. Soc*, 10:855–859, 1959.

[Sch85] M.R. Schroeder. *Number Theory in Science and Communication*. Springer-Verlag, Berlin, second edition, 1985.

[Sha57] H.S. Shapiro. Extremal problems for polynomials and power series. Sc.M. thesis, Massachusetts Institute of Technology, 1957.

[Zyg59] A. Zygmund. *Trigonometric Series*. Cambridge University Press, second edition, 1959.

ANTENNA DIAGNOSIS BY MICROWAVE HOLOGRAPHIC METROLOGY

Y. RAHMAT-SAMII

Jet Propulsion Laboratory
California Institute of Technology
Pasadena, CA 91109

1. INTRODUCTION

Reflector antenna performance can be improved by identifying the location and the amount of the surface distortions and then by correcting them. Similarly, the performance of array antennas can be improved by determining the location of the defective radiating elements in the array. For example, Fig. 1 shows how the surface distortion can deteriorate the gain performance of large antennas used for Deep Space exploration, which may result in the loss of valuable science data. A review of mechanical and optical surface measurement techniques for radio telescopes may be found in [1]. Many of these methods can become time-consuming, especially when repeated measurements are required. Recently, considerable attention has been given to using alternative approaches either based on the photogrammetric concept [2] or based on microwave holographic techniques [3-10], which have proven to be convenient and quick with acceptable accuracy. Strictly speaking the word "holography" means "total recording" and should not be confused with the more common term in laser holography.

In applying microwave holographic metrology, one measures the complex (amplitude and phase) far-field (or Fresnel zone) pattern of the antenna and then applies the Fourier transform relationship that exists between the radiation pattern and the function related to the current distribution on the reflector surface (or equivalent aperture field). From the constructed phase data one can then determine the deviations of the surface from its ideal geometry. The technique has already been used successfully by many organizations worldwide. It is worthwhile to mention that the holographic reconstruction (diagnosis) can be applied to the complex far-field data, either directly measured or constructed from near-field measurements. This latter application is very useful for the organizations who utilize near-field measurement techniques. Fig. 2 depicts a block diagram of utilization of the microwave holographic reconstruction. As mentioned earlier, the procedure can also be used for the detection of defective radiating elements in arrays.

In this chapter, an attempt is made to provide an overview of the mathematical basis of the technique, numerical

17

B. de Neumann (ed.), Electromagnetic Modelling and Measurements for Analysis and Synthesis Problems, 17–50.
© 1991 Kluwer Academic Publishers. Printed in the Netherlands.

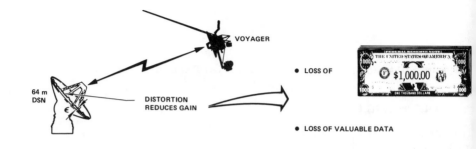

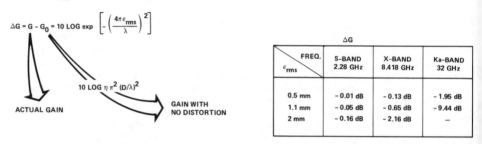

ϵ_{rms} = rms VALUE OF SURFACE DISTORTION

$$\Delta G = G - G_0 = 10 \ LOG \ exp \left[-\left(\frac{4\pi\epsilon_{rms}}{\lambda} \right)^2 \right]$$

ACTUAL GAIN

10 LOG $\eta \ \pi^2 \ (D/\lambda)^2$

GAIN WITH NO DISTORTION

		ΔG		
ϵ_{rms}	FREQ.	S–BAND 2.28 GHz	X–BAND 8.418 GHz	Ka–BAND 32 GHz
0.5 mm		– 0.01 dB	– 0.13 dB	– 1.95 dB
1.1 mm		– 0.05 dB	– 0.65 dB	– 9.44 dB
2 mm		– 0.16 dB	– 2.16 dB	–

FIG. 1. Estimated gain loss due to surface distortions.

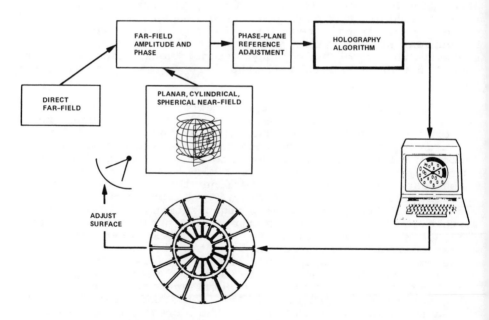

FIG. 2. Methodology of applying microwave holographic diagnosis to complex (amplitude and phase) far-field data, either directly measured or constructed from near-field measurements.

considerations, measurement aspects, and the results of measurements using both the direct far-field and near-field data. Sample cases are shown to demonstrate how effectively surface distortions can be identified.

2. FOURIER TRANSFORM RELATIONSHIP

In this section a summary of mathematical developments in describing the fundamentals of the holographic technique is given. The concept of physical optics is used to derive the required expressions. The geometry of a parabolic reflector with diameter $D = 2a$ (radius a) and focal length F is shown in Fig. 3. It is assumed that the reflector is illuminated by a feed located at the focal point (the off-focused feeds can be similarly discussed) and that the reflector surface may have some irregularities. Using the construction of the physical optics method , one can show that the radiated field can be expressed as

$$\vec{E} = -jk\eta \frac{e^{-jkr}}{4\pi r} \left(T_\theta \hat{\theta} + T_\phi \hat{\phi} \right) \tag{1}$$

where $j = \sqrt{-1}$, $k = 2\pi/\lambda$ (λ is the wavelength), $\eta = 120\pi$ (η free space impedance), and

$$\vec{T}(\theta,\phi) = \int_S \vec{J}(\vec{r}') \, e^{jk\vec{r}' \cdot \hat{r}} \, dS' \quad . \tag{2}$$

In (2), J is the induced surface current defined as

$$\vec{J} = 2\hat{n} \times \vec{H}^i \tag{3}$$

where n is the surface unit normal, S is the reflector surface, and H^i is the incident magnetic field generated by the feed.

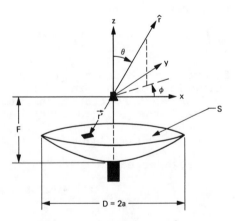

FIG. 3. Geometry of a parabolic reflector.

Integration in (2) is performed on the curved surface S with integration parameters defined on it. As shown in [11], this integral can be performed in terms of the aperture coordinates (ρ', ϕ') or (x', y') by using the concept of the surface projection Jacobian. This allows one to express (2) as

$$\vec{T}(\theta,\phi) = \int_S \vec{J}(\vec{r}') \; e^{jk\vec{r}'\cdot\hat{r}} \; J_s \; dx'dy' \qquad (4)$$

where the Jacobian transformation J_s is

$$J_s = \sqrt{1 + \left(\frac{\partial f}{\partial x'}\right)^2 + \left(\frac{\partial f}{\partial y'}\right)^2} \; , \qquad (5)$$

and f describes the reflector surface

$$z = f(x,y) \; , \qquad (6)$$

and s designates the area of the projection of the reflector surface S onto the plane x-y (for symmetric reflectors this is a circular area).

Equation (4) can further be simplified by employing the following definitions and identities:

$$\begin{cases} \vec{\underset{\sim}{J}}(x',y') = \vec{J}(\vec{r}') \; J_s = 2\vec{N} \times \vec{H}^i \\[2ex] \vec{r}'\cdot\hat{r} = z'\cos\theta + ux' + v\,y' \end{cases} \qquad (7)$$

where

$$\begin{cases} u = \sin\theta\sin\phi \\ v = \sin\theta\cos\phi \; . \end{cases} \qquad (8)$$

By substituting (7) into (4), one obtains

$$\vec{T}(u,v) = \int_S \vec{\underset{\sim}{J}}(x',y') \left[e^{jkz'\cos\theta} \right] e^{jk(ux'+vy')} \; dx'dy' \qquad (9)$$

For the flat reflector, $z' = $ const, the term in the bracket will not be a function of the integration variable, and therefore T will be an exact Fourier transform of the induced current. However, for the curved surfaces, due to the appearance of the bracket term in (9), T is not, in general, simply a two-dimensional Fourier transform of J.

If, however, (9) is reexpressed as

$$\vec{T}(u,v) = \int_S \vec{\underset{\sim}{J}}(x',y') e^{jkz'} \left[e^{-jkz'(1-\cos\theta)} \right]$$

$$\cdot e^{jk(ux'+vy')} \; dx'dy' \qquad (10)$$

it can then be expanded in terms of the Taylor series for small values of θ, namely,

$$\vec{T}(u,v) = \sum_{p=0}^{P \to \infty} \left[-jk(1-\cos\theta)\right]^p \vec{T}_p \tag{11}$$

where

$$\vec{T}_p = \int_S z'^p \vec{\tilde{J}}(x',y') e^{jkz'} e^{jk(ux'+vy')} dx' dy' \tag{12}$$

Notice that (11) is now a summation of Fourier transforms with its dominant term expressed as

$$\vec{T}(u,v) = \int_S \vec{\tilde{J}}(x',y') e^{jkz'} e^{jk(ux'+vy')} dx' dy' . \tag{13}$$

The contribution of higher-order terms in the series expansion of (11) becomes significant for wide-angle observations (large θ) and laterally displaced feed [11]. However, as far as the scope of this chapter is concerned, the only important term is T_0 (renamed T in (13) for simplicity).

Once T is determined, the far-field pattern can then be obtained from (1). In many cases, T may be determined by its Cartesian components T_x, T_y, and T_z, which can be transformed to the spherical components by

$$\begin{Bmatrix} T_\theta \\ T_\phi \end{Bmatrix} = \begin{pmatrix} \cos\theta\cos\phi & \cos\theta\sin\phi & -\sin\theta \\ -\sin\phi & \cos\phi & 0 \end{pmatrix} \begin{Bmatrix} T_x \\ T_y \\ T_z \end{Bmatrix} . \tag{14}$$

Furthermore, application of Ludwig's third definition [12] allows the definition of the copolar and cross-polar components of the far-field patterns

$$\begin{Bmatrix} T_{co-pol} \\ T_{cross-pol} \end{Bmatrix} = \begin{pmatrix} \sin\phi & \cos\phi \\ \cos\phi & -\sin\phi \end{pmatrix} \begin{Bmatrix} T_\theta \\ T_\phi \end{Bmatrix} . \tag{15}$$

In (15) it is assumed that the radiated field is predominantly y-polarized. For the x-polarized field case, the rows of the square matrix in (15) must be interchanged. From (14) and (15) one can show that for the cases where θ is small, the following holds

$$\begin{Bmatrix} T_{co-pol} \\ T_{cross-pol} \end{Bmatrix} \overset{\sim}{=} \begin{Bmatrix} T_y \\ T_x \end{Bmatrix} . \quad \text{for small } \theta \tag{16}$$

The application of the reciprocity theorem implies that the transmitting and receiving patterns are identical. Therefore if the reflector antenna is impinged by a plane wave, the output at the feed port will be proportional to the antenna receiving pattern for a given direction of incidence and plane wave polarization. For the sake of simplicity, it is assumed that the dominant polarization is in the y direction, and therefore, by using (16) and (13), the following expression for the pattern is obtained

$$T = \int_{S} \tilde{J}(x',y') e^{jkz'}\, e^{jk(ux'+vy')} dx'dy' \; . \tag{17}$$

It is noted that the vector notations are removed in (17), in order to emphasize that only one polarization is considered.

In summary, it has been shown that, in general, (1) can be represented in terms of a summation of many Fourier transforms. However, under certain conditions where the reflector is large in terms of wavelength and when the angular region of the required pattern data is small, an exact Fourier transform relationship results. Similar interpretation can also be made by using aperture field method.

One may also use the concept of the so-called "slicing or focusing" approach, as employed in tomography, to establish a Fourier transform relationship in (1) for each fixed value of $z' = z_f$. When the inverse Fourier transform is performed, it can provide the surface profile on the reflector at locations sliced where $z = z_f$. In practice, one may only use a limited number of these slices and then construct the entire surface profile from them. For deep dishes, this "focusing" approach improves the quality of the reconstruction. It is noticed that this procedure can be simply applied by changing the phase reference plane of the far-field data. As the phase center reference becomes closer to the reflector surface, better surface reconstruction is obtained (see Fig. 4).

3. SURFACE PROFILE CONSTRUCTION

Once both the amplitude and phase of the reflector pattern are measured, the surface profile can be determined via an inverse Fourier transform. In order to demonstrate that the surface profile information can be extracted from (17), let it first be assumed that the phase center of the feed is located at the focal point and that the surface irregularities are described by function $\epsilon(x, y)$ in the normal direction, namely,

$\epsilon(x, y)$ = surface irregularity in the normal direction

Using the geometrical relationship shown in Fig. 5 and making the assumption that distortion is a small fraction of the wavelength, one can then establish the following expressions for the surface distortion in the normal direction to the surface

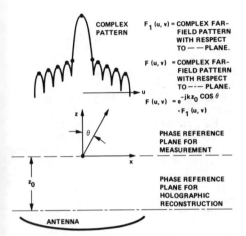

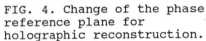

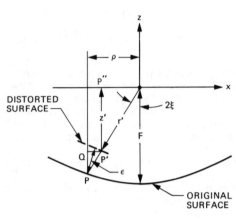

FIG. 4. Change of the phase reference plane for holographic reconstruction.

FIG. 5. Relationships among the ray paths and surface deviation in the normal direction.

$$P'P + PQ = \frac{\epsilon}{\cos\xi} + \frac{\epsilon}{\cos\xi} \cos2\xi = 2\epsilon\cos\xi \tag{18}$$

where for a parabolic reflector,

$$\cos\xi = \frac{1}{\sqrt{1 + \frac{\rho2}{4F^2}}} = \frac{1}{\sqrt{1 + \frac{x^2+y^2}{4F^2}}} . \tag{19}$$

The term inside (17) may now be expressed as

$$\tilde{J}(x',y')e^{jkz'} = |\tilde{J}(x',y')|e^{-jkr'}e^{jkz'} . \tag{20}$$

Note that the phase term exp(-jkr') is obtained from (3) because the phase center of the feed is located at the focal point (one may have to perform special corrections for the cases when the phase front of the feed is not spherical in the subtended angle of the reflector). The exponent in (20) can be written as

$$-r' + z' = -r' - P'P'' = -r' - P'P'' - P'P - PQ + P'P + PQ . \tag{21}$$

For a parabolic reflector,

$$-r' - P'P'' - P'P - PQ = 2F \tag{22}$$

Which finally result in

$$-r' + z' = -2F + 2\epsilon\cos\xi . \tag{23}$$

If the distortion phase error is defined as

$$\delta = 4\pi \frac{\epsilon}{\lambda} \cos\xi , \tag{24}$$

(17) can then be written as

$$T(u,v) = e^{-j2kF} \iint_S |\overset{\curvearrowright}{J}(x',y')| \, e^{j\delta} \, e^{jk(ux'+vy')} dx'dy' \quad (25)$$

The constant phase term $\exp(-j2kF)$ in (25) is the result of positioning the center of the coordinate system at the focal point. Different phase constants can result if the center of the coordinate system is displaced from this focal point. In general, it is preferred to bring the center of the coordinate system as close as possible to the reflector surface. It should be noticed that, in general, both copolar and cross-polar patterns can be used to determine δ from (25).

Once both the amplitude and the phase of the reflector far-field (or Fresnel zone) are measured (or simulated), the surface distortion can be determined from (25) via an inverse Fourier transform, namely,

$$|\overset{\curvearrowright}{J}(x,y)| e^{j\delta} = e^{j2kF} \mathscr{F}^{-1}\left[T(u,v)\right] \quad , \quad (26)$$

where $\mathscr{F}$ designates the Fourier transform.

From (26), (24), and (19), the surface distortion $\epsilon(x, y)$ in terms of the wavelength is finally constructed, namely,

$$\frac{\epsilon(x,y)}{\lambda} = \frac{1}{4\pi} \sqrt{1 + \frac{x^2+y^2}{4F^2}} \; \text{Phase} \left\{ e^{j2kF}\mathscr{F}^{-1}\left[T(u,v)\right] \right\} . \quad (27)$$

In constructing $\epsilon(x,y)$ from (27), the constant and linearly dependent phase terms in the braces must first be extracted. The former accounts for the ambiguity in defining the constant phase reference and the latter accounts for the beam displacement due to the lateral feed displacement. It is also possible to identify the best fit quadratic dependence that could be used to remove the feed axial defocusing by readjusting the feed position. In doing so, one must first extract the non-spherical phase front variation of the feed. Similar expressions can also be derived for dual reflector systems in which one should also consider the effects of subreflector displacements and its diffraction. If complex patterns in the Fresnel zone are used, one must also extract the quadratic terms due to the Fresnel zone dependance.

4. DFT, FFT, AND SAMPLING THEOREMS

The numerical evaluation of (27) can be performed by applying different algorithms and the choice is dictated by the nature of the measured (or simulated) data. For example, if the data is not distributed on a uniform rectangular grid in u-v space, one may either employ an interpolation procedure to regularize the data and then apply FFT or use the data as it is and perform the inverse Fourier transform by using the DFT algorithm in brute force manner. In the latter case, one can

use a DFT algorithm to evaluate the inverse Fourier transform in (27) by

$$t(x,y) = \mathscr{F}^{-1}[T(u,v)] = \sum_i \sum_j T(u_i, v_j) e^{-jk(u_i x + v_j y)} \qquad (28)$$

where (u_i, v_i) designates the measured (or simulated) sample points.

Depending on the antenna drive mechanism, the measured data can be collected in a variety of ways such as raster scan, etc. In Appendix A, a procedure is suggested to allow the far-field pattern measurement on regularized u-v distribution by properly controlling the motion of the antenna under the test for a given antenna mount mechanism.

If, however, the data are provided in a regularized rectangular distribution in u-v space, one can then apply an FFT algorithm to evaluate the inverse Fourier transform in (27). For large numbers of sampled points, FFT has the advantage of being very fast compared to brute force application of DFT.

Since the integrand of (17) has a finite support (reflector surface), its Fourier transform T is an analytic function which extends to infinity. Conversely, since the transform of T is a function with a finite support, the sampling theorem can be invoked to express T(u, v) by only its values at the sampling points. This concept can also be generalized to nonuniform sampling technique [13]. For a reflector with diameter D, the sampling interval is

$$\text{Largest sampling interval} = \frac{1}{D/\lambda} \qquad (29)$$

which is closely related to reflector beam width (uniform illumination case). However, since the Fast Fourier transform (FFT) algorithm is used to evaluate (27), intervals smaller than the largest sampling interval must be used to overcome aliasing problem. For these cases, the sampling intervals are typically chosen to be

$$\Delta u = \Delta v = \frac{\kappa}{D/\lambda} . \qquad (30)$$

where D is the reflector diameter and $0.5 < \kappa < 1$ is a factor to ensure the required sampling criterion.

The value $\epsilon(x, y)/\lambda$ from (27) can be determined almost exactly, provided the pattern T(u, v) is known in its entirety (infinite extended) and with acceptable level of SNR (Signal-to-Noise-Ratio). In practice, this is not the case, and T(u,v) is measured or simulated only in a finite range. If the total number of measured (simulated) data in each of the u and v directions is designated by N_{msr}, then the total number of measured data is

$$\text{The total number of measured data} = N_{MSR} \cdot N_{MSR} = (2N+1) \cdot (2N+1) . \qquad (31)$$

The variable N is introduced in (31) to denote the number of measured data to one side of the boresight (see Fig. 6). Similarly, the total number of points used to perform FFT can be defined as

$$\text{total number of points for FFT} = N_F N_F \qquad (32)$$

For the cases when $N_F > N_{msr}$, however, the domain of the measured pattern must be extended, as will be discussed in the next section.

Since (27) is evaluated using FFT, the values of $\epsilon(x, y)$ are determined at the intervals of

$$\Delta x = \frac{D}{\kappa(N_F - 1)} \qquad \Delta y = \frac{D}{\kappa(N_F - 1)} \qquad (33)$$

However, as far as the actual surface resolution for identifying the surface errors are concerned, the parameter N_{msr} is a more dominant factor than N_F.

5. AN ITERATIVE SCHEME

There are many cases in which it may be necessary to extend the domain of the measured data T(u,v) in order to possibly improve the achievable resolution. This may also be required to improve the extended zero-in domain used in applying the FFT. Many algorithms are available to accomplish this; however, it has been found that an iterative algorithm similar to the one used in [14] is simple to apply and provides improved results. The procedure basically follows the steps shown in Fig. 7, and relies on the fact that the induced current has bounded support extended to the diameter of the reflector antenna. This allows one to introduce a truncation operator in step 2 of Fig. 7 and then continue the process of iteration by improving the tail end of the pattern in step 3. The procedure, typically, converges rapidly and reduces the amplitude artifacts outside the reflector domain considerably. Attempts have also been made to use similar and other iterative techniques to perform surface diagnosis without the knowledge of the phase data [15]. This is an area that demands more investigation and verification of the achievable accuracy.

The steps of this iterative procedure as applied to microwave holography are presented here. First, (17) is reexpressed as

$$T(u,v) = \mathscr{F}[Q] = \iint_S Q(x',y')e^{jk(ux'+vy')}dx'dy' \quad . \qquad (34)$$

Since Q is zero outside the reflector boundary (assuming neither a ground reflection nor any outside obstacles), the domain of integration in (34) can be extended to infinity; hence, the Fourier transform relationship holds. The function Q is constructed from T using the relation

$$Q(x,y) = \mathscr{F}^{-1}[T] \quad . \qquad (35)$$

PATTERN

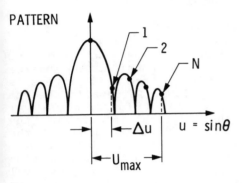

REFLECTOR

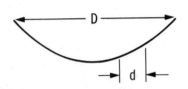

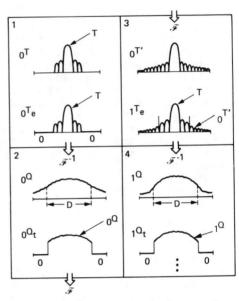

FIG. 6. Sampling points of far-field patterns. Surface resolution is $d \approx D/(2N\kappa)$, and error in determining ϵ is $2N\lambda/(4\pi SNR)$ where SNR is the signal to noise ratio at the boresight.

FIG. 7. Steps of the iterative scheme. T is the measured (or simulated) complex far-field pattern.

Since T is only given in the measurement domain, the iterative procedure is used to extend this domain, with the steps depicted in Fig. 7. The procedure starts by filling in zeros in the desired extended domain to give the zeroth iterated extended pattern $_0T_e$ such that it takes the same value as T inside the original domain and zero value outside, namely,

$$_0T_e = \begin{cases} T & \text{in the original domain} \\ 0 & \text{in the extension} \end{cases} \tag{36}$$

Next, the inverse Fourier transform (using FFT) is used to find $_0Q(x, y)$, the zeroth iteration of $Q(x, y)$. Obviously, since $_0T_e$ is a truncated function, its Fourier transform will extend to infinity. However, it is known from the physics of the problem that $_0Q(x, y)$ should only be extended within the region where the antenna surface exists. This information is next used to truncate $_0Q(x, y)$ to the domain of the antenna (radius D/2). Once the truncated version of $_0Q$, namely $_0Q_t$, is constructed, then a forward transform is taken to obtain $_0T'$. From T and $_0T'$ the first iterated form of T, namely $_1T_e$, is constructed such that

$$1^T e = \begin{cases} T & \text{in the original domain} \\ 0^{T'} & \text{in the extension} \end{cases} \tag{37}$$

It is clear that $1^T e$ is an improvement over $0^T e$, since its extended domain is filled with a nonzero function. The inverse Fourier transform of $1^T e$ is denoted by 1^Q, and this process is followed for higher orders of iterations. Parseval's theorem can be used to show that the tail end of function $Q(x, y)$ gradually diminishes outside the physical domain of the antenna, which is a manifestation of the convergence of the procedure. Once the nth iterated value of Q is determined, $\epsilon(x, y)$ can then be constructed from (27). In the next section the use of this procedure will be demonstrated by numerical examples. It is worthwhile to mention that this iterative procedure can become more effective for smaller values of parameter κ.

6. NUMERICAL SIMULATIONS

In order to assess how accurately the reflector profile can be determined from a set of measured data, it becomes necessary to develop simulation algorithms capable of providing all possible measurement scenarios. This is, especially, important to establish meaningful accuracy bounds on the reconstructed surface data for the cases when the measured data is subject to a variety of systematic and random errors. Fig. 8 depicts a schematic block diagram for interdisciplinary simulation studies, which considers all aspects of the microwave holography technique. Simulations have been performed using both aperture models and reflector models.

6.1 Aperture Models

Aperture simulation models are simple to apply and provide an insight into the application of the holography technique. In these models, aperture and phase distribution are typically defined and then far-field data are constructed. In general an integration or FFT scheme may be used to obtain the far field. However, for certain special aperture distributions, such as those which are circularly symmetric, closed form expressions can be used. In the following the steps of this aperture model is presented.

The geometry of a circular aperture with different annular regions is shown in Fig. 9, where the radii of these regions are designated by $a_0, a_1, a_2, \ldots, a_N = a$ (a_0, in this figure, is used to designate the central blockage region). Let us assume that the amplitude and phase distributions across the aperture are given by Q and δ, respectively, and that these functions are circularly symmetric. The far-field pattern of this aperture distribution is constructed using the Fourier transform in polar coordinates to obtain

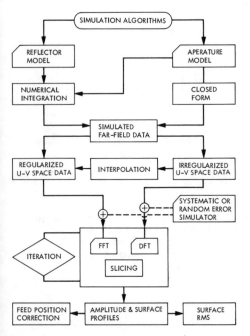

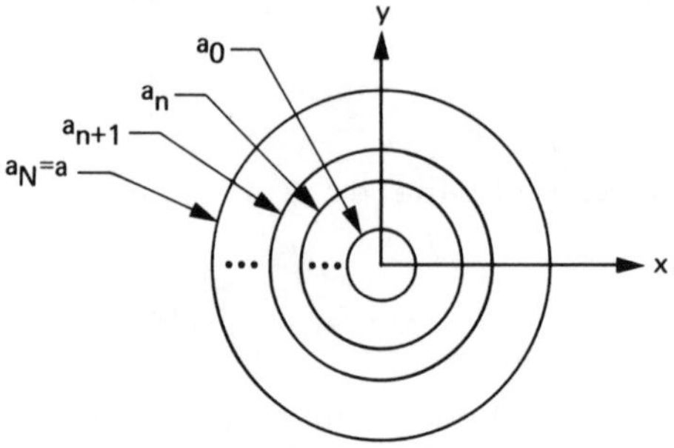

FIG. 9. A circular aperture with annular phase distortions.

FIG. 8. Block diagram of simulation studies for microwave holography.

$$T = \int_{a_0}^{a} \int_{0}^{2\pi} Q(\rho')e^{j\delta(\rho')} \; e^{jk\rho'\sin\theta\cos(\phi-\phi')}\rho'd\rho'd\phi' \; . \tag{38}$$

with the inner region having radius a_0 considered as the blockage region ($a_0 = 0$, no blockage).

It is further assumed that $\delta(\rho')$ takes the constant value δ_n in the nth annular region, which can be interpreted as the phase irregularity (surface irregularity) in the nth annular region. With this assumption, (38) may be expressed as

$$T = \sum_{n=1}^{N} T_{n,n-1} \; e^{j\delta_n} \tag{39}$$

with

$$T_{n,n-1} = T_n - T_{n-1} \tag{40}$$

$$T_n = 2\pi \int_{0}^{a_n} Q(\rho') \; J_0(k\rho'\sin\theta)\rho'd\rho' \; . \tag{41}$$

with J_0 being the zero-order Bessel function.

The aperture amplitude distribution is defined as

$$\begin{cases} Q(\rho) = B + C\left[1-\left(\frac{\rho}{a}\right)^2\right]^P \\ \\ B + C = 1 \end{cases} \qquad (42)$$

where parameter B is used to control the edge taper, namely,

$$\text{Edge taper} = 20 \log B . \qquad (43)$$

It has been found for values of $1 \le P \le 2$, (42) is an adequate representation of the aperture amplitude distribution for many typical reflectors. In this work, the values of $P = 1$ and 2 are used, as (41) can be integrated in a closed form [16]. To arrive at this, the following identity is applied

$$\begin{cases} I_p(\alpha) = 2 \int_0^1 x^{2p+1} J_0(\alpha x)\,dx = \frac{2}{\alpha} J_1(\alpha) - p\left(\frac{2}{\alpha}\right)^2 J_2(\alpha) + p(p-1)\left(\frac{2}{\alpha}\right)^3 J_3(\alpha) - \ldots \\ \\ I_p(0) = \frac{1}{p+1} . \end{cases} \qquad (44)$$

Substituting (42) into (41), suing (44) and introducing the notation

$$\dot{u}_n = ka_n \sin\theta , \qquad (45)$$

one finally arrives at

$$P = 1 : \begin{cases} T_n = \pi a_n^2\left[B\frac{2}{u_n} J_1(u_n) + C\left\{\frac{2}{u_n} J_1(u_n)\right.\right. \\ \\ \left.\left. -\frac{a_n^2}{a^2}\left[\frac{2}{u_n} J_1(u_n) - \left(\frac{2}{u_n}\right)^2 J_2(u_n)\right]\right\}\right] \\ \\ T_n(0°) = \pi a_n^2\left[B + C\left(1 - \frac{a_n^2}{2a^2}\right)\right] \end{cases} \qquad (46)$$

$$P=2 : \begin{cases} T_n = \pi a_n^2 \Bigg[B \frac{2}{u_n} J_1(u_n) + C \Bigg\{ \frac{2}{u_n} J_1(u_n) - \frac{2a_n^2}{a^2} \cdot \\ \\ \qquad \Bigg[\frac{2}{u_n} J_1(u_n) - \frac{4}{u_n^2} J_2(u_n) \Bigg] + \frac{a_n^4}{a^4} \Bigg[\frac{2}{u_n} J_1(u_n) \\ \\ \qquad -2\left(\frac{2}{u_n}\right)^2 J_2(u_n) + 2\left(\frac{2}{u_n}\right)^3 J_3(u_n) \Bigg] \Bigg\} \Bigg] \\ \\ T_n(0°) = \pi a_n^2 \Bigg[B+C\left(1 - \frac{a_n^2}{2} + \frac{a_n^4}{3a^4}\right) \Bigg] . \end{cases} \qquad (47)$$

Expressions (46) and (47) can be used to construct the far field from (39). In order to further relate the aperture phase error to the reflector surface distortion, the following value for δ_n is used

$$\delta_n = 4\pi \frac{\epsilon n}{\lambda} \frac{1}{\sqrt{1 + \frac{\tilde{a}_n^2}{4F^2}}} \qquad (48)$$

where ϵ_n and a_n are the surface error in the normal direction and the average radius of the nth zone, respectively, namely,

$$\tilde{a}_n = \frac{a_n + a_{n-1}}{2} , \qquad (49)$$

and F is the focal length of the parabola as shown in Fig. 3.

This model has been used and many useful representative data have been generated. For example, the following reflector parameters are considered

a/λ radius, equal to 243.3;
F/λ equivalent focal length, equal to 1030.56;
a_0/λ blockage radius, equal to 25.5; $\qquad$ (50)
ET edge taper, equal to -10 dB;
$P = 1$

A computer program has been developed based on the mathematical formulations as presented before. In one case, far-field patterns are constructed for the reflector with no surface profile errors and, in another case, with surface profile errors as follows

$\varepsilon_1/\lambda = 0 \qquad 25.5 < \rho/\lambda < 80$

$\varepsilon_2/\lambda = 0.02 \qquad 80 < \rho/\lambda < 100$

$\varepsilon_3/\lambda = 0 \qquad 100 < \rho/\lambda < 150 \qquad (51)$

$\varepsilon_4/\lambda = 0.10 \qquad 150 < \rho/\lambda < 180$

$\varepsilon_5/\lambda = 0 \qquad 180 < \rho/\lambda < 243.3$

It is noticed that the above surface profile designates two annular rings of width 20λ and 30λ with surface errors of λ/50 and λ/10, respectively. Fig. 10 shows the simulated amplitude and phase far-field patterns of a reflector antenna subject to two annular ring surface distortions. Once these patterns are constructed using the developed computer programs, they are then used in the microwave holographic reconstruction computer program to determine the simulated surface profile distortions. Results of Fig. 11 clearly indicate that for the employed number of sampling and FFT points (i.e., N_{msr} = 63, and N_F = 128 in this case), the simulated surface profile has been recovered very well. Notice that in this figure the application of the previously mentioned iterative approach has provided cleaner reconstruction.

FAR-FIELD PATTERNS

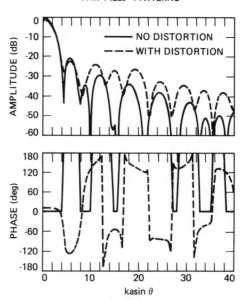

FIG. 10. Far-field patterns of a reflector with and without distortions.

As yet another example, an antenna with dimensions and surface distortions as shown in Fig. 12 is considered. In accordance with (43), both -10 dB and -25 dB edge tapers have been used to generate the amplitude and phase far-field patterns as displayed in Fig. 13. The reconstructed surface profiles using N_{msr} = 101 and N_F = 128 are shown in Fig. 14 for both edge taper cases. The reconstructed profiles match the simulated distortions of Fig. 12 very well. It is worthwhile to mention that in this simulation it has been assumed a perfect signal-to-noise-ratio for both cases. However, in real situation due to the existence of noise it will be more difficult to reconstruct the edge panels for the -25 db taper case than the -10 dB taper case.

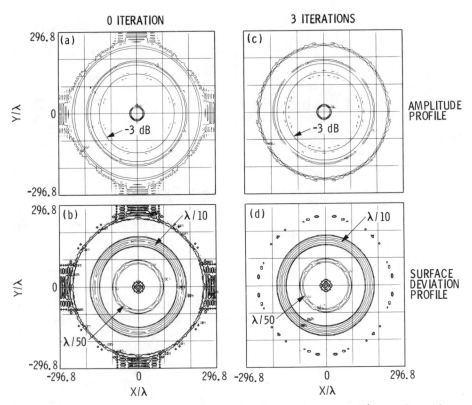

FIG. 11. Constructed amplitude and surface profiles for the simulated distortions. (a-b) Zero iteration. (c-d) Three iterations.

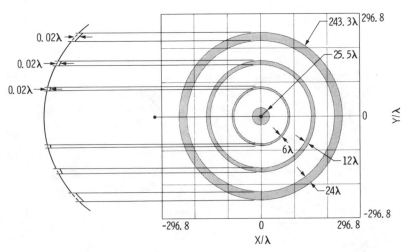

FIG. 12. Geometry of a reflector antenna with three annular rings of surface distortions and a central blockage region.

34

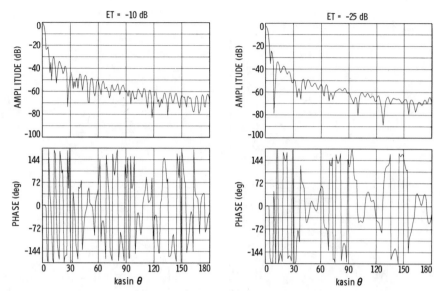

FIG. 13. Amplitude and phase far-field patterns for the reflector of Fig. 12 with -10 dB and -25 dB edge tapers.

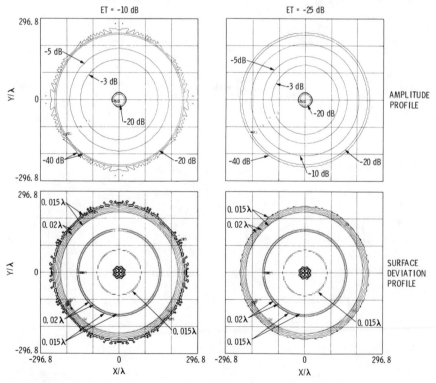

FIG. 14. Constructed amplitude and surface profiles based on far-field patterns of Fig. 13.

6.2 Reflector Models

Because most of the surfaces of large antennas are adjusted by properly setting surface panels, it is important to be able to simulate the effects of the panels which are displaced. The simplified aperture model presented earlier is not capable enough for the panel studies. In this section attention is focused on the reflector analysis model. The vector diffraction analysis for both single and dual reflectors is based on the methods developed in [11,17]. These methods use Physical Optics and Geometrical Theory of Diffraction (GTD) formulations for the efficient determination of the far-field patterns. In particular, reflectors with displaced panels are analyzed (see Fig. 15). For the sake of simplicity, the panels are modeled as regions which, when projected into the aperture, result in a radial and circumferential boundary as shown in Fig. 15. The panel displacement is achieved by axially moving the panel along the z-axis. Under these conditions, the parabolic reflector surface with n displaced parabolic panels may be represented by

$$
\begin{cases}
z = z_{parabola} \text{ or } z_{Panel} & x,y \in s \\[2ex]
z_{parabola} = -F + \dfrac{x^2+y^2}{4F^2} & x,y \in s - \displaystyle\sum_{n=1}^{N} s_n \\[3ex]
z_{panel} = \left(-F + \epsilon_n + \dfrac{x^2+y^2}{4F^2}\right) & x,y \in s_n
\end{cases}
\qquad (52)
$$

where s_n designates the area of the nth panel and ϵ_n is the axial displacement of the nth panel.

As an example, results are shown for a representative case with reflector dimensions of (see Fig. 16)

$$a/\lambda = \text{radius} = 897.7$$

$$F/\lambda = \text{focal length} = 760.4$$

$$f = \text{frequency} = 8.415 \text{ GHz}$$

$$ET = \text{edge taper} = -10 \text{ dB}$$

(53)

Note that, as shown in Fig. 16, although the panel is relatively large, it can be regarded as a combination of many panels with equal displacements. In addition, the panel is chosen such that it is not symmetric with respect to the x-axis as shown in Fig. 16. Figs. 17 and 18 depict the far-filed patterns in two principal planes for both the non-displaced (perfect) and displaced panel cases. The far-field data are generated for regularized u-v space grids with intervals $\Delta u = \Delta v = 0.000278$ which uses $\kappa = 0.5$ in accordance with (30). These data are then employed in an FFT/iterative scheme to determine how well the displaced panel can be recovered. Results are shown in Figs. 19 and 20 for different

36

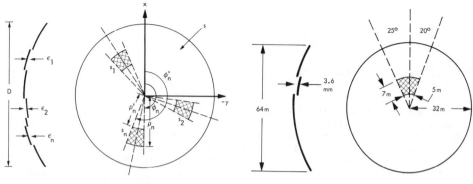

FIG. 15. Geometry of a
reflector antenna with
displaced surface panels.

FIG. 16. A parabolic reflector
with an axially displaced
panel used for numerical
simulation (λ = 3.57 cm).

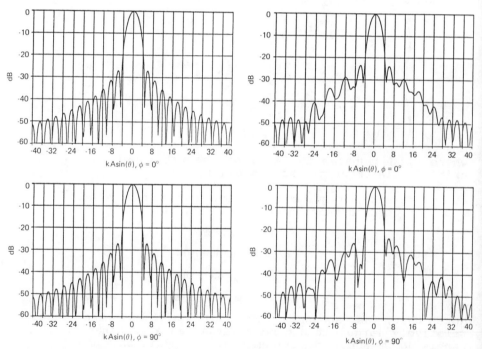

FIG. 17. Reflector far-field
patterns with no displaced
panel.

FIG. 18. Reflector far-field
patterns with displaced panel.

numbers of simulated far-field data points. It is clear from
these results that, in order to resolve the narrow dimension
of the panel with an acceptable degree of accuracy, a
sufficient amount of far-field data is needed. The numerical
simulation has shown that roughly two to four points per
smallest dimension is required to properly resolve the panel.

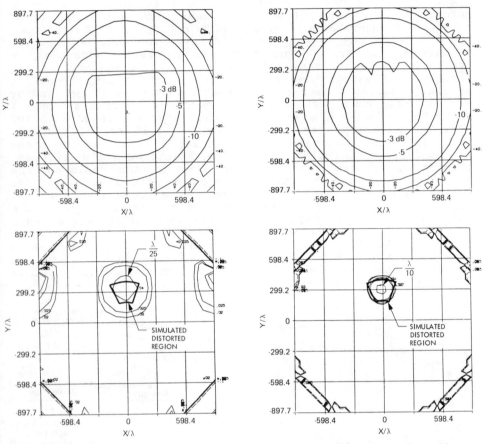

FIG. 19. Amplitude and surface profiles for 32 x 32 sampling points and with κ = 0.5. The panel displacement is $\lambda/10$ which is not recovered.

FIG. 20. Amplitude and surface profiles for 64 x 64 sampling points and with κ = 0.5. the panel displacement is $\lambda/10$ which is recovered.

In summary, these kind of simulations allow one to select proper number of measured sample data for a desired resolution and to investigate the needed signal-to-noise-ratio and overall phase stability in the course of the measurement for achieving specified accuracy tolerances.

7. FAR-FIELD MEASUREMENTS

Representative holographic reconstruction results based on the direct far-field measurements are summarized in this section. Results are shown for one of the three NASA/JPL DSN 64m-antenna as depicted in Fig. 21. Several measurements were performed on these antennas and the reader is referred to [7,9,18] for details. Only some preliminary results are discussed here. To perform a holography measurement, which necessitates the amplitude and phase measurements, in

38

practice, a second antenna is needed to provide a phase reference. This can often be a small and low-cost design antenna. In the case of many radio astronomy installations, the reference may already exist if the antenna to be measured belongs to an array. The test and reference antennas are typically illuminated at a single frequency using terrestrial, satellite-borne, or celestial radiation sources of small angular diameter. Different sources have been used by various researchers to measure a number of antennas, from 1 meter to 100 meters, using frequencies over the range 2 to 100 GHz. The accuracy attainable depends on a number of factors, including the signal-to-noise-ratio, tropospheric phase fluctuations, pointing stability, instrumental effects, truncation effects, etc. Fig. 22 depicts the basis of the microwave holographic metrology using far-field measurements. Notice that in this setup the reference antenna always tracks the source in order to provide the reference signal. The amplitude and phase data can be determined in variety of ways based on the architecture of the receiver, including the connected-element interferometer procedure. It is anticipated that amplitude and phase accuracies better than 0.1 of dB and a few degrees can be achieved. This can result in surface accuracy reconstruction better than $\lambda/100$.

FIG. 21. NASA/JPL 64-m Deep Space Network (DSN) antenna.

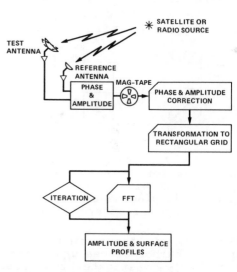

FIG. 22. Schematic of holographic measurement using a reference antenna.

Originally, a low resolution holographic measurement (matrix of 11 x 11) was performed on one of the DSN antennas using a radio source at 2.28 GHz [7,8]. The results of this measurement clearly identified that most of the necessary computer programs have been developed and the overall measurement approach is sound. Recently, a high resolution measurement (matrix of 189 x 189) has been performed at 11.45 GHz, using a linearly polarized beacon from a geosynchronous-

orbit satellite as the illumination source. This is necessary in order to obtain the required signal-to-noise-ratio for performing the high resolution holographic reconstruction. The detailed description of this measurement can be found in [9,18]. The recorded data was corrected for satellite motion, phase drift, pointing offset errors, baseline phase errors, and azimuth/elevation to rectangular u-v space interpolation [18]. This 189 x 189 matrix of complex numbers of the far-field was then processed using the holographic reconstruction computer program by utilizing a 256 x 256 FFT operation. These results were then processed to display two effective maps: effective surface error map (derived from the phase) and effective surface current map (amplitude). Figs. 23 and 24 are black-and-white presentations of the 14-color computer display [9]. Additional processing may be needed to remove nonspherical phase variation of the feed phase pattern, feed and subreflector displacements, subreflector diffraction effects, etc. For these reasons the maps are called "effective maps". It is noticed that panels as small as 1 meter have been resolved.

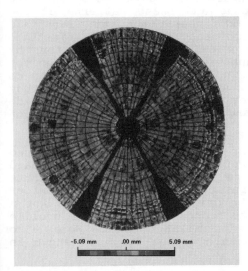

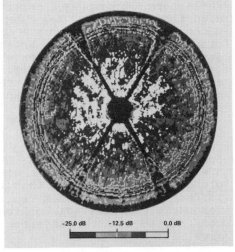

FIG. 23. Effective surface error map (black and white presentation of 14 color computer display).

FIG. 24. Effective surface current map (amplitude). N_{msr} = 189 and N_F = 256.

8. NEAR-FIELD MEASUREMENTS

As mentioned earlier, microwave holographic reconstruction may be used in conjunction with near-field measurements. This provides one with a useful diagnosis tool. This diagnosis approach has been used by several researchers using planar near-field facilities. For example, at JPL the plane-polar near-field facility is utilized to perform surface diagnosis on reflector antennas [19]. In the summer of 1986, this author

had the occasion to spend four months as a Guest Professor at the Technical University of Denmark. Among different projects conducted there, a study was performed to use their spherical near-field facility and conduct a microwave holographic diagnosis on some existing antennas. A synopsis of this investigation is described here and the reader is referred to [20] for more details.

This hybrid approach of combining spherical near-field measurement and holographic reconstruction would allow an in-depth evaluation of the accuracy of the holographic technique because the spherical near-field measurement provides accurate amplitude and phase far-field pattern for the entire angular range. The steps of this hybrid approach are depicted in Fig. 25. Under controlled conditions, two sets of spherical near-field measurements were performed on a 156-cm reflector at 11.3 GHz. In the first measurement, the antenna was measured in its existing condition (see Fig. 26a), while in the second measurement, four bumps of different sizes and heights were attached at several locations to the reflector surface (see Fig. 26b). These measured near-field data were used to generate the far-field amplitude and phase patterns of the reflectors using a spherical near-field to far-field algorithm. Figs. 27a-b depict the constructed far-field patterns from the measured spherical near-field data for the original antenna and the antenna with attached bumps. Then the steps of the block diagram shown in Fig. 25 were employed to generate the surface profiles. After the far-field amplitude and phase were determined, they were interpolated to provide the far-field data on rectangularly distributed u-v data points.

Similar steps were followed for both the original and bumped reflectors and the resulting profiles are shown in Figs. 28a-d for the original antenna. These figures are black-and-white presentations of 16-color computer displays and also contour plots. Due to the fact that the original reflector was not a perfect parabolic reflector, its reconstructed surface profile demonstrated many distortions. Similar results were also constructed for the antenna with attached bumps as shown in Figs. 29a-d.

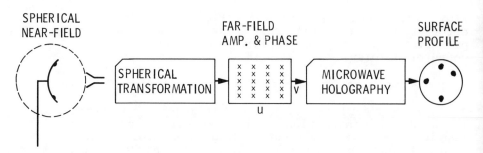

FIG. 25. Steps in performing microwave holographic diagnosis using spherical near-field measurements.

FIG. 26a. Original antenna in the spherical near field chamber.

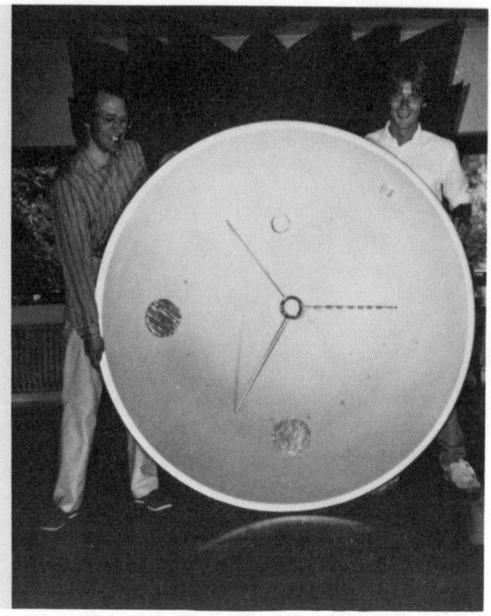

FIG. 26b. Antenna with four attached bumps of different sizes, locations and thicknesses.

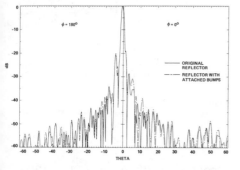

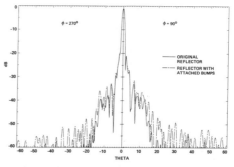

FIG. 27a. Far-field patterns constructed from spherical near-field measurements. $\phi = 0$ and 180 deg. planes.

FIG. 27b. Same as Fig. 27a except for $\phi = 90$ and 270 deg. planes.

However, in order to demonstrate how successfully the bumps were recovered, the results of original and bumped reflectors were used to obtain the final result and to remove the contamination due to the reflector's original imperfection. To this end, the generated holographic surface profiles of the two measurements (i.e., Figs. 28b and 29b) were subtracted to essentially remove the effects due to the original surface imperfection, struts diffraction, feed misalignments,

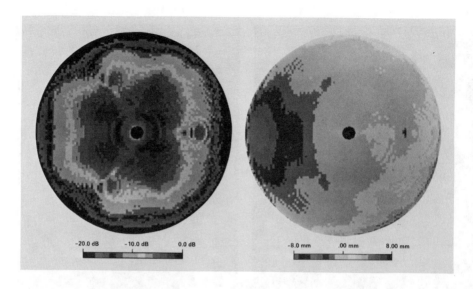

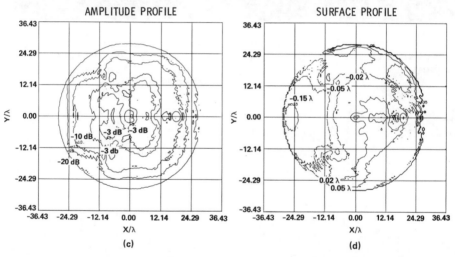

FIG. 28. Results of holographic reconstruction at 11.3 GHz on the original reflector. (a) Amplitude profile. (b) Surface distortion profile. (c) Contour plot of the amplitude profile. (d) Contour plot of surface distortion profile.

nonspherical phase pattern of the feed, and other factors. Results are shown in Figs. 30a-b using both the black-and-white and contour plot presentations, which clearly demonstrate how well the four attached bumps are recovered. To generate these results a matrix of 127 x 127 complex far-field data with u-v spacings of 0.0136 were used.

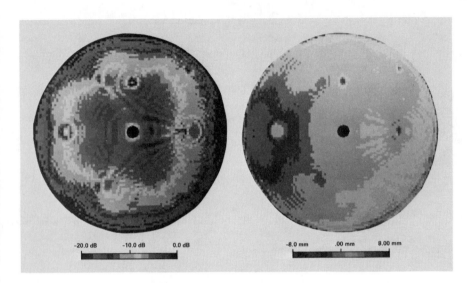

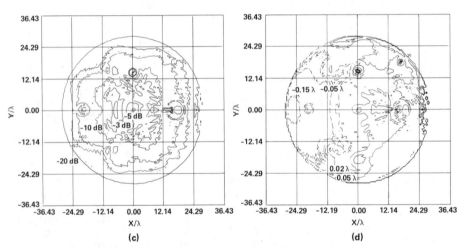

FIG. 29. Results of holographic reconstruction at 11.3 GHz on the reflector with attached bumps. (a) Amplitude profile. (b) Surface distortion profile. (c) Contour plot of the amplitude profile. (d) Contour plot of surface distortion profile.

9. CONCLUSIONS

In this chapter, an attempt has been made to familiarize the reader with the application of the microwave holographic metrology, which is currently being used at many organizations worldwide. The technique is a very powerful diagnosis tool for identifying the reflector surface distortions and defective radiating elements of an array. Although important measurement improvements, algorithm developments, error evaluations, and

44

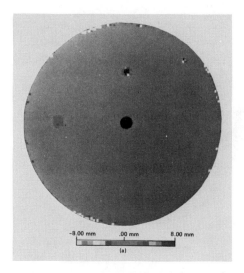

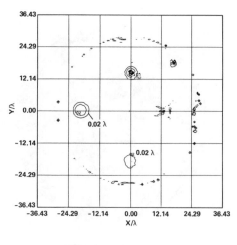

(b)

FIG. 30 **Reconstructed** surface distortion profile after subtraction of the surface profiles of the original and bumped reflectors. (a) Black-and-white representation of 16-color computer display. (b) contour plot representation.

display methodologies have been reported in recent years, there are still several areas remaining for further improvements by reducing the complexity of the overall implementation of the technique and by assessing the ultimate achievable accuracies.

APPENDIX A: RELATIONSHIP BETWEEN AZ-EL AND U-V SPACES

In performing far-field holographic measurements, one must move the test antenna with respect to the illuminating source in order to measure the amplitude and phase data. Where the data is taken depends on the relative motion of the antenna with respect to the source. As discussed in previous sections, it is advantageous to measure the data in a regularized u-v space which would allow a direct application of FFT.

The antenna motion is typically controlled by the azimuthal and elevation (or hour angle and declination) angles depending on the antenna mount mechanism. Most of the mounts used for large antennas are elevation over azimuth (or declination over hour angle) as shown in Fig. 21. It is the purpose of this Appendix to demonstrate the relationship between AZ-El and u-v spaces for a generalized mount configuration and provide a procedure to obtain the regularized u-v space data distribution.

The geometry of coordinate systems which specifies the angular orientation of the reflector antenna with respect to the source is shown in Fig. 31. It is assumed that the motion of

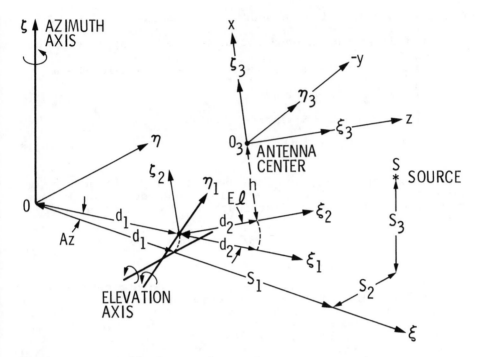

FIG. 31. Coordinate systems for a generalized antenna mount. Antenna motion is controlled by rotations about the azimuthal and elevation axes.

the antenna is controlled by azimuthal and elevation angles about their corresponding axes, as shown in the figure. Furthermore, for the sake of generality, it is assumed that these axes do not intersect and are separated by distance d_1 (this situation is important for the hour-angle and declination antenna mounts). The azimuthal and elevation based coordinates are designated by (ξ, η, ζ) and (ξ_1, η_1, ζ_1), respectively. The reflector coordinate is defined by (ξ_3, η_3, ζ_3) or (x, y, z), as shown in Fig. 31. It is noticed that the receiving point is positioned at the origin of the (x, y, z) system which is located at the height h from (ξ, η) plane when the azimuthal and elevation angles are zero.

The first objective is to determine the u-v coordinates of the source point with respect to the (x, y, z) antenna coordinates. this is achieved by successive coordinate transformations, as shown in Fig. 31. the final result may be expressed by

$$\begin{cases} u = \sin\theta\cos\phi = x/r \\ \\ v = \sin\theta\sin\phi = y/r \end{cases} \qquad \text{(A-1)}$$

where

$$r = \sqrt{x^2 + y^2 + z^2}$$

46

One can then show that a source with coordinates (S_1, S_2, S_3) in the mount coordinate (ξ, η, ζ) will have the following coordinates in the antenna coordinate (x, y, z) namely,

$$
\begin{cases}
x = S_3 \cos El - (S_1 \cos Az + S_2 \sin Az - d_1 \cos Az) \sin El - h \\
y = -S_2 \cos Az + S_1 \sin Az \\
z = (S_1 \cos Az + S_2 \sin Az - d_1) \cos El + S_3 \sin El - d_2
\end{cases} \tag{A-2}
$$

where (Az, El) are the azimuthal and elevation angles, respectively.

From (A-2) and (A-1), it can be seen that, in general, for uniformly spaced (Az, El) angles, one does not obtain a uniformly spaced u-v distribution. For example, for the parameters

$$d_1/\lambda = 100, \quad d_2/\lambda = 300, \quad h/\lambda = 800$$
$$S_1/\lambda = 10^6, \quad S_2/\lambda = 0, \quad S_3/\lambda = 10^6 \tag{A-3}$$

Fig. 32 demonstrates that the u-v space distribution is non-rectangular. The amount of deviation depends on the angular range, source location, and other geometrical parameters. In most practical cases, S_1 and S_3 are much larger than their values in (A-3); hence, less distortion may be observed. In order to demonstrate the effects of the non-rectangular u-v space distribution in performing FFT, the -25 db taper case of Fig. 13 is considered with the non-rectangular u-v distribution of Fig. 32. The resulting profile reconstruction based on this non-rectangular distribution is shown in Fig. 33 which clearly demonstrates how wrong it is.

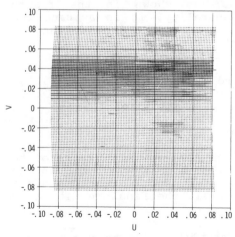

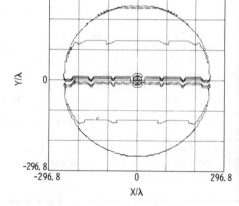

FIG. 32. Non-rectangular (u,v) distributions resulting from rectangular (Az, El) distributions.

FIG. 33. An erroneous surface profile reconstruction using non-rectangular (u, v) distributions with FFT.

The next important problem is to determine the appropriate (Az, El) which gives a uniform (u, v) distribution. This requires that the following coupled transcendental equations be solved, viz.,

$$
\begin{cases}
u = \dfrac{S_3 \cos El - (S_1 \cos Az + S_2 \sin Az - d_1)\sin El - h}{r} \\[2mm]
v = \dfrac{-S_2 \cos Az + S_1 \sin Az}{r}
\end{cases}
\tag{A-4}
$$

In general, it is very difficult to solve the above coupled equations for a specified value of (u, v) due to the appearance of r in the denominator. However, if one makes the assumption that r is large and does not change significantly, one can then express (A-4) as

$$
\begin{cases}
u = A \cos El + B \sin El - h/r \\[2mm]
v = C \cos Az + D \sin Az
\end{cases}
\tag{A-5}
$$

where

$$
A = S_3/r
$$

$$
B = -(S_1 \cos Az + S_2 \sin Az - d_1)/r
$$

$$
C = -S_2/r
$$

$$
D = S_1/r
$$

$$\tag{A-6}$$

It is noticed that B is a function of the Az angle. Fortunately, in (A-5), v is only a function of Az angle, which allows determination of Az once v is specified. The final result may be expressed as

$$
Az = \cos^{-1}(v/\sqrt{C^2+D^2}) + \tan^{-1}(D/C)
$$

$$
El = \cos^{-1}[(u+u_0)/\sqrt{A^2+B^2}] + \tan^{-1}(B/A)
$$

$$\tag{A-7}$$

where

$$
u_0 = h/r, \quad r = \sqrt{x^2 + y^2 + z^2}
\tag{A-8}
$$

Since, in general, r varies for different values of Az and El angles, one can employ an iterative scheme to improve the value of r from (A-8) for different values of Az and El determined from (A-7). Typically, the number of iterations is very limited for sources far away from the antenna. Some representative numerical data are discussed next.

For the parameters specified in (A-3), the values of Az and El are determined to result in a uniform rectangular grid distribution in (u, v) space. These results are plotted in Fig. 34, which clearly exhibits the non-rectangular distribution in (Az, El) space. Their corresponding (u, v) are then determined to show how well the desired rectangular

48

distribution is obtained. These results are depicted in Fig. 35. A computer program has been developed to allow the generation of the required Az and El which, with some modifications, could be used to derive the antenna controller mechanism.

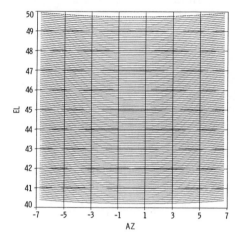

FIG. 34. Required non-rectangular (Az, El) distributions resulting in rectangular (u, v) distributions.

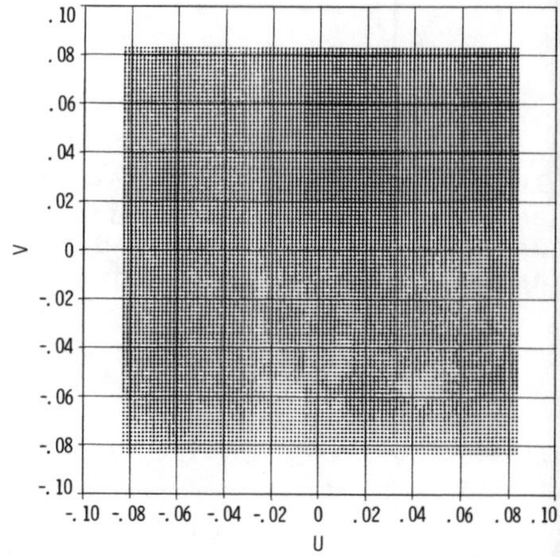

FIG. 35. Resultant rectangular (u, v) distributions based on the non-rectangular (Az, El) distributions.

ACKNOWLEDGMENTS

The research described in this chapter were carried out at the Jet Propulsion Laboratory, California Institute of Technology, under contract with the National Aeronautics and Space Administration. The author would like to express thanks for contributions and useful discussions from D. Rochblatt, J. Mumford, M. Gatti, B. Seidel, and D. Bathker of JPL and J. Lemanczyk and K. Brown of the Technical University of Denmark.

REFERENCES

1. Findlay J. W., "Filled-aperture antennas for radio astronomy," Annu. Rev. Astron. Astrophys., vol. 9, pp.271-192, 1971.

2. Merrick W. D., F. L. Lansing, F. W. Stoller and V. B. Lobb, "Precision photogrammetric measurements of NASA-JPL 34-m antenna reflectors," in Proc. of MONTECH'86, pp. 286-290, Montreal, Canada, 1986.

3. Bennet J. C., A. P. Anderson, P. A. McInnes, and A. J. T. Whitaker, "Microwave holographic metrology of large reflector antennas," IEEE Trans. Antennas Propagat., AP-24, pp. 295-303, 1976.

4. Scott P. F. and M. Ryle, "A rapid method for measuring the figure of a radio telescope reflector," <u>Royal Astronom. Soc. Monthly Notices</u>, vol. 178, pp. 539-545, 1977.

5. Godwin M. P., A. J. T. Whitaker, and A. P. Anderson, "Microwave diagnostics of the Chilbolton 25 m antenna using OTS satellite," in <u>Proc. Inst. Elec. Eng. Int. Conf.</u>, York, England, 1981, pp.232-236.

6. Mayer C. E., J. H. Davis, W. L. Peters, and W. J. Vogel, "A holographic surface measurement of the Texas 4.9-meter antenna at 86 GHz," <u>IEEE Trans. Instru. Meas.</u>, vol. IM-32, pp. 102-109, 1983.

7. Rahmat-Samii Y., "Surface diagnosis of large reflector antennas using microwave holographic metrology -- An iterative approach," <u>Radio Science</u>, vol. 13, pp. 1205-1217, 1984.

8. Rahmat-Samii Y., "Microwave holography of large reflector antennas -- Simulation algorithms," <u>IEEE Trans. Antennas Propagat.</u>, vol. AP-33, pp. 1194-1203, 1985 (see minor corrections in vol. AP-34, pp. 853, 1986).

9. Rochblatt D. J., Y. Rahmat-Samii, and J. H. Mumford, "DSN microwave antenna holography," JPL/TDA Progress Report 42-87, pp.92-97, 1986.

10. Godwin M. P., E. P. Schoessow, and B. H. Grahl, "Improvement of the Effelsberg 100 meter telescope based on holographic reflector surface measurement," <u>Astron. Astrophys.</u>, vol 167, pp. 390-394, 1986.

11. Rahmat-Samii Y. and V. Galindo-Israel, "Shaped reflector antenna analysis using the Jacobi-Bessel series, " <u>IEEE Trans. Antennas Propagat.</u>, vol. 28, pp. 425-435, 1980.

12. Ludwig A. C., "The definition of cross polarization," <u>IEEE Trans. Antennas Propagat.</u>, vol. 21, pp. 116-119, 1973.

13. Rahmat-Samii Y. and R. Cheung, "Nonuniform sampling techniques for antenna applications," <u>IEEE Trans. Antennas Propagat.</u>, vol. 35, pp. 268-279, 1987.

14. Papoulis A., "A new algorithm in spectral analysis and band-limited extrapolation, " <u>IEEE Trans. Circuits Syst.</u>, vol. 22, pp. 735-742, 1975.

15. Anderson A. P. and S. Sali, "New possibilities for phaseless microwave diagnostics," <u>IEE Proc.</u>, vol. 132, pp. 291-298, 1985.

16. Rahmat-Samii Y., "An efficient computational method for characterizing the effects of random surface errors on the average power pattern of reflectors," <u>IEEE Trans. Antennas Propagat.</u>, vol. 31, pp. 92-98, 1983.

17. Rahmat-Samii Y. and V. Galindo-Israel, "Scan performance of dual offset reflector antennas for satellite communications," Radio Science, vol. 16, pp. 1093-1099, 1981.

18. Godwin M. P., E. P. Schoessow and P. J. Richards, "Final report on holographic tests at S-band and K-band on the DSS-63 64-meter antenna," Prepared for the Jet Propulsion Laboratory under Contract No. 956984, 1986.

19. Gatti M. and Y. Rahmat-Samii, "FFT applications to plane-polar near-field antenna measurements," Int. IEEE/AP-S Symp., Blacksburg, VA, June 15-19, 1987.

20. Rahmat-Samii Y. and J. H. Lemanczyk, "Microwave holographic diagnosis of antennas using spherical near-field measurements," Int. IEEE/AP-S Symp., Blacksburg, VA, June 15-19, 1987.

LARGE ANTENNA EXPERIMENTS ABOARD THE SPACE SHUTTLE -- APPLICATION OF NONUNIFORM SAMPLING TECHNIQUES

Y. RAHMAT-SAMII

Jet Propulsion Laboratory
California Institute of Technology
Pasadena, CA 91109

1. INTRODUCTION

It is very likely that antennas in the range of 20 meters or larger will be an integral part of future satellite communication and scientific spacecraft payloads. For example, Fig. 1 depicts the conceptual evolution of the Land Mobile Satellite System which is anticipated to evolve from utilizing approximately 6-9 meter reflectors to 55 meter reflectors in the era spanning the late 1980's to early 2000. In order to commercially use these large, low sidelobe and multiple-beam antennas, a high level of confidence must be established as to their performance in the 0-g and space environment. Certain ground (1-g) testing can be performed to validate the workability of different segments of such large structures; however, it will be a formidable task to characterize the performance of the entire structure on the ground. For this

FIG. 1. Evolution of the proposed Land Mobile Satellite System (LMSS).

51

B. de Neumann (ed.), Electromagnetic Modelling and Measurements for Analysis and Synthesis Problems, 51–60.
© 1991 Kluwer Academic Publishers. Printed in the Netherlands.

reason, a conceptual study has been initiated with the intention to describe an experiment aboard the Space Shuttle to demonstrate the reliability deployment of the antenna structure, to measure thermal and dynamic structural characteristics, and to verify performance specification under all expected conditions. In particular, special consideration is being given to the RF far-field pattern measurements which should provide the ultimate characterization for the antenna performance (Fig. 2).

Several potential scenarios have been considered and the relative merit of each of them are shown in qualitative manner in Fig. 3. Among all these possibilities the application of the scenario shown in the last row appeared most feasible. The RF experiment is anticipated to be performed on a 20-meter offset reflector at L-band using the Remote Mini-Flyer, a NASA-developed reusable and retrievable spacecraft (a modified Spartan), as the carrier for an RF beacon. This beacon is used to illuminate the antenna in a similar fashion as one does in the ground-based far-field ranges using transmit illuminators. An artist's rendition of this spaced-based experiment is depicted in Fig. 4.

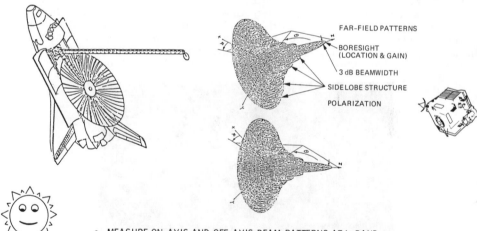

FAR-FIELD PATTERNS
BORESIGHT (LOCATION & GAIN)
3 dB BEAMWIDTH
SIDELOBE STRUCTURE
POLARIZATION

- MEASURE ON-AXIS AND OFF-AXIS BEAM PATTERNS AT L-BAND
  - MEASURE UNDER VARIOUS THERMAL CONDITIONS
  - MEASURE AFTER ON-ORBIT SURFACE RECONFIGURATION
- DEMONSTRATE IN-SPACE RF MEASUREMENT TECHNOLOGY
- CORRELATE MEASURED RF PERFORMANCE WITH MEASURED SURFACE AND FEED ALIGNMENT
- VERIFY AND UPDATE MATHEMATICAL AND COMPUTER MODELS OF RF PERFORMANCE ANALYSIS AND PREDICTION

FIG. 2. Spaced-based RF experiment objectives aboard the Space Shuttle.

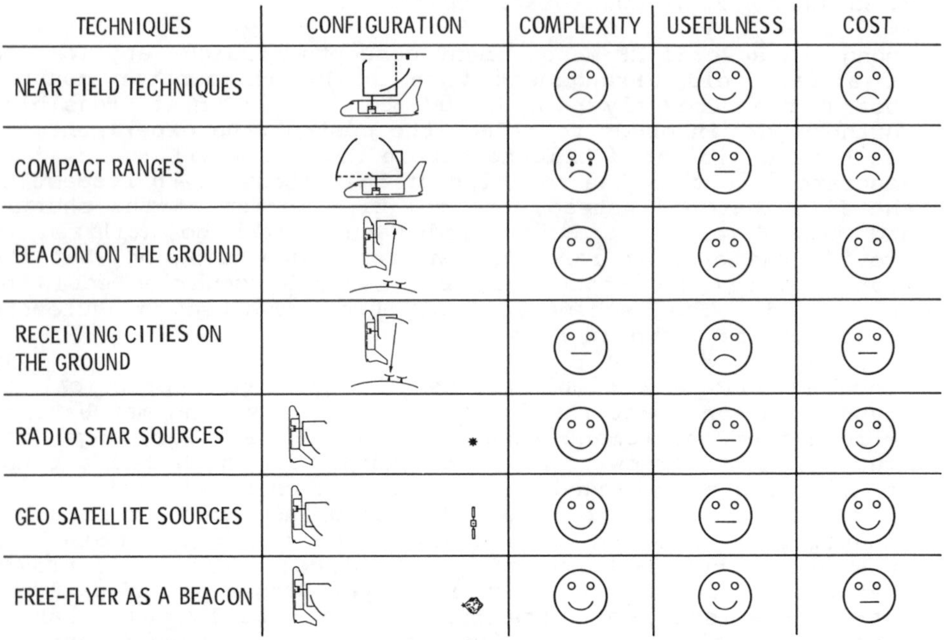

TECHNIQUES	CONFIGURATION	COMPLEXITY	USEFULNESS	COST
NEAR FIELD TECHNIQUES		:-(	:-)	:-(
COMPACT RANGES		:-(	:-|	:-(
BEACON ON THE GROUND		:-|	:-(	:-|
RECEIVING CITIES ON THE GROUND		:-|	:-(	:-|
RADIO STAR SOURCES		:-)	:-|	:-)
GEO SATELLITE SOURCES		:-)	:-|	:-)
FREE-FLYER AS A BEACON		:-)	:-)	:-|

FIG. 3. A qualitative comparison of different measurement techniques aboard the Space Shuttle.

FIG. 4. An artist's rendition of the proposed large antenna Shuttle experiment.

54

2. RF MEASUREMENT CONSIDERATIONS

Among the several RF measurement concepts studied, application of a far-field arrangement with an RF illuminator mounted aboard the free-flyer appeared to be the most feasible. Furthermore, in order to reduce the cost of the experiment, it has been anticipated that no gimble mechanism will be used to accurately control the position of the antenna with respect to the illuminator. Instead, the relative motion of the shuttle and free-flier in a controlled manner will be utilized to provide the angular range of interest. Depending on what the exact covered u-v space will be, several scenarios could be considered. The schematics of two possible measurement scenarios are shown in Figs. 5 and 6.

Since without any gimble system it will be impractical to measure antenna patterns in specified ϕ cuts, one may have to perform the measurement in a specified u-v angular range (for example, ±5 degrees) in a time period in which the antenna structure is not changed appreciably for an RF viewpoint. Fig. 5 shows the possibility of measuring very dense but nonuniformly measured data points from which the needed ϕ-cuts or contour patterns can be constructed. If it is proven that the possibility of measuring a very dense set of data may not be realistic, an alternate scheme should be available. This alternate scheme is depicted in Fig. 6, which assumes that the measured (amplitude and phase) are obtained at

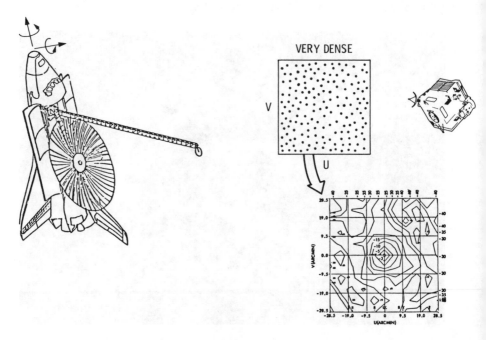

FIG. 5. Schematic of a nonuniform and very densely measured sampled points.

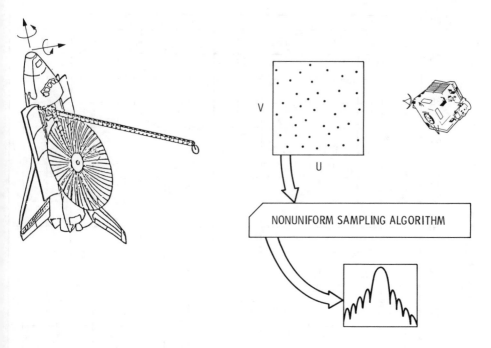

FIG. 6. Schematic of a nonuniform and relatively sparsely measured sampled points.

relatively sparse and nonuniformly distributed u-v points. The question is, then, whether or not one can construct the ϕ-cuts from a set of nonuniformly distributed measured data points?

Recently, Rahmat-Samii and Cheung [1] have demonstrated that a two-dimensional nonuniform sampling technique which utilizes irregularly spaced samples (amplitude and phase) can be used to generate the far-field patterns. The mathematical developments of this two-dimensional nonuniform sampling technique have been detailed in [1]. Additionally, a powerful simulation algorithm has then been developed to test the applicability of this sampling technique for a variety of reflector measurement configurations [1]. For example, Figs. 7 and 8 show the simulated nonuniform sample points and the reconstructed far-field patterns in specified ϕ-cuts for a 20 meter offset reflector antenna with a defocused feed operating at L-band [1]. In these figures the solid curves are reconstructed co-polar and cross-polar patterns using the nonuniform sampling technique. It is noted that even though no sample points are captured in these cuts, the reconstructed patterns agree well with the ideal patterns in the angular range where the nonuniform sample points have been generated, i.e., ±3.2 degrees. Many tolerance studies have also been performed to demonstrate the required measurement accuracies in applying the nonuniform sampling technique [1].

56

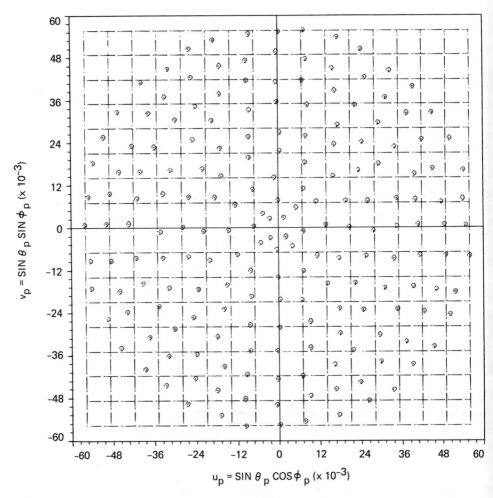

FIG. 7. One hundred ninety two nonuniform sampled point distribution in (u,v) coordinates which covers an angular region of $\theta = \pm 3.2$ degrees.

To validate the accuracy of nonuniform sampling technique for antenna pattern construction, several measurements have been performed as reported in [2]. In one of the measurements JPL's 1200-ft far-field range was used, where a 1.47-m circularly polarized Viking reflector antenna [3] (Fig. 9) was measured at X-band (8.415 GHz) using a corrugated horn as the illuminating antenna. The far-field amplitude and phase were measured in the directions shown in Fig. 10 which consisted of 585 nonuniformly distributed sampled points in (u,v) coordinates.

The total system errors for amplitude measurement have been estimated to be less than 0.2 dB at -40 dB level relative to

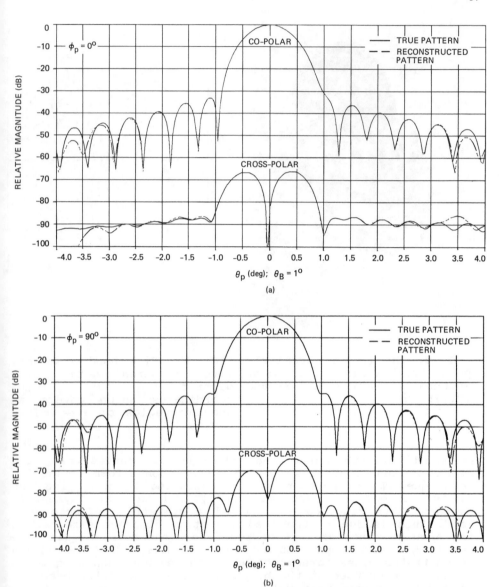

FIG. 8. Far-field patterns of a defocused (tilted beam) offset reflector reconstructed from 192 sampled points using nonuniform sampling technique. (a) ϕ = 0 degree cut, (b) ϕ = 90 degree cut.

the boresight power level and the phase measurement error has been within ±5 degrees. The pointing accuracy has been estimated to be better than 0.1 degrees. The co-polar far-field patterns for ϕ = 0 and 90 degrees are depicted in Figs. 11(a) and 11(b), respectively. The solid curves are the

58

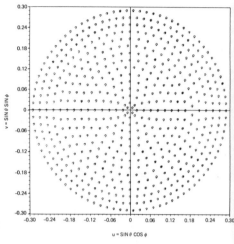

FIG. 9. The 1.47-m circularly polarized Viking reflector antenna operating at X-band (8.415 GHz).

FIG. 10. Five hundred eighty five measured data point distribution in (u,v) coordinates which covers an angular region of $\theta = \pm 16.8$ degrees.

standard azimuth cuts and the dashed curves are the reconstructed patterns using the nonuniform sampling technique. These patterns are constructed by utilizing the window concept as discussed in [1,2]. Note that asymmetric patterns have resulted even though a symmetric reflector was used. this is due to the feed and strut blockage effects. In the angular range of 16.8 degrees where the measured data are available , the comparison between the solid and dashed curves demonstrates close agreement. Note that even though no sampled point has been used at the boresight, the peak of the beam has been reconstructed very well.

ACKNOWLEDGEMENTS

The research described in this chapter were carried out at the Jet Propulsion Laboratory, California Institute of Technology, under contract with the National Aeronautics and Space Adminstration. The Author would like to thank R. Cheung for his assistant during the development and testing of the two-dimensional nonuniform sampling algorithm.

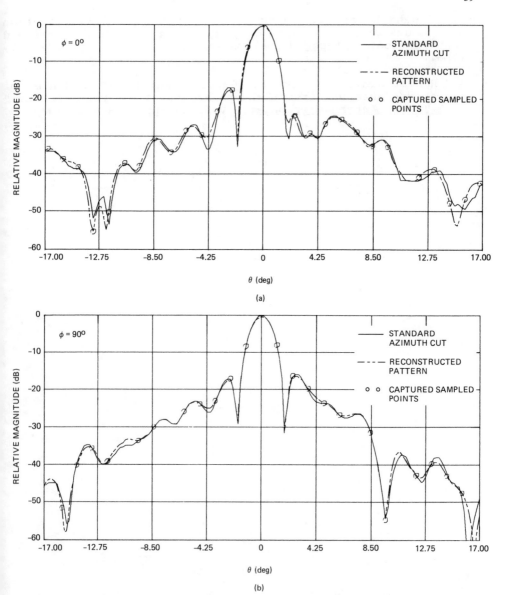

FIG. 11. Reconstructed co-polar far-field patterns of the Viking reflector at X-band using nonuniformly distributed measured data. (a) θ = 0 degree cut, (b) θ = 90 degree cut.

REFERENCES

1.   Rahmat-Samii Y. and R. Chueng, "Nonuniform sampling techniques for antenna applications," IEEE Trans. Antennas Propagat., vol. 35, pp. 268-279, 1987.

2.   Chueng R. and Y. Rahmat-Samii, "Experimental verification of   nonuniform   sampling   technique   for   antenna   far-field construction," <u>Electromagnetics</u>, vol. 6, no. 4, pp.277-300, 1986.

3.   Rahmat-Samii   Y.   and   M.   Gatti,   "Far-field   patterns   of spaceborne antennas from plane-polar near-field measurements," <u>IEEE Trans. Antennas Propagat.</u>, vol. 33, pp. 638-648, 1985.

OPTIMAL ANGULAR COORDINATE MEASUREMENTS BY OPTICAL LOCATION

Lev B. Levitin

Boston University, USA

ABSTRACT. The measurement of the angular coordinate of a point object by use of a scanning optical location system is considered, with quantum properties of light taken into account. Both the signal and the noise photons flows are assumed to be independent Poisson random processes with angle-dependent expected values. The maximum likelihood estimate of the angular coordinate is found, and the asymptotic variance of the estimate is calculated. The exact distribution of the estimate and other explicit results are obtained for the case of Gaussian angular distribution. A numerical example is given.

KEY WORDS: optical location; quantum effects in laser location; angular coordinate measurements; estimation in Poisson process; quantum detection and estimation.

1. INTRODUCTION

The fundamental limitation of the accuracy of optical measurements are imposed by the dualistic wave-corpuscular nature of light. In contrast with particles of non-zero rest mass, the light quanta - photons - behave in the classical limit case as a continuous wave field, while the corpuscular properties of light are of essentially quantum nature. Owing to this fact, the effects of the wave properties of electromagnetic radiation were investigated during more than a hundred years (the Abbe theory of diffraction in optical instruments, etc.), while the quantum limitations attracted the attention of researchers comparatively recently. (See the excellent book [1] by C.W. Helstrom for a comprehensive review and bibliography up to 1976.) In particular, in spite of a great number of works devoted to various aspects of the optical location, the problem of the measurement of the angular position of an object under the quantum regime was treated in a few papers only (e.g. [2]), where a method based on the use of photon correlations in two different points of the space was studied. Measurements by use of a scanning locator beam, which is of practical importance, apparently were not yet considered. This method is discussed in the present paper for the case when the radiation registered by the locator should be described as a discrete folow of photons.

It is known that from the standpoint of classical physics the angular position of an object of a known form (e.g., a point object) can be determined with arbitrary high accuracy. But the quantum properties of light change the situation in principle.

2. FORMULATION OF THE PROBLEM

Consider the procedure of measuring the angular coordinate of a point object by a

B. de Neumann (ed.), Electromagnetic Modelling and Measurements for Analysis and Synthesis Problems, 61–65.

scanning optical location system. The words "point object" imply, as usual, that the details of the object cannot be resolved, i.e., that the angle of view at which the object is seen does not exceed the diffraction angle of the receiver. The object is illuminated by the locator beam and scatters a portion of the light irradiated by the locator. A portion of this scattered light is registered in its turn by the receiver of the location system.

Suppose that the angular distribution (directivity diagram) of the intensity of the light irradiated by the locator is given by $f(\beta)$, where β is the angle referred to the normal to the locator antenna plane $(0 \leq \beta < 2\pi; f(0) = 1)$. Let the angular distribution of the additive background noise radiation reaching the reciever of the locator be given by $g(\alpha)$, where α is the coordinate angle of the locator $(0 \leq \alpha < 2\pi)$, and $g(\alpha_0) = 1$, where α_0 is the angular coordinate of the object. Denote the mean number of signal photons registered by the receiver in unit time if the locator is directed exactly towards the object by N_s, and the mean number of noise photons under the same condition by N_n. Obviously, the amount of light scattered by the object and reaching the receiver is proportional to the intensity of the light irradiated by the locator in the direction of the object. Hence, the mean number of photons registered by the receiver in unit time is given by

$$N(\alpha) = N_s f(\alpha - \alpha_0) + N_n g(\alpha) \tag{1}$$

We assume that the signal and the noise photon flows are independent Poisson random processes (as was assumed for other purposes in [3,4]). Then all the photons registered by the receiver form a Poisson random process with angle dependent expected value $N(\alpha)$. We assume also that the numbers of photons registered in different angle positions of the locator are independent random variables. These assumptions are valid for the coherent and wide-band thermal radiation.

Our problem is to estimate the angular coordinate α_0 of the object on the base of registerd photons and to evaluate the accuracy of such a measurements.

3. THE MAXIMUM LIKELIHOOD ESTIMATE

According to the Dugue theorem ([5], p. 500) under certain general conditions, the optimal estimate can be found by Fisher's maximum likelihood method. The estimate is asymptotically unbiased, asymptotically normal, and asymptotically efficient, with asymptotic variance $\sigma_{as}^2(\hat{\alpha}_0)$ given by

$$\sigma_{as}^2(\hat{\alpha}_0) = \frac{1}{I} \tag{2}$$

where $\hat{\alpha}_0$ is the estimate of the unknown parameter α_0 and I is Fisher's information.

In order to find the maximum likelihood estimate it is convenient to suppose in the beginning that the direction of the locator changes in a discrete way, so that the locator remains during a time interval τ in a position α_i and then jumps to the position $\alpha_{i+1} = \alpha_i + \Delta\alpha(i = 1, 2, ...)$. Then the mean number of photons registered in the position α_i is

$$\bar{n}_i = [N_s f(\alpha_i - \alpha_o) + N_n g(\alpha_i)]\tau \tag{3}$$

The likelihood function can be written in the form:

$$p(n_1, n_2, ...n_i.../\alpha_0) = \prod \frac{\bar{n}_i^{n_i} e^{-\bar{n}_i}}{n_i!} \tag{4}$$

and Fisher's information is given by

$$I = \sum_{n_1=0}^{\infty} \sum_{n_2=0}^{\infty} ...p(n_1, n_2, .../\alpha_0) \left[\frac{\partial \ln p(n_1, n_2, .../\alpha_0)}{\partial \alpha_0} \right]^2 \tag{5}$$

Accomplishing calculations, we obtain:

$$I = \sum_i \frac{N_s^2 \tau (\frac{df}{d\alpha})_{\alpha=\alpha_i-\alpha_0}^2}{N_s f(\alpha_i - \alpha_o) + N_n g(\alpha_i)} \tag{6}$$

In order to obtain a result for a continuously moving locator, we shall pass to the limit $\tau \to 0, \Delta\alpha \to 0$, keeping $\frac{\Delta\alpha}{\tau}$, where ω is the angular velocity of the locator rotation while scanning. It leads to the following expression for the asymptotic variance of the estimate:

$$\sigma_{as}^2(\hat{\alpha}_0) = \frac{\omega}{N_s^2} \left[\int_0^{2\pi} \frac{(\frac{df(\alpha)}{d\alpha})^2 d\alpha}{N_s f(\alpha) + N_n g(\alpha + \alpha_0)} \right]^{-1} \tag{7}$$

The maximum likelihood estimate $\hat{\alpha}_0$ is to be found from the equation

$$\frac{\partial \ln p(n_1, n_2, ...n_i, .../\hat{\alpha}_0)}{\partial \hat{\alpha}_0} = 0 \tag{8}$$

which gives finally

$$\sum_k \frac{N_s (\frac{df}{d\alpha})_{\alpha=\alpha_k-\hat{\alpha}_0}}{N_s f(\alpha_k - \hat{\alpha}_0) + N_n g(\hat{\alpha}_k)} = 0 \tag{9}$$

where α_k is the angle at which the k-th photon has been registered.

If there is no additive noise, the equation for the maximum likelihood estimate becomes simpler:

$$\sum_k \left(\frac{d \ln f(\alpha)}{d\alpha} \right)_{\alpha=\alpha_k-\hat{\alpha}_0} = 0 \tag{10}$$

4. THE GAUSSIAN ANGULAR DISTRIBUTION

The most understandable and simplest explicit results are obtained in a case when there is no additive noise and the angular distribution of the light irradiated by the locator is Gaussian:

$$f(\alpha) = exp\left(-\frac{a^2}{2\pi^2\lambda^2}\alpha^2 \right) \tag{11}$$

Here λ is the wavelength of the light and a is a parameter characterizing the effective diameter of the radiator of the location system. (This assumption implies, of course, that the directivity diagram of the locator is sufficiently narrow, so that the intergration with respect to the angle can be formally performed from $-\infty$ to ∞).

The equation (10) proves to be linear in this case, and the optimal estimate $\hat{\alpha}_0$ is equal to the arithmetical average of all the angles where photons have been registered.

$$\hat{\alpha}_0 = \frac{1}{N} \sum_{k=1}^{N} \alpha_k \tag{12}$$

Here N is the total number of registered photons.

The asymptotoic variance is given by

$$\sigma_{as}^2(\hat{\alpha}_0) = \frac{\omega\lambda\sqrt{2\pi}}{2aN_s} \tag{13}$$

The variance is proportional to the angular velocity of scanning ω to the diffraction angle of the radiator $\frac{\lambda}{a}$ and inversely proportional to the intensity of radiation, which is of transparent physical meaning.

It is interesting to compare the asymptotic result (13) with the exact value of the variance. This is possible in this case, due to the simple form of the estimate (12). After rather tedious calculations we obtain explicitly the exact probability density function of the estimate

$$p(Z/\alpha_0) = \frac{a}{\pi\sqrt{2\pi}\lambda} \left[exp\left(-\frac{\pi\sqrt{2\pi}N_s\lambda}{a\omega} \right) \right] \sum_{k=0}^{\infty} \left[\frac{\pi\sqrt{2\pi}N_s\lambda}{a\omega} \right] \cdot \frac{\sqrt{k}}{k!} exp\left(-\frac{a}{2\pi\lambda}kZ^2 \right) \tag{14}$$

(Here Z is the value taken on by the random variable $\hat{\alpha}_0 - \alpha_0$).

The exact variance is calculated by use of (14):

$$\sigma^2(\hat{\alpha}_0) = \frac{\omega\lambda\sqrt{2\pi}}{2aN_s^2} + \frac{\omega^2}{2\pi N_s^2} + O\left(\frac{\omega^3 a}{\lambda N_s^3} \right) \tag{15}$$

The results shows that the asymptotic expression (13) is valid under the condition

$$\frac{\lambda N_s}{\omega a} \gg 1 \tag{16}$$

The meaning of the criterion (16) is very simple: the expected value of hte number of photons which have been registerd during the time of observation must be greater than one.

Consider now the effect of the additive noise, assuming that the noise is isotropic: $g(\alpha) = 1$.

Then the asymptotic variance of the estimate is a function of two parameters: the signal intensity N_s and the noise-to-signal ratio $r = \frac{N_n}{N_s}$.

$$\sigma_{as}^2(\alpha_0) = \frac{\pi\omega\lambda}{aN_s} \left[\int_{-\infty}^{\infty} \frac{x^2 exp(-x^2)dx}{exp(\frac{-x^2}{2}) + r} \right]^{-1} \tag{17}$$

For $r \gg 1$ (large noise) it yields:

$$\sigma_{as}^2(\alpha_0) \simeq \sigma_{as}^2(\alpha_0)_{r=0} 2\sqrt{2}r \tag{18}$$

the variance is proportional to the noise-to-signal ratio. For $r = 1$ the numerical evaluation shows that the variance is approximately four times greater than in the absence of

noise.

5. A NUMERICAL EXAMPLE

Consider a numerical example for the case, when formula (13) is valid. Let us take the following values of the parameters:

$\lambda = 7 \cdot 10^{-7} m$ —the wavelength of light

$a = 10^{-1} m$ —the effective diameter of the light radiator

$P = 1W$ —the power irradiated by the locator

$l = 10^4 m$ —the distance between the object and the locator

$\omega = 1 rad/sec$ —the quantum efficiency of the receiver

Suppose that the object crosses the whole light beam, and the reflectivity of the object surface is 1. Then the quantum limit of accuracy given by the standard deviation is $\sigma(\alpha_0) \simeq 1.8 \cdot 10^{-5} rad$, which is several times greater than the diffraction limit of resolution $\frac{\lambda}{a} \simeq 7 \cdot 10^{-6} rad$. Thus the quantum effects can be essential for some practical situations.

REFERENCES

1. C.W. Helstrom, "Quantum Detection and Estimation Theory", Academic Press, 1976.

2. V.V. Paddubny, B.E. Trivozhenko, "On the Potential Accuracy of the Measurements of the Angular Position of a Photon Source", Proc. of the I National Conference on Information Transmission by Use of Laser Radiation, Kiev, 1968 (in Russian).

3. I. Bar-David, "Communication Under the Poisson Regime", IEEE TRans. Inform. Theory, IT-15, pp. 31-37, Jan. 1969.

4. P.N. Misra and H.W. Sorenson, "Parameter Estimation in Poisson Process", IEEE Trans. Inform. Theory, IT-21, pp. 87-90, Jan. 1975.

5. H. Cramer, Mathematical Methods of Statistics, Princeton University Press, Princeton, 1966.

GENERAL COVARIANCE OF THE ELECTROMAGNETIC FIELD APPLIED TO E.M. FIELD
TRANSFORMATIONS

S. CORNBLEET

DEPARTMENT OF PHYSICS, UNIVERSITY OF SURREY,
GUILDFORD, SURREY, GU2 5XH, ENGLAND.

1.   INTRODUCTION

The application of the theory of functions of a complex variable to
two dimensional electrostatic and hydrodynamic problems is well known.
This uses the conformal property of the mapping to transform the
boundaries and fields and the appropriate boundary conditions of a
canonical problem into a second system whose field solution is that
required.   In electrostatics the fields concerned are the lines of
electric and magnetic force and their orthogonality and boundary
conditions are retained by the conformal property of the transformation as
expressed by Cauchy-Riemann equations.   These equations also ensure that
the fields obey the two dimensional transverse Laplace equation.   This
article is concerned with the possible extension of this method to
problems concerning radiating fields and scattering surfaces.   This
therefore requires a transformation of a four-dimensional field and it is
proposed to investigate in this respect functions of a four-dimensional
hypercomplex variable.   Since the Laplacian of the two dimensional case
has to be extended to a D'Alembertian, the appropriate algebra required is
that which gives the proper signature $(-,+,+,+)$ for this operator.   The
four-dimensional complex variable theory of quaternions was established by
Hamilton in the mid nineteenth century [1].   However this is deficient in
the provision of the indefinite signature and has to be extended by the
inclusion of the complex imaginary $\sqrt{-1}$ which will be referred to
throughout as the Argand imaginary.   This then results in an eight
dimensional complex algebra related to the octonions of Hamilton or more
recently to the Clifford algebra.   Much has been written recently on the
applications of Clifford algebra to physical problems, but, for the
purposes of this introductory exposition it can be reduced to the more
familiar four dimensions provided at least one element remains Argand
imaginary.   In conformity with relativity theory, this imaginary will in
general denote the temporal component of four-component entities giving
the usual 1 + 3 dimensions of space-time algebra.

Functions of quaternion variables are studied in much the same manner
as functions of a complex variable by Scheffers [2] and Fuerter [3].   A
recent study by Imaeda [4] applies these to the formulation of classical
electrodynamics and the analysis presented here draws heavily on these
methods.   Many other hypercomplex algebras can be postulated and it has
become necessary at the outset to select the one giving results most
consistent with the equations of the electromagnetic field as they arise.
A comparison of these and further justification for the choice made and
the notation decided upon, would be, as their investigation proved, an
unduly tedious exercise and is therefore omitted.   Functions of this
hypercomplex variable are shown to generate space-time functions

B. de Neumann (ed.), Electromagnetic Modelling and Measurements for Analysis and Synthesis Problems, 67–88.
© 1991 Kluwer Academic Publishers. Printed in the Netherlands.

68

satisfying the wave equation and potential functions from which electromagnetic field can be derived. The Cauchy-Riemann conditions are extended and a "regularity" condition applied which proves to be a derivation of the Maxwell field equations. This generation of Maxwell equations by purely algebraic methods is itself of considerable interest.

The conformality of the electromagnetic field under space-time transformation has hitherto been completely concerned with the covariance of the field equations under a Lorentz transformation in the restricted theory of Relativity. In this transformation the signature is obtained by the definition of the metric tensor. Lorentz transformations as a quaternion similarity operation was performed at a very early time [5]. The Argand imaginary in that treatment, is subsumed into a separate symbol and its significance is not apparent thereafter. The algebra selected for use in this article conforms with this analysis and is the main reason for adapting and translating the previous work of Fuerter and Imaeda. It has been shown, however, that other transformations of the electromagnetic field, and specially of the field in its geometrical optics limit, exist [6] which would require a more general covariance. It is apparent that in order for this general situation to encompass the earlier transformations a similar hypercomplex method needs to be sought.

Such transformations, for identical reasons, must needs be conformal mappings. It so happens that both are inversions which are the only conformal mappings of the three dimensional space and the four-dimensional real space into themselves. Transformations of this nature were the major study of Bateman [7] who, specifying that the wave equation and eikonal equation had to remain covariant, found the result to be basically inversions. This author derived similar inversions to Bateman's by considering the differentiability condition for a function of a quaternion variable [8] and the extended set of Cauchy-Riemann equations that result. Bateman also used a specification of the electromagnetic field in terms of two spatial scalar variables. These can be extended to space-time scalars and will be termed Bateman potentials. They are closely related to the Clebsch potentials [9] and will be shown to be solutions of a wave equation. As such they can be derived by the process of Fuerter and Imaeda from hypercomplex variable theory.

It has to be admitted that the theory is, at this stage, not totally complete. The inclusion of boundary conditions is unduly elaborate and the move from free space conditions into a variable isotropic medium as required for one of the ray inversion theorems is yet to be completed. The evidence presented shows how it is anticipated these problems may be approached.

2. THE BIQUATERNION ALGEBRA

We establish an algebra over the reals with basic elements the quaternion elements of Hamilton 1, α, β, γ where

i) $\alpha^2 = \beta^2 = \gamma^2 = -1$

ii) The base elements anti-commute $\alpha\beta = -\beta\alpha$

iii) $\alpha\beta\gamma = -1$ and hence $\alpha\beta = \gamma$ etc. cyclically

iv) An inverse exists given by $\alpha\alpha^{-1} = \alpha^{-1}\alpha = 1$
A quaternion is a four element entity

(1)

$$q = q_0 + \alpha q_1 + \beta q_2 + \gamma q_3 \qquad (q_i \text{ real}) \tag{1}$$

This can be variously written as

$$q = \{q_0, q_1, q_2, q_3\} \qquad \text{or}$$

$$q = \{q_0 + \underset{\sim}{q}\} \qquad \text{where } q_0 \text{ is the "scalar" part of } q \text{ and } \underset{\sim}{q} \text{ the "vector"}$$

part.

We now include the Argand imaginary i as an element which commutes with the basis elements, that is $i\alpha = \alpha i$ etc. $i^2 = -1$.

A biquaternion is then the eight component entity

$$Q = A + iB \qquad \text{where A and B are quaternions of the form given by}$$
(1).

For the purposes of this article we mainly concern ourselves with the reduced biquaternion

$$Q = ia_0 + \alpha b_1 + \beta b_2 + \gamma b_3 \tag{2}$$

that is $= (0, b_1, b_2, b_3) + i(a_0, 0, 0, 0)$

The reasons for this choice out of all other possibilities will be apparent later.

The biquaternions are an extension of the field of real numbers and the complex numbers since the real $a_0 \equiv \{a_0, 0, 0, 0\} + i\{0, 0, 0, 0\}$ and

$$x + iy \equiv \{x, 0, 0, 0\} + i\{y, 0, 0, 0\}$$

The basis elements $i\alpha$, $i\beta$, $i\gamma$ are isomorphic to the Pauli spin matrices $-\sigma_3$, $-\sigma_2$, $-\sigma_1$.

v) Addition and multiplication are associative and distributive but multiplication is in general non-commutative i.e. $Q_1 Q_2 \neq Q_2 Q_1$

vi) We define an inner product

$$\underset{\sim}{a} \odot \underset{\sim}{b} = a_1 b_1 + a_2 b_2 + a_3 b_3 \tag{3}$$

and a cross product

$$\underset{\sim}{a} \otimes \underset{\sim}{b} = (a_2 b_3 - a_3 b_2)\alpha + (a_3 b_1 - a_1 b_3)\beta + (a_1 b_2 - a_2 b_1)\gamma \tag{4}$$

where $\underset{\sim}{a}$, $\underset{\sim}{b}$ are the vector parts of the biquaternion

$$A = (ia_0, a_1, a_2, a_2) \text{ and } B = (ib_0, b_1, b_2, b_3) \tag{5}$$

The product $AB = -a_0 b_0 + i(a_0 \underset{\sim}{b} + b_0 \underset{\sim}{a}) - \underset{\sim}{a} \odot \underset{\sim}{b} + \underset{\sim}{a} \otimes \underset{\sim}{b}$

vii) We define two conjugation processes for the biquaternion

$$Q = iq_0 + \alpha q_1 + \beta q_2 + \gamma q_3$$

a) the quaternion conjugate

$$\overline{Q} = iq_0 - \alpha q_1 - \beta q_2 - \gamma q_3 \tag{6}$$

and b) the complex conjugate

$$Q* = -iq_0 + \alpha q_1 + \beta q_2 + \gamma q_3$$

Hence $\overline{Q}^* = \overline{Q*}$ (= -Q for this special case)

These conjugations satisfy $(\overline{A\ B}) = \overline{B}\ \overline{A}$

$$(A\ B)* = B*\ A*$$

viii) The norm of a biquaternion Q is then

$$Q\ \overline{Q} = \overline{Q}\ Q = -q_0^2 + q_1^2 + q_2^2 + q_3^2 = |Q|^2$$

ix) The inverse of a biquaternion Q, whose norm is not zero, is

$$Q^{-1} = \overline{Q}/|Q|^2 \tag{7}$$

The inverse obeys the rules $(AB)^{-1} = B^{-1}A^{-1}$

Biquaternions are related to four vectors as will be specified later. The basic coordinate four-vector of Minkowski space becomes the coordinate biquaternion (ict, αx, βy, γz). From the product rules given in vii) above it can be seen that all vector operations in the cartesian basis $\hat{i}$, $\hat{j}$, $\hat{k}$ obey similar rules and the algebra of the quaternion basis is that of the vector cross product. In fact, by eliminating the scalar complex element, the Heaviside-Gibbs notation of ordinary vector algebra is immediately derived [10]. Likewise with differential operators of the biquaternion form, all the relations of vector calculus are immediately translated into biquaternion form.

x) We thus can define the differential operator

$$\boxed{} = (\frac{1}{ic}\frac{\partial}{\partial t}, \frac{\partial}{\partial x}, \frac{\partial}{\partial y}, \frac{\partial}{\partial z}) \tag{8}$$

$$= (\frac{1}{ic}\frac{\partial}{\partial t} + \alpha\frac{\partial}{\partial x} + \beta\frac{\partial}{\partial y} + \gamma\frac{\partial}{\partial z})$$

$$= (\frac{1}{ic}\frac{\partial}{\partial t}, \nabla)$$

the biquaternion gradient incorporating the quaternion gradient ∇

xi) Properties, analogous to ordinary gradient, divergence and curl, now derive from the algebraic rules of the biquaternion algebra

$$\Box \, \underline{Q} = \Box \, \{q_0, \, \underline{q}\}$$

$$= \{\frac{1}{ic} \frac{\partial q_0}{\partial t} - \nabla \odot \underline{q} \, , \quad \frac{1}{ic} \frac{\partial \underline{q}}{\partial t} + \nabla \, q_0 + \nabla \otimes \underline{q}\} \tag{9}$$

where $\nabla \, q_0 = \alpha \frac{\partial q_0}{\partial x} + \beta \frac{\partial q_0}{\partial y} + \gamma \frac{\partial q_0}{\partial z}$

$$\nabla \odot \underline{q} = \frac{\partial q_1}{\partial x} + \frac{\partial q_2}{\partial y} + \frac{\partial q_3}{\partial z}$$

and $\nabla \otimes \underline{q} = \alpha(\frac{\partial q_3}{\partial y} - \frac{\partial q_2}{\partial z}) + \beta(\frac{\partial q_1}{\partial z} - \frac{\partial q_3}{\partial y}) + \gamma(\frac{\partial q_2}{\partial z} - \frac{\partial q_1}{\partial y})$

Hence all the major first order derivatives are incorporated in the single algebraic operation equation 9.

xii)   The D'Alembertian is obtained as usual as the norm of the Laplace-Beltrami operator in equation 8.

$$\Box \, \overline{\Box} = \frac{-1}{c^2} \frac{\partial^2}{\partial t^2} + \frac{\partial^2}{\partial x^2} + \frac{\partial^2}{\partial y^2} + \frac{\partial^2}{\partial z^2} \tag{10}$$

Physical biquaternions are analogous to the four vectors of relativity theory.   We shall be mainly concerned with space time transformations of the coordinate biquaternion

$$X = \{ict, \, \underline{r}\} = \{ict, \, x, \, y, \, z\} \tag{11}$$

(using real (x, y, z) rather than involve unnecessary suffices) and the field four-potential

$A = \{\frac{\phi}{ic} \, , \, \underline{A}\},$ where ϕ and $\underline{A}$ are the quarternion 'scalar' and vector

potentials.
   A more comprehensive treatment applied to general physics theory would include the biquaternion representations of velocity, momentum, force and wave vectors.   These will be obvious where they arise.

Example 1.   Applying the operator of equation 9 to the potential

$A = \{\frac{\phi}{ic} \, , \, \underline{A}\}$

$$\Box \, A = \{\frac{-1}{c^2} \frac{\partial \phi}{\partial t} - \nabla \odot \underline{A}, \, \frac{1}{ic} \frac{\partial \underline{A}}{\partial t} + \frac{1}{ic} \nabla \, \phi + \nabla \otimes \underline{A}\} \tag{12}$$

If the field satisfies the Lorentz condition

$$\frac{1}{c^2} \frac{\partial \phi}{\partial t} + \nabla \odot \underline{A} = 0 \tag{13}$$

then $\quad \mathbf{A} = \{0, \dfrac{i\underset{\approx}{E}}{c} + \underset{\approx}{B}\}$ $\qquad\qquad$ (14)

where $\quad \underset{\approx}{B} = \underset{\approx}{\nabla} \otimes \underset{\approx}{A}$

and $\quad \underset{\approx}{E} = \dfrac{-\partial \underset{\approx}{A}}{\partial t} - \underset{\approx}{\nabla} \phi$ $\qquad\qquad$ (15)

<u>Example 2.</u> $\quad$ Specify a biquaternion $\Phi = \{-i\ \phi_0,\ \phi_1,\ \phi_2,\ \phi_3\}$

and $\quad \underset{\text{\rlap{—}Q}}{} = \{\dfrac{-i}{c} \dfrac{\partial}{\partial t}, \dfrac{\partial}{\partial x}, \dfrac{\partial}{\partial y}, \dfrac{\partial}{\partial z}\}$

then $\quad \text{\rlap{—}Q}\ \phi = 0$ results in

$$- \frac{1}{c} \frac{\partial \phi_0}{\partial t} - \frac{\partial \phi_1}{\partial x} - \frac{\partial \phi_2}{\partial y} - \frac{\partial \phi_3}{\partial z} = 0$$

$$+ \alpha\{-i\frac{\partial \phi_0}{\partial x} - \frac{i}{c} \frac{\partial \phi_1}{\partial t} + \frac{\partial \phi_3}{\partial y} - \frac{\partial \phi_2}{\partial z}\}$$

$$+ \beta\{-i \frac{\partial \phi_0}{\partial y} - \frac{i}{c} \frac{\partial \phi_2}{\partial t} - \frac{\partial \phi_3}{\partial x} + \frac{\partial \phi_1}{\partial z}\}$$

$$+ \gamma\{-i \frac{\partial \phi_0}{\partial z} - \frac{i}{c} \frac{\partial \phi_3}{\partial t} + \frac{\partial \phi_2}{\partial x} - \frac{\partial \phi_1}{\partial y}\} = 0 \qquad (16)$$

These conditions will be referred to as "left regularity"
{"right regularity" is the condition $\Phi\ \text{\rlap{—}Q} \equiv \text{\rlap{—}Q}\ \bar{\Phi} = 0$} $\qquad$ (17)

We put $\Phi = \mathbf{A} + i\mathbf{B} = \{0,\ a_1,\ a_2,\ a_3\} + i\{0,\ b_1,\ b_2,\ b_3\}$

$\qquad\qquad \equiv \alpha\{a_1 + ib_1\}\ \beta\{a_2 + ib_2\} + \gamma\{a_3 + ib_3\}$

Then equations 16 yield

$$\underset{\approx}{\nabla} \odot \underset{\approx}{a} = 0 \qquad \underset{\approx}{\nabla} \odot \underset{\approx}{b} = 0$$

$$- \frac{1}{c} \frac{\partial \underset{\approx}{a}}{\partial t} + \underset{\approx}{\nabla} \otimes \underset{\approx}{b} = 0$$

$$\frac{1}{c} \frac{\partial \underset{\approx}{b}}{\partial t} + \underset{\approx}{\nabla} \otimes \underset{\approx}{a} = 0 \qquad (18)$$

These are source free Maxwell equations, the derivation being solely algebraic and due to the skew symmetric or symplectic nature of the algebra.
Putting $\underset{\approx}{a} \equiv \underset{\approx}{E}$ and $\underset{\approx}{b} \equiv c\underset{\approx}{B}$ we obtain

$$\text{curl } \underset{\approx}{E} = - \frac{\partial \underset{\approx}{B}}{\partial t}$$

$$\text{curl } \underset{\approx}{B} = \frac{1}{c^2} \frac{\partial \underset{\approx}{E}}{\partial t}$$

(curl being the quaternion operator $\underset{\approx}{\nabla} \otimes$) $\qquad\qquad$ (19)

Sources can be included by equating $\square\Phi$ to a source biquaternion S

$$S = \{\tfrac{\rho}{\epsilon}, \underset{\sim}{j}_m\} + ic \{\rho_m, \mu\underset{\sim}{j}\} \tag{20}$$

ρ, ρ_m, $\underset{\sim}{j}$ and $\underset{\sim}{j}_m$ being electric and magnetic charge and current densities.

The complete set of symmetrised Maxwell equations result[11]

$$\nabla \otimes \underset{\sim}{E} = -\frac{\partial \underset{\sim}{B}}{\partial t} - \mu \underset{\sim}{j}_m \;\; ; \;\; \nabla \otimes \underset{\sim}{B} = \mu\underset{\sim}{j} + \epsilon\frac{\partial \underset{\sim}{E}}{\partial t}$$

$$\nabla \odot \underset{\sim}{B} = \rho_m \;\;\; ; \;\;\; \nabla \odot \underset{\sim}{E} = \frac{\rho}{\epsilon} \tag{21}$$

[Note that A, B are specified to have zero scalar components. Imaeda terms this a "vector condition" on the free choice of such biquarternions]

A + iB is then an electromagnetic field $\underset{\sim}{E} + i c \underset{\sim}{B}$

Finally, since $\square \, \Phi = S$ does not uniquely determine Φ any left regular biquaternion K can be added. That is with $\square K=0:\square\Phi=\square(\Phi+K)$ constituting a gauge transformation.

Example 3. An extension to example 1 can be made with the inclusion of the electric vector potential A_e and scalar magnetic potential ϕ_m.

Hence with A = $\{\dfrac{-i\phi}{c} , \underset{\sim}{A}\} + i\{-i\phi_m, cA_e\}$

then $\square A = \underset{\sim}{B} + \dfrac{i}{c} \underset{\sim}{E}$ results in

$$\text{div}\underset{\sim}{A} + \frac{1}{c^2} \frac{\partial \phi}{\partial t} = 0 \;\; ; \;\; \text{div}\underset{\sim}{A}_e + \frac{1}{c^2} \frac{\partial \phi_m}{\partial t} = 0$$

$$\underset{\sim}{E} = -\frac{\partial \underset{\sim}{A}}{\partial t} - \nabla\phi - \nabla\otimes\underset{\sim}{A}_e \tag{22}$$

$$\underset{\sim}{B} = \nabla\otimes\underset{\sim}{A} + \nabla\phi_m + \frac{\partial \underset{\sim}{A}_e}{\partial t}$$

The Lorentz condition and symmetrised potential description of the E.M. field. The apparently arbitrary definition of the regularity condition of equation 16 is given full justification by Imaeda[4] in terms of functional derivatives in a four dimensional complex space.

Example 4.
Let F be a force biquaternion F = $\left\{-\dfrac{if_o}{c} , \dfrac{\underset{\sim}{f}}{c}\right\}$

and U a velocity biquaternion U = $\{-iu_o, \underset{\sim}{u}\}$

then if A is a field biquaternion A = $\left\{0, \underset{\sim}{B} + \dfrac{i\underset{\sim}{E}}{c}\right\}$

F = $-\tfrac{1}{2}\,\{\text{AU}-\text{UA*}\}$

gives $f_0 = \dfrac{\underset{\sim}{u} \cdot \underset{\sim}{E}}{c}$; $\underset{\sim}{f} = \left\{ \dfrac{\underset{\sim}{E}}{c} + \underset{\sim}{u} \times \underset{\sim}{B} \right\}$

the Lorentz force and temporal component of the four force

3. FUNCTIONS OF A BIQUATERNION VARIABLE

As stated in the introduction, for the two dimensional electrostatic case, solutions of the transverse Laplace equation can be derived from the real and imaginary parts of the function $w = f(z) = f(x+iy) = u+iv$. There are similar methods whereby solutions of the wave equation can be derived from functions of a biquaternion variable in the form

$$F(ict, \underset{\sim}{r}) = (icu, \underset{\sim}{v}) \tag{23}$$

However whereas all functions can be considered in the two dimensional conformal mapping of the Argand complex planes, the mappings of equation 23 are subject to constraints and a priori conditions which severely limit the functions available. If, for example, the same method of defining differentiability is used as for ordinary complex variable theory a set of twelve extended Cauchy-Riemann equations result[12]. Since functions that are left regular as shown in equation 16 can be used to derive solutions of Maxwell's equations, methods are required that can produce such functions. We note at the outset that for source free field solutions as shown in equations 17-19 the function Φ required has to have zero scalar component in both the real and imaginary parts of the biquaternion.

3.1 Fuerter Polynomials

Fuerter[3] introduced polynomials derived by a generating function

$$F_n(X, \underset{\sim}{t}) = (-itx_0 + \underset{\sim}{t} . \underset{\sim}{x})^n \tag{24}$$

where $\underset{\sim}{t} = \{0, t_1, t_2, t_3\}$ is a parameter

and $x = \{-ix_0, x_1, x_2, x_3\}$ a biquaternion

Thus $F_n(X, \underset{\sim}{t}) = \{x_1 t_1 + x_2 t_2 + x_3 t_3 - i\alpha x_0 t_1 - i\beta x_0 t_2 - i\gamma x_0 t_3\}^n$ (25)

The coefficients of the powers and products of t_i in this expansion are polynomials of the same degree in x_i. These polynomials, $P_{n_i}(i=1,2,3)$, are the "Fuerter polynomials" and obey the regularity condition $\Box P = 0$. i.e.

$$F_n(X, \underset{\sim}{t}) = \sum_{\substack{n_1 n_2 n_3 \\ n_1 + n_2 + n_3 = n}} n! P_{n_1 n_2 n_3}(X) \; t_1^{n_1} t_2^{n_2} t_3^{n_3} \tag{26}$$

This is best illustrated by a simple example

$$\{(x_1t_1+x_2t_2+x_3t_3)-i\alpha x_0t_1-i\beta x_0t_2-i\gamma x_0t_3\}^2 \qquad (27)$$

$$= \quad t_1^2\{x_1^2+x_0^2 - 2i\alpha x_0x_1\} \quad (=2t_1^2\frac{\{x_1-i\alpha x_0\}^2}{2})$$

$$+ \quad t_2^2\{x_2-i\beta x_0\}^2$$

$$+ \quad t_3^2\{x_3-i\gamma x_0\}^2$$

$$+ \quad 2t_1t_3(x_1x_2-i\alpha x_0x_2-i\beta x_0x_1\}$$

$$+ \quad 2t_2t_3\{x_2x_3-i\beta x_0x_3-i\gamma x_0x_2\}$$

$$+ \quad 2t_1t_3\{x_1x_3-i\alpha x_0x_3-i\gamma x_0x_1\}$$

corresponding to Fuerter polynomials

$$P_{200}t_1^2+P_{020}t_2^2+P_{002}t_3^2+P_{110}t_1t_2+P_{011}t_2t_3+P_{101}t_1t_3 \qquad (28)$$

the sum of the suffices equalling the degree of the polynomials.

It is to be noted that only the vector part of the parametric biquaternion $T_n=(t_0,\underline{t})$ has been used in this derivation. The resulting

functions in equation 28 all include powers of x_0. If we label this series $T_n^{(0)}$ other series of polynomials $T_n^{(1)}$ can be derived by the same method by taking any three of the four t_i components of T_n (in order) as the parametric basis. Totally spatial P functions result in some cases, for example

$$P_{001}^{(1)} = z+\alpha x \quad ; \quad P_{001}^{(0)} = yz-\beta xy+\gamma xz \qquad (29)$$

These are suitable for the solution of static problems. A complete algebra of Fuerter polynomials results including recurrence relations, completeness and orthogonality properties, series expansions addition formulae and interdependence relations between the four classes of polynomial sets.

As shown by example 2 of the previous section the real and (Argand) imaginary parts of the $P_n(X)$ are independently potential functions of electric and magnetic fields since they all obey $\Box P = 0$. However for source-free solutions it is essential that the scalar part of these functions are reduced to zero. Such solutions can be derived from the P functions by a variety of methods. One such is to form a series of functions of the same kind and of the same degree, for example

$$a\alpha P_{201}^{(0)} + b\beta P_{111}^{(0)} + c\gamma P_{102}^{(0)}$$

and equating the coefficients of the scalar components to zero. Other conditions, boundary of initial value conditions can then be applied to completely determine the coefficient values.

A more powerful method is to form $P(X)A(X)$ where $A(X)$ is a constant biquaternion. This is best illustrated by an example.

Example 5. Selecting at random P_{110} in equation 28 (and transferring to t,x,y,z, notation) we form

$$\Pi(X) = (xy-i\alpha cty-i\beta ctx)(-ia_0+\alpha a_1+\beta a_2+\gamma a_3)$$

resulting in a complex vector

$$\pi = \alpha\{a_1 xy + a_0 cty\} + \beta\{a_2 xy + a_0 ctx\} + \gamma\{a_3 xy\}$$
$$+ i\{(a_0 xy + a_1 cty + a_2 ctx) - \alpha a_3 ctx + \beta a_3 cty$$
$$+ \gamma(a_1 ctx - a_2 cty)\}$$

then $\Box\pi = 0$ and equating π to $\underset{\sim}{E} + ic\underset{\sim}{B}$ gives a field which satisfies

$$\text{curl } \underset{\sim}{E} = - \frac{\partial \underset{\sim}{B}}{\partial t} \quad \text{etc.} \quad \text{e.g. putting } a_0 = a_1 = a_2 = 0$$

$$\underset{\sim}{E} = a_3 xy\hat{k}$$

$$\underset{\sim}{B} = -a_3 t(x\hat{i} - y\hat{j}) \text{ in ordinary vector notation.}$$

From equations 24 and 26 we can expand

$$\exp\{-itx_0 + \underset{\sim}{t}.\underset{\sim}{x}\} \equiv \exp\{-i(\underset{=}{t}x_0 - i\underset{=}{t}.\underset{=}{x})\}$$

$$= \sum_{n=0}^{\infty} \frac{(-i)^n}{n!} (\underset{\sim}{t}x_0 - \underset{=}{t}.\underset{\sim}{x}')^n \qquad (30)$$

$$= \sum_{n=0}^{\infty} \frac{(-i)^n}{n!} F_n^o(X',t)$$

where X' is the biquaternion $\{-ix_0', ix_1', ix_2', ix_3'\}$ replacing X in equation 25.

Each term in the expansion remains left regular and hence the exponential function is left regular. Fourier representation, Fourier integrals and plane wave expansions then have analogous representations in the biquaternion space.

A system of polynomials $Q_{n_1 n_2 n_3}^{(i)}$ X $\{i=0,1,2,3\}$ can be derived from

inverse powers of X. These too are regular functions and are orthogonal to the P functions. The completeness of the series enables expansions of arbitrary functions to be made as a series of either form.

3.2 Solutions of the wave equation

Since all the Fuerter polynomials $P(X)$ obey the left regularity condition, $\boxed{Q}P = 0$ it follows that they all obey the wave equation

$$\boxed{Q}^2 P = \boxed{Q}\overline{\boxed{Q}}P \equiv \overline{\boxed{Q}}\boxed{Q}P = 0$$

Other solutions may be obtained from regular functions in the manner of equation 23 provided u and v obey Cauchy-Riemann like relations. That is for the form $\underset{\sim}{}$

$$F(ict,\underset{\sim}{r}) = u(ict,\underset{\sim}{r})+(\alpha x+\beta y+\gamma z)v(ict,\underset{\sim}{r})$$

then $\dfrac{u}{|\underset{\sim}{r}|}$ and $\dfrac{v}{|\underset{\sim}{r}|}$ satisfy the wave equation if and only if

$$\frac{\partial u}{c\partial t} = \frac{\partial v}{\partial \underset{\sim}{r}} \quad \text{and} \quad \frac{\partial v}{c\partial t} = \frac{\partial u}{\partial \underset{\sim}{r}} \tag{31}$$

Unfortunately this severely restricts the functions F to which this can be usefully applied. Even simple powers of $X = \{ict,\underset{\sim}{r}\}$ do not qualify

completely. It appears to have gone unnoticed that, for such powers of X,

$U = \dfrac{u}{|\underset{\sim}{r}|}$ does, in fact, always obey the wave equation independently of

the relations in equation 31. For the first five powers of X^n these functions are

$$U^1 = \frac{ct-r}{r} \qquad\qquad [r=(x^2+y^2+z^2)^{\frac{1}{2}}]$$

$$U^2 = -\frac{c^2t^2-r^2}{r} \;\; ; \;\; U^3 = \frac{-ct(c^2t^2+3r^2)}{r}$$

$$U^4 = \frac{c^4t^4+6r^2c^2t^2+r^4}{r} \;\; ; \;\; U^5 = \frac{c^5t^5+10r^2c^3t^3+5r^4ct}{r}$$

[These are alternately real and imaginary; in conjunction with the missing $\dfrac{v}{|\underset{\sim}{r}|}$ the pattern of binomial coefficients becomes apparent and the series

is simple to continue].

However in such cases where $\Box^2 X^n \neq 0$, the remainder is itself a regular function that is if

$$\Box^2 X^n = Y \;\; ; \;\; \Box Y = 0 \tag{32}$$

4. FIELD DESCRIPTION BY SCALAR POTENTIALS

In three dimensional vector calculus the field is derived from a vector potential $\underset{\sim}{A}$ obtained from

$$\text{div } \underset{\sim}{B} = 0 \;\; ; \;\; \underset{\sim}{B} = \text{curl } \underset{\sim}{A}$$

A second solution exists for div $\underset{\sim}{B}=0$ namely

$$\underset{\sim}{B} = \nabla\sigma \times \nabla\tau \tag{33}$$

for scalar fields $\sigma(x,y,z)$ and $\tau(x,y,z)$. These were first introduced by Bateman [13] in searching for field transformations which left the wave equation and the eikonal equation invariant (up to a scale factor). We

shall therefore designate them Bateman potentials.

Hence
$$\Box \underset{\sim}{A} = \tfrac{1}{2}[\tau\nabla\sigma - \sigma\nabla\tau] \tag{34}$$

$$\Phi = \tfrac{1}{2}\left[\sigma\,\frac{\partial\tau}{\partial t} - \tau\,\frac{\partial\sigma}{\partial t}\right] \tag{35}$$

$$\underset{\sim}{E} = \frac{\partial\sigma}{\partial t}\,\nabla\tau - \frac{\partial\tau}{\partial t}\,\nabla\sigma \tag{36}$$

Maxwell equations are obeyed but with rather complex sources

$$\mu\underset{\sim}{j} = \nabla\times(\nabla\tau\times\nabla\sigma) + \frac{1}{c^2}\frac{\partial}{\partial t}\left[\frac{\partial\tau}{\partial t}\,\nabla\sigma - \frac{\partial\sigma}{\partial t}\,\nabla\tau\right]$$

$$\frac{\rho}{\epsilon} = \Delta\tau.\frac{\partial}{\partial t}\,\nabla\sigma - \nabla\sigma.\frac{\partial}{\partial t}\,\nabla\tau + \frac{\partial\sigma}{\partial t}\,\nabla^2\tau - \frac{\partial\tau}{\partial t}\,\nabla^2\sigma. \tag{37}$$

The Bateman potentials satisfy the wave equation, but most interestingly, if they are considered to be invariant under Lorentz transformation, the transformation of the differential operators alone provides all the correct relations for the transformation of the E.M field. From equations 34 and 35 it is possible to derive Hertz vectors and the Lorentz condition directly since $\underset{\sim}{A} = \frac{\partial\underset{\sim}{\pi}}{\partial t}$ and $\phi = c^2\nabla.\underset{\sim}{\pi}$ automatically satisfy both.

Bateman regarded σ and τ as orthogonal stream and velocity potentials in the hydrodynanic analogy.
The Poynting vector is

$$\underset{\sim}{E}\times\underset{\sim}{B} = \frac{\partial\sigma}{\partial t}\,\{\nabla\sigma|\nabla\tau|^2 - \nabla\tau(\nabla\sigma.\nabla\tau)\} - \frac{\partial\tau}{\partial t}\,\{\nabla\tau|\nabla\sigma|^2 - \nabla\sigma(\nabla\tau.\nabla\sigma)\}$$

and if σ = constant and τ = constant are orthogonal surfaces $\nabla\sigma.\nabla\tau=0$ and ray directions are given by

$$\underset{\sim}{E}\times\underset{\sim}{B} = \frac{\partial\sigma}{\partial t}\,\nabla\sigma|\nabla\tau|^2 - \frac{\partial\tau}{\partial t}\,\nabla\tau|\nabla\sigma|^2 \tag{38}$$

4.1 Biquaternion representation

We analytically continue into the complex four space by operating with the biquaternion operator $\underset{\sim}{Q} = \{\frac{i}{c}\frac{\partial}{\partial t}\,,\,\underset{\sim}{\nabla}\}$

Equations 34 and 35 are combined in

$$\underset{\sim}{A} = \tfrac{1}{2}[\tau\underset{\sim}{Q}\sigma - \sigma\underset{\sim}{Q}\tau] \equiv \left\{\frac{\phi}{ic}\,,\,\underset{\sim}{A}\right\}$$

where $\underset{\sim}{A} = \tfrac{1}{2}[\tau\underset{\sim}{\nabla}\sigma - \sigma\underset{\sim}{\nabla}\tau]$ and $\phi = \tfrac{1}{2}\left[\tau\,\frac{\partial\sigma}{\partial t} - \sigma\,\frac{\partial\tau}{\partial t}\right]$

Then $\frac{1}{2}[\underset{\sim}{Q}\sigma\overline{\underset{\sim}{Q}}\tau - \underset{\sim}{Q}\tau\overline{\underset{\sim}{Q}}\sigma] = \{0, \dfrac{iE}{c} + \underset{\sim}{B}\}$

where $\underset{\sim}{E} = \dfrac{\partial\sigma}{\partial t}\,\nabla\tau - \dfrac{\partial\tau}{\partial t}\,\nabla\sigma$ and $\underset{\sim}{B} = \nabla\sigma\otimes\nabla\tau$

and $\underset{\sim}{Q}\sigma\otimes\underset{\sim}{Q}\tau$ is an antisymmetric tensor (cf $\nabla\sigma\times\nabla\tau$)

$$F^{\mu\nu} = \dfrac{\partial\sigma_\mu}{\partial x}\dfrac{\partial\tau_\nu}{\partial x} - \dfrac{\partial\tau_\mu}{\partial x}\dfrac{\partial\sigma_\nu}{\partial x}$$

$$F^{\mu\nu} = \begin{bmatrix} 0 & \dfrac{i}{c}\left(\dfrac{\partial\sigma}{\partial t}\dfrac{\partial\tau}{\partial x} - \dfrac{\partial\tau}{\partial t}\dfrac{\partial\sigma}{\partial x}\right) & \dfrac{i}{c}\left(\dfrac{\partial\sigma}{\partial t}\dfrac{\partial\tau}{\partial y} - \dfrac{\partial\tau}{\partial t}\dfrac{\partial\sigma}{\partial y}\right) & \dfrac{i}{c}\left(\dfrac{\partial\sigma}{\partial t}\dfrac{\partial\tau}{\partial z} - \dfrac{\partial\tau}{\partial t}\dfrac{\partial\sigma}{\partial z}\right) \\ & 0 & \dfrac{\partial\sigma}{\partial x}\dfrac{\partial\tau}{\partial y} - \dfrac{\partial\tau}{\partial x}\dfrac{\partial\sigma}{\partial y} & \dfrac{\partial\sigma}{\partial x}\dfrac{\partial\tau}{\partial z} - \dfrac{\partial\tau}{\partial x}\dfrac{\partial\sigma}{\partial z} \\ & & 0 & \dfrac{\partial\sigma}{\partial y}\dfrac{\partial\tau}{\partial z} - \dfrac{\partial\tau}{\partial y}\dfrac{\partial\sigma}{\partial z} \\ & & & 0 \end{bmatrix} \tag{39}$$

$$= \begin{bmatrix} 0 & \dfrac{i}{c}E_x & \dfrac{i}{c}E_y & \dfrac{i}{c}E_z \\ & 0 & B_z & -B_y \\ & & 0 & B_x \\ & & & 0 \end{bmatrix} \tag{40}$$

the electromagnetic field tensor.

Forming the dual tensor $*F^{\mu\nu}$ either by the transformation $E \rightarrow H$, $H \rightarrow E$, $\mu \longleftrightarrow \epsilon$, or the product with the antisymmetric tensor density $\epsilon_{ij\mu\nu}$, results as usual in the form of Maxwell's equations given by

$$F^{\mu\nu}{}_{,\nu} = 0 ; \qquad *F^{\mu\nu}{}_{,\nu} = -J^\mu \tag{41}$$

$$\underset{\sim}{Q}^2 A = \underset{\sim}{Q}\,\overline{\underset{\sim}{Q}}A = J = \dfrac{1}{ic}\,\{\dfrac{\rho}{\epsilon} , ic\mu\underset{\sim}{j}\}$$

and both $\underset{\sim}{A}$ and ϕ obey the separated wave equations

$$\nabla^2 \phi - \frac{1}{c^2} \frac{\partial^2 \phi}{\partial t^2} = - \frac{\rho}{\epsilon} \; ; \; \nabla^2 \underset{\approx}{A} - \frac{1}{c^2} \frac{\partial^2 \underset{\approx}{A}}{\partial t^2} = \mu \underset{\approx}{j}$$

and Maxwell's equations have the biquaternion form

$$\nabla \underset{\approx}{\otimes} E = - \frac{\partial \underset{\approx}{B}}{\partial t} \; ; \qquad \nabla \underset{\approx}{\odot} E = \frac{\rho}{\epsilon}$$

$$\nabla \underset{\approx}{\otimes} B = \frac{1}{c^2} \frac{\partial \underset{\approx}{E}}{\partial t} + \mu \underset{\approx}{j} \; ; \qquad \nabla \underset{\approx}{\odot} B = 0 \; ; \qquad \nabla \underset{\approx}{\odot} j + \frac{\partial \rho}{\partial t} = 0$$

$\square A = 0$ is the Lorentz condition.

Finally the Lorentz condition requires both potentials σ, τ to obey the wave equation $\square^2 \sigma = 0$ solutions of which are obtainable by the quaternion methods of section 3. Since the vector field contains the product, it is expedient to take only one function as time dependent in the frame of observation. [Since the potentials themselves are invariants, zero time dependence in one frame does not guarantee time independence in a co-moving frame.]

5. TRANSFORMATIONS OF PHYSICAL BIQUATERNIONS

Biquaternions transform under similarity transformations of the kind

$$X' = Q*XQ \tag{42}$$

where Q is a biquaternion and $Q*$ its Argand complex conjugate. We list without proofs the following

a) $Q = (a_0 + i \underset{\approx}{b} | / (a_0^2 + \underset{\approx}{b}.\underset{\approx}{b})^{\frac{1}{2}}$ then Q has unit norm and equation 42 represents a spatial rotation with axis $\underset{\approx}{b}$ through the origin of coordinates and magnitude

$$\omega = 2\tan^{-1} \left[\frac{|\underset{\approx}{b}|}{a_0} \right] \tag{43}$$

b) $Q = \frac{1}{\sqrt{2}} (\sqrt{\lambda+1} - i \frac{\underset{\approx}{v}}{v} \sqrt{\lambda-1}) \tag{44}$

$$\underset{\approx}{v} = \alpha v_x + \beta v_y + \gamma v_z \; ; \qquad \lambda = \frac{1}{\sqrt{1 - \dfrac{v^2}{c^2}}} \qquad \text{norm } Q = 1$$

Equation 42 then represents a pure Lorentz transformation [13]. That is if

$$X' = ix_0' + \underset{\sim}{x}'$$

$$x_0' = \lambda\left[x_0 - \frac{\underset{\sim}{X}.\underset{\sim}{V}}{c}\right] \; ; \quad \underset{\sim}{x}' = \underset{\sim}{x} + \frac{\underset{\sim}{V}}{c}\left[\underset{\sim}{x}.\underset{\sim}{V}(\lambda-1)+\lambda x_0\right] \tag{45}$$

the standard form.

Transformations may be combined by the product rule with elementary transformations such as translations and magnifications to give more general linear transformations [14].

It has been proved by Maxwell and subsequently by Forsyth [15] that the only conformal mapping of a space of three dimensions onto itself is an inversion (again omitting the trivial translations, rotations, reflections and magnifications). Following the analysis of Forsyth it is possible to arrive at the same conclusion for a real space of four dimensions and the process of continuing this to a complex space of four dimensions is continuing. Hence in our pursuit of a conformal mapping technique, it is apparent that the processes of inversion be more closely examined. This is borne out by the two optical theorems to be presented in the following section.

Under conformal mappings angles remain invariant. Hence coordinates such as $x = r\sin\theta\cos\phi$ etc. transform under inversion to $x' = \frac{1}{r}\sin\theta\cos\phi$ or space inversions take the form

$$\underset{\sim}{r}' = \frac{1}{r^2}\underset{\sim}{r} \tag{46}$$

Solutions of the wave equation for which angles are invariant become solutions of the one dimensional radial wave equation which take the form of Euler's wave function $\frac{f(ct\pm r)}{r}$. Now a polynomial function of a biquaternion variable $Q = \{iq_0, \underset{\sim}{q}\}$ can be put into the form

$$f(Q) = [f_1(q_0+q)+f_2(q_0-q)] \; ; \quad \frac{i\underset{\sim}{q}}{2q}[f_1(q_0+q)-f_2(q_0-q)] \tag{47}$$

[since $f(ict+r)/r$ is not a radial wave function] where f_1 and f_2 are arbitrary functions of their arguments and $q = |\underset{\sim}{q}|$. The relationship between equation 47 and simple powers of $(ict, \underset{\sim}{r})$ as given in equations 31 and 32 is obvious. Hence choosing Q to be the basic coordinate biquaternion $Q = \{ict, \underset{\sim}{r}\}$ we can regard equation 47 as a transformation and obtain

$$
\begin{aligned}
ct' &= \tfrac{1}{2}[f_1(ct+r) + f_2(ct-r)] \\
x' &= \frac{x}{2r}[f_1(ct+r) - f_2(ct-r)] \\
y' &= \frac{y}{2r}[f_1(ct+r) - f_2(ct-r)] \\
z' &= \frac{z}{2r}[f_1(ct+r) - f_2(ct-r)]
\end{aligned}
\tag{48}
$$

For example choosing $f_1(\zeta) \equiv f_2(\zeta) = \frac{1}{\zeta}$ we have [16]

$$t' = \frac{t}{s^2} \; ; \; x' = \frac{x}{s^2} \; ; \; y' = \frac{y}{s^2} \; ; \; z' = \frac{z}{s^2}$$

where $s^2 = c^2 t^2 - r^2$.

It can readily be verified that if $f(x,y,z,t)$ is a wave function then

$$\frac{1}{s^2} f\left[\frac{x}{s^2}, \frac{y}{s^2}, \frac{z}{s^2}, \frac{t}{s^2}\right] \tag{49}$$

is also a wave function and that for t=0 [t'≠0] this becomes the space inversion required by equation 46.

A further transformation is given as

$$t' = \frac{s^2 + 1}{2c(z - ct)} \; ; \; x' = \frac{x}{z - ct} \; ; \; y' = \frac{y}{z - ct} \; , \; z' = \frac{s^2 - 1}{2c(z - ct)} \tag{50}$$

and this is "equivalent to a conformal transformation of a space of four dimensions which was discovered by Cremona". It has not proved possible as yet to obtain this transformation by biquaternion methods. Most use is made by Bateman and Cunningham of the transformation given in equation 49[17], the former for optical transformations and the latter for electromagnetic field transformations.

6. TRANSFORMATIONS OF OPTICAL SYSTEMS

The several transformations of optical systems which are known are listed below and the relevance of inversions is apparent.

6.1 Damien's theorem [18]

If a ray from a source O meets a refractive surface g at a point M, it is refracted, according to Snell's law, at the point of intersection into a new direction MS say. If MS is the distance marked off **in reverse** along

the direction of the refracted ray by MS = $-\frac{1}{n}$ OM (n being the refractive

index), then points M on the refracting surface are mapped into points S on a surface h, termed the zero distance phase front. As a phase front all rays are normal to it and it acts as if it were the source in a medium of infinite extent with the same refractive index.

Damien's theorem then states "If a surface g and a source O produce a zero distance phase front h, and if the inverse of g in a circle centered at O is g', and the inverse of h is h', then g' is the zero distance phase front of h' with respect to the same source and for the same refractive index." The peculiar nature of this transformation is that it inverts surfaces into phase fronts and vice versa. It has found application in deriving the second surface of an astigmatic lens from any given source and first surface. For a reflector n is taken to be -1 and the surface h is then on the opposite side of g from O and is in fact the pedal curve of g.

6.2 Cornbleet's inversion for spherical symmetric media[19]

A spherically symmetrical nonhomogeneous medium with refractive index $n(r)$ can be transformed into a second medium with index $n'(r)$ by putting

$$rn(r) = f(r) \text{ and } rn'(r) = f\left(\frac{1}{r}\right) \tag{51}$$

that is the inverse applies to rn and **not** to n alone. Rays in the first with equation $r = f(\theta)$ invert directly into rays in the second with equation

$$\frac{1}{r} = f(\theta) \tag{52}$$

Maxwell's fish eye is invariant under the transformation.

For the transverse cross-section of a cylindrically symmetric medium the transform can be generalised to

$$n'(r) = \frac{1}{r} \, n\left(\frac{1}{r}n\right) \qquad \text{and rays}$$

$$r = f(\theta) \text{ become } \frac{1}{r}n = f(\theta) \tag{53}$$

6.3 The Legendre transform[20]

If surfaces $S(x,y,z) =$ constant are phase fronts in a medium with refractive index $n(x,y,z)$, and $T(\lambda,\mu,\nu) =$ constant are phase fronts in a medium with refractive index $n(x,y,z)$ then

$$\left(\frac{\partial S}{\partial x}\right)^2 + \left(\frac{\partial S}{\partial y}\right)^2 + \left(\frac{\partial S}{\partial z}\right)^2 = n^2$$

$$\tag{54}$$

$$\text{and } \left(\frac{\partial T}{\partial \lambda}\right)^2 + \left(\frac{\partial T}{\partial \mu}\right)^2 + \left(\frac{\partial T}{\partial \nu}\right)^2 = n^2$$

and they are connected by a Legendre transformation if

$$\frac{\partial S}{\partial x} = \lambda \; ; \; \frac{\partial s}{\partial y} = \mu \; ; \; \frac{\partial S}{\partial z} = \nu$$

$$\tag{55}$$

$$\text{and } \frac{\partial T}{\partial \lambda} = x \; ; \; \frac{\partial T}{\partial \mu} = y \; ; \; \frac{\partial T}{\partial \nu} = z$$

This enables a transformation to be made between the two media, for example Maxwell's fish-eye is transformed into the Eaton lens[21]. It appears from known examples to be equivalent to the transformation $r \rightarrow n$, $n \rightarrow r$.

6.4 Budden's reciprocity[22]

In an inhomogeneous, and not necessarily isotropic, medium, the

vectors tangent to a curved ray can be mapped into radius vectors from an origin with the same direction and of magnitude equal with refractive index at the corresponding point on the ray. The result will be a surface which has been termed the "refractive surface" by Budden or the "surface of components" by Hamilton[23]. The actual phase front is Fresnel's wave surface or the "surface of normal slowness" or simply the ray surface. Budden shows that the normals of each of these surfaces are each parallel to the corresponding radius vectors of the other. Hamilton [loc cit] states that an inversion of one surface is connected to the pedal surface of the other.

6.5 Batemans' inversion[16]

This is based on the result given in equation 49. The transformation, he states, "makes a standard wave surface t=0 (the zero distance phase front possibly) in the original system correspond to a standard wave surface t=0 the second. (Here Damien's theorem differs - see section 7) - also since the equations of the surfaces are

$$F(x,y,z,0) = 0 \text{ and } F\left(\frac{x}{r^2}, \frac{y}{r^2}, \frac{z}{r^2}, 0\right)$$

respectively, it is clear that one is the inverse of the other with regard to a unit sphere whose centre is the origin." He goes on to prove that the laws of refraction and reflection remain the same, that the refractive index is unaltered - and applies the theory to the transformation of a centred lens with two spherical surfaces.

The connective tissue between these various transformations of optical systems are manifest. It is quite obvious that a non-uniform medium could not support a zero distance phase front. However this latter is the source for an infinite uniform medium - it's place is therefore taken by the Fresnel wave surface - a continuously varying phase front in a non-uniform non-isotropic medium. The inversion of refractive index means that a uniform medium is inverted into a medium with refractive index

proportional to $\frac{1}{r^2}$.

$$\text{For} n(r) = f(r) = r \; ; \; f\left(\frac{1}{r}\right) = \frac{1}{r} = rn'(r) \; \therefore \; n'(r) = \frac{1}{r^2} \tag{56}$$

If we apply Bateman's transform (with t=0) to the outgoing spherical wave function $\frac{e^{ikr}}{r}$ by putting $r' = \frac{1}{r}$ we obtain $r'^2\left(\frac{e^{ik'r'}}{r'}\right)$ where $k' = \frac{k}{r^2}$.

This in effect is a wave in a medium with a pseudo refractive index of $\frac{1}{r^2}$ (and of decreasing amplitude).

Buchdahl[24] has shown the analogy between optical rays and particle trajectories, in particular the relation between Maxwell's fish-eye and Kepler orbits. It is well known that the exterior problem of a spherical

mass in gravitation theory is the Kepler orbit (potential $\propto \frac{1}{r}$) and the

inverse, the interior problem is the harmonic oscillator (potential $\propto r^2$).

Finally Stavroudis[25] illustrates the geometrical fact that if a curved ray in a non-uniform medium has its tangent, normal and binormal vectors labelled by the refractive index at every point, the resultant triad of vectors obey Maxwell-like equations.

7. SUMMARY AND DISCUSSION

From the point of view of scattering and diffraction the most applicable theory is that of Damien's inversion as it applies to perfectly conducting surfaces, although, of course, any comprehensive theory must encompass the other transformations. There are limitations involved at the outset. Inversion alone gives a very small choice of cross sectional profiles and the theory is applied to a point source of radiation. Other illumination functions require extended sources, which by the nature of the inversion, cannot be included. A plane wave illumination can be achieved by removing the source to infinity, but then inversion becomes a plane reflection and even more elementary in form. It has to be noted that Damien's theorem has only been proved in the plane. As such it applies to the cross section of a two dimensional problem or that of a rotationally symmetrical one. In fact more general transformations exist for cylindrical problems and one could expect to deal with scattering and diffraction from cylinders of any cross section by some form of extended conformal mapping and even Schwarz-Christoffel methods.

The generation of field variables by hypercomplex methods likewise appears to give rise only to elementary solutions. For a full harmonic analysis one has to establish an infinite series of "plane-wave" powers such as terms in equation 49. This Imaeda does but the resulting analysis would appear to be a complicated version of ordinary Fourier expansions.

To use Fuerter functions to obtain Bateman potentials is comparatively straightforward. It is noteworthy that Bateman's work on transformations of electromagnetic fields was effectively concluded with publication of his book ref 16 in 1914 whereas his analysis involving potentials such as σ and τ was published in the 1920's. Bateman potentials themselves have one major advantage. The form $\underset{\sim}{B} = \nabla\sigma \times \nabla\tau$ means that, if one potential is

time independent, it can be taken to be the equation of a conducting

surface. Then, say, with $\frac{\partial \tau}{\partial t} = 0$; if $\tau(x,y,z)$ = constant is a conductor,

B and E automatically obey the boundary conditions at the surface. Taking

$\sigma(x,y,z,t)$ = constant to be a phase front it will be of the form $\sigma(x,y,z)$ = at for different values of a. Presumably t=0 would refer to a zero distance phase front. There is a difference in emphasis between a being zero and t=0. Damien's inversion implies that under inversion $\sigma \to \tau$ and $\tau \to \sigma$ in some manner, as given in the quotation from Bateman's analysis in section 6.5.

All such procedures depend on an a priori specification of the potentials. A method has to be derived whereby the potentials can be found to suit the field problem being posed. Alternatively a body of canonical problems can be established which will enable suitable potentials to be predicted. Then the possibility of actually conformally

mapping (cylindrical) problems may be addressed. The potentials of a known problem **may**, under the conformal mapping appropriate to hypercomplex functions, produce the potentials and hence the fields of the conformally mapped cross section. It is to be hoped that this would not merely occur in the optical range. The problem is that ray optics forms a static geometry in space and the essential time dependence of real fields is suppressed. In all the adaptations of the hypercomplex theory yet produced, static or quasi-static examples have been given[26].

Other conformal transformations of the E.M. field have been applied usually involving the introduction of a non-uniformity in the refractive index throughout the transformed space[27].

The Fuerter series relies on integer (positive and negative) powers of the hypercomplex variable. It is possible that fractional powers will produce diffractive effects in the same way that the cut Riemann plane does in ordinary complex variable theory[28]. This implies a higher order Riemann theory, a two dimensional cut of a three dimensional Riemann surface for example.

The problem of non-uniform media is more complex still. Only few electromagnetic analyses have been made and most for the linear stratified case. However C.T. Tai[29] quotes the same limitation for the complete spherical problem as against the cylindrical. In dealing both with non-uniform and with moving media, generalized wave functions are used which incorporate the refractive index, setting up a refractive space in which direction cosines are multiplied by refractive index. Cunningham's paper shows that moving media (media under Lorentz transformation) and media under inversion have the same transformation laws for the constitutive relations. Thus one should look for "refractive potentials" where in the manner of the function $rn(r)$ (section 6.2) the coordinates are transformed separately from the product of coordinate and refractive index.

Again only basic problems, spherical or cylindrical, can be investigated. This author has proved, for example, that a confocally elliptically stratified non-uniform medium cannot be designed to refocus a point source on its surface in the manner of an elliptical Maxwell fish-eye. But then the mapping is both non-conformal and non-inversion. Outstanding problems of this nature are currently being investigated[30] with applications to inverse scattering problems and the radiation from conformal arrays.

ACKNOWLEDGEMENT

The author acknowledges with gratitude the assistance received from the advanced copy of the comprehensive study by Imaeda into the application of biquaternion theory to electromagnetic problems.

REFERENCES

1. Halbertstam M and Ingram R E: The Collected Works of Sir W R Hamilton, vol 3, Algebra Cambridge University Press, 1967.
2. Scheffers G: "Verallgemeinerung der Grundlagen der gewohnlich complexen Functionen" Rep. Trans. Roy Soc. Sci., Leipzig, vol 45, p828, 1893.
3. Fuerter R: Commentarii Mathematici Helvetica, Vol 16, P19, 1943, "On the analytic representation of regular functions of a quaternion variable", Vol 7, p307, 1934-5, also Vol 8, p371, and Vol 9, 1936-37.

4. Imaeda K: "Quaternionic Formulation of Classical Electrodynamics and Theory of Functions of a Biquaternion Variable", Report FPL-1-83-1, The Fundamental Physics Laboratory, Department of Electronic Science, Okayama University of Science, Okayama, Japan 1983, also Nuovo Cimento, March 1976.
5. Silberstein L: The Theory of Relativity, MacMillan, New York, 1924.
6. Cornbleet S: "Microwave and Optical Ray Geometry", Wiley & Sons, England, 1984.
7. Bateman H: "The Transformations of coordinates which can be used to transform one physical problem into another", Proc London Math Soc, Vol 7, p70, 1908.
8. Cornbleet S: "Microwave Optics", pp341-344, Academic Press, 1976.
9. Rund H: "Generalized Clebsch Representation on Manifolds" in Topics in Differential Geometry, Ed. H Rund, Academic Press, New York, 1976.
10. Crowe M J: "A History of Vector Analysis", Notre Dame University Press, 1967.
11. Papas C H: "The Theory of Electromagnetic Wave Propagation", McGraw Hill, 1956.
12. Ref 4, p232.
13. Bateman H: "On lines of Electric Induction and the Conformal Transformations of a Space of Four Dimensions", Proc Lond Math Soc, Ser 2, Vol 21, p256, Nov. 1920.
14. Ref 4, p50.
15. Maxwell J C: The Scientific Papers of J C Maxwell, Dover Press, New York, 1956.
    Forsyth A R: "Differential Geometry", Cambridge University Press, p410, 1912.
16. Bateman H: "The Mathematical Analysis of Electrical and Optical Wave Motion", (1914), Dover Publications, p31, 1955.
17. Cunningham E: "The Principle of Relativity in Electrodynamics and an Extension thereof", Proc Lond Math Soc, Vol 8, p469, 1910.
18. Damien R: "Theoreme sur les Surfaces d'Ondes en Optique Geometrique", Gauthier-Villars, Paris, 1955.
19. Cornbleet S and Jones M C: "The transformation of spherical nonuniform lenses", Proc I.E.E. pt H, Vol 129, no 6, p321, 1982.
20. Luneburg R K: "The Mathematical Theory of Optics", p102, University of California Press, 1964.
    Sommerfeld: "Lectures on Theoretical Physics", Vol IV Optics, Academic Press, p139, 1964.
21. Eaton J E: "On Spherically Symmetric Lenses", Trans I.R.E., Antennas and Propagation AP-4, p66, 1952.
22. Budden K G: "Radio Waves in the Ionosphere", Cambridge University Press, p252, 1961.
23. Conway A W and Synge J L: "The Collected Works of Sir W R Hamilton", Vol 1 Optics, Cambridge University Press, 1931.
24. Buchdahl H A: "Conformal transformations and conformal invariance of optical systems", Optik Vol 43, p259.
    "The Kepler Problem and Maxwell's Fish-eye", Amer Jour Phys Vol 46, p840, 1978.
25. Stavroudis O N: "The Optics of Rays, Wavefronts and Caustics", Academic Press, New York, pp194-198, 1972.
26. Evans D D: "Complex Variable Theory Generalized to Electromagnetics", PhD Thesis, University of California, 1976.

88

27. Myazaki Y: "Propagation properties of optical signal waves in perturbed dielectric waveguide by a conformal mapping technique", Topical Meeting on Integrated Optics, Las Vegas, Nevada, February 1972.
    Tischer F J: "Conformal mapping in Waveguide Considerations", Proc IEEE, Vol 51, p1056, July 1963.
28. Baker H F and Copson E T: "The Mathematical Theory of Huygen's Principle", Oxford University Press, 1953.
29. Tai C T: "Dyadic Green's Functions in Electromagnetic Theory", Intext Educational Publishers, p193, 1971.
30. Cornbleet S: "Geometrical Optics Reviewed", Proc IEEE, Vol 71, no 4, p471, April 1983.

COMPLEX-SOURCE-POINT THEORY AND THE OPEN RESONATOR

A.L. CULLEN

UNIVERSITY COLLEGE LONDON

1. INTRODUCTION

The theory of the open resonator is now established, at least so far as scalar wave functions are adequate to describe the field distribution. In this treatment we shall pay particular attention to the rather subtle modifications to the theory that arise when a complete vector description of the electromagnetic field is undertaken. The complex-source-point method, so powerfully applied to related problems by Felsen (1), is the basis of the present theory. We shall, however, omit the complicating effects of diffraction, and we begin by justifying this apparently serious defect. We then outline briefly the main results of the scalar theory of open resonatiors, and follow this by describing the procedures and results when the vector theory is applied. In particular, we shall show, following Yu and Luk (2), that the linearly-polarised higher-order modes of conventional theory do not exist. The situation is analogous in some ways to the erroneous assumptions that higher-order TE and TM modes exists in lossy rectangular waveguides, though in the present case the curvature of the mirrors rather then ohmic losses is the crucial factor. In a final section, some experimental aspects of the problem are considered. Experimental evidence is given confirming the vector theory of the higher-order modes, and applications of the open resonator to microwave measurements are briefly indicated.

2. DIFFRACTION

Weinstein (3) gives results for the loss per pass in an open resonator due to diffraction for the case of a confocal resonator. It is reasonable to assume that for resonators which do not depart too greatly from the confocal configuration Weinstein's results will at least be a good guide to the order of magnitude of the diffraction loss. In Table 1 below the loss per pass due to diffraction is compared with the loss per pass due to ohmic loss in the mirrors for the special case of copper mirrors at a frequency of 40 GHz; this loss will be almost independent of the parameter ka^2/D when this parameter is large enough, and is here assumed to be completely independent in the range quoted.

B. de Neumann (ed.), Electromagnetic Modelling and Measurements for Analysis and Synthesis Problems, 89–102.

Table 1

Parameter	Loss per pass	
ka^2/D	Diffraction	Ohmic
6	0.001	0.0005
7	0.0001	0.0005
8	0.00002	0.0005
9	0.000003	0.0005

In the above table, a is the radius of the mirrors, (assumed identical). The radius of curvature is of course equal to the mirror separation D in the confocal case. The free-space TEM phase coefficient is denoted by k as usual. Typical values for a millimeter-wave open resonator might be: a = 4cm, D = 14cm, and ka^2/D then has the value 9 and Table 1 shows that diffraction losses are truly negligible in comparison with ohmic losses in this case. We shall in fact assume that diffraction is negligible in what follows.

3. SCALAR WAVE THEORY

Fig 1 below shows the geometry of an open resonator with spherical mirrors symmetrically opposed on either side of a central plane. We can imagine a wave setting off from this central plane and travelling along the z-axis, spreading by diffraction as it travels. This wave is then re-focused by the concave mirror on reflection, first converging and then diverging as it approaches the opposing mirror, where the same process takes place. Resonance occurs if the round-trip phase change is an integral multiple of 2π. The central plane becomes the focal plane in this symmetrical case, the case on which we shall concentrate here. We now put this qualitative description into formal mathematical language. We begin by looking for a solution of the scalar wave equation corresponding to a narrow beam of radiation travelling mainly along the axis of the mirror system; the z-axis in the chosen co-ordinate system. We take out the factor exp(-jkz), which we may regard as the zeroth approximation to the required solution, and assume a solution of the form

$$u(x,y,z) = \psi(x,y,z)\exp(-jkz) \tag{1}$$

Substituting this formula for u into the three-dimensional wave equation, recognising that ψ may be assumed to be a slowly-varying function of z, and so neglecting its second derivative with respect to z, we find the following approximate wave equation for ψ:

$$\nabla_t^2 \psi - 2jk\frac{\partial \psi}{\partial z} = 0 \tag{2}$$

This equation is of the same form as the time-dependent Schrödinger equation for two space dimensions, z now playing the role of time. The solutions are well known, and may be used in (1) to give the following solutions for the wave function u:

$$u(\rho,\theta,z) = \left(\sqrt{2}\frac{\rho}{w}\right)^{\ell} L_{p}^{\ell}\left(\frac{2\rho^2}{w^2}\right)\frac{w_0}{w} \exp\left\{-j(kz - \Phi_{p\ell})\right.$$

$$\left. -\rho^2\left(\frac{1}{w^2} + j\frac{k}{2R}\right)\frac{\cos}{\sin}\ell\theta \right. \tag{3}$$

The wave function u can be identified with x- or y- component of the electric field, and this is the conventional procedure in the basic theory of Gaussian beams. Note that in this approximation linearly polarized beams are envisaged. It is of course possible to combine two orthogonally polarized beams to produce other field structures; for example, with $\ell = 1$, wholly radial or wholly circumferential electric field lines are possible.

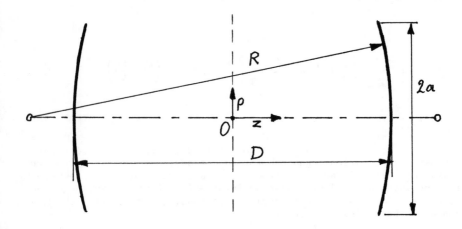

Fig 1 Open resonator geometry and notation

We can, moreover, combine forward and backward travelling beam waves to form a standing wave pattern of the kind needed to satisfy the boundary conditions in an open resonator of the kind shown in fig.1. If the radius of curvature of the mirrors is sufficiently large, it will be reasonable to put the transverse component of the electric field equal to zero over each mirror surface, though this is only strictly correct on the axis. When this is done, the scale radius w of the Gaussian beam can be calculated in terms of the mirror spacing D and the radius of curvature R of the mirrors. In the central (focal) plane its value is given by:

$$w_o^2 = \frac{\lambda}{2\pi} \sqrt{D(2R - D)} \qquad (4)$$

At each mirror, however, it has a larger value wl, where

$$w_1^2 = \frac{\lambda R}{\pi} \sqrt{\frac{D}{2R - D}} = w_o^2 \left\{ \frac{2R}{2R - D} \right\} \qquad (5)$$

In the confocal case R=D, and (5) becomes

$$w_1^2 = \frac{\lambda D}{\pi} = 2 w_o^2 \qquad (6)$$

The parameter ka^2/D introduced in the discussion of diffraction can be written $2a^2/w_1^2$. It is reasonable to assume that in the non-confocal case this latter form will be more appropriate for use in Table 1. The resonant frequencies of the various modes are given by:

$$f_{p\ell q} = \frac{c}{2D} \left\{ q + 1 + \frac{2p + \ell + 1}{\pi} arc\, cos \left(1 - \frac{D}{R} \right) \right) \qquad (7)$$

The theory as outlined above is remarkably successful in optical and microwave applications, and has formed the basis of a number of valuable measurement techniques. Nevertheless, there are a number of approximations involved, and of course no information about the vector field structure is given in the foregoing equations. Specifically, the theory contains the following defects:

(i)   The second derivative of ψ with respect to z has been neglected.
(ii)   The nodal surfaces of the Gaussian beam standing wave are parabolic so that the assumed boundary condition cannot be met over the whole surface of a spherical mirror.
(iii)   The boundary condition is actually incorrect; the electric field, if finite at the mirror must be normal to it. This implies a finite transverse component except on the axis.

4.   COMPOSITE MODES

As mentioned earlier in connection with equation (3) above, the function u can be identified with either Ex or Ey. We can superimpose these two polarizations of waves of the same p and ℓ values to obtain more complex field patterns than either polarization alone. The patterns so obtained are intrinsically interesting, and have the rather

surprising merit of being more fundamental than the linearly polarized modes from which they are constructed, as we shall see. There are two types of composite mode which are important in what follows; both are included in the following equations for the transverse electric field:

$$
\left.
\begin{aligned}
E_x &= \left(\sqrt{2}\,\frac{\rho}{w}\right)^{\ell} L_p^{\ell}\!\left(\frac{2\rho^2}{w^2}\right) \frac{w_0}{w}\, exp\left\{-j(kz - \Phi)\right. \\
&\qquad\qquad \left. -\rho^2\!\left(\frac{1}{w^2} + \frac{jk}{2R}\right)\right\} \cos \ell\theta \\[2ex]
E_y &= \pm\left(\sqrt{2}\,\frac{\rho}{w}\right)^{\ell} L_p^{\ell}\!\left(\frac{2\rho^2}{w^2}\right) \frac{w_0}{w}\, exp\left\{-j\left(kz - \Phi\right)\right. \\
&\qquad\qquad \left. -\rho^2\!\left(\frac{1}{w^2} + \frac{jk}{2R}\right)\right\} \sin \ell\theta
\end{aligned}
\right\} \quad (8)
$$

The two types of mode arise from the choice of sign in the second equation of (8). For the present, following Yu and Luk (2), we designate those with the negative sign as Series A modes and those with the positive sign as Series B modes. Obviously these "composite" modes can be regarded as basic, and the more familiar linearly polarized modes obtained by appropriate superposition of these Series A and Series B modes; as we shall see, in the open resonator context the Series A and Series B modes are to be preferred. Fig 2. shows the electric field line patterns for some of these modes. In the approximate formulation we are using, the electric (and magnetic) fields are entirely transverse, and the usual $TEM_{p\ell}$ mode designation is used for the present. Equation (7) shows that modes having the same values of $2p+\ell$ have the same resonant frequency. Thus, for example, the TEM_{20} mode has the same resonant frequency as the TEM_{12} mode, though their field patterns are quite different. Moreover, in the present approximation, Series A and Series B modes of the same mode numbers have the same resonant frequencies. In what follows, we shall find by applying a more accurate theory that neither of these statements is strictly true.

5.   COMPLEX SOURCE-POINT METHOD

Surprisingly, the starting point of the complex source-point method is the Hertz potential of an infinitesimal dipole. Consider an electric dipole located at the origin of a Cartesian system of co-ordinates and oriented along the x-axis. The corresponding Hertz potential is:

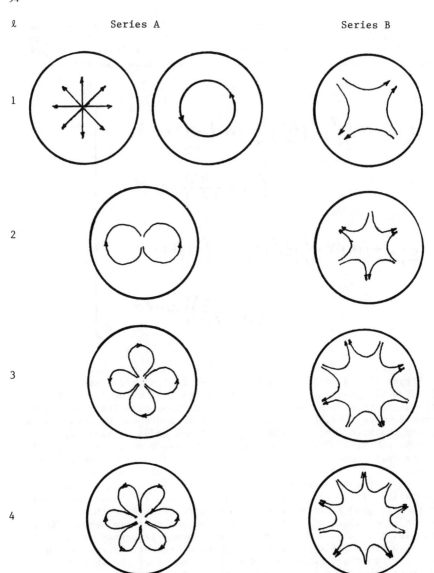

Fig.2. Field patterns for the $TEM_{0\ell}$ modes.

$$\Pi_x = \frac{1}{r} \exp\left(-jkr\right) \tag{9}$$

where

$$r = \left(x^2 + y^2 + z^2\right)^{1/2} \tag{10}$$

The entire electromagnetic field can be found from this potential. Now suppose that the dipole is moved a distance z_0 along the z-axis. (9) is still valid if (10) is replaced by (11) below:

$$r = \left\{x^2 + y^2 + (z + z_0)^2\right\}^{1/2} \tag{11}$$

Physically, the reason is obvious; we have not changed the field by moving the dipole, but we have moved it, and this must be taken into account, as it is when (10) is replaced by (11). The new form of (9) is an exact solution of Maxwell's equations, just as the original form was. Mathematically, the reason for this is that z_0 is a constant; it happens to be a real constant, but it need not be if the sole criterion is that a solution of Maxwell's equation should result. This key observation leads the way to an electromagnetic theory of Gaussian beams, as we shall see. The procedure is to use a purely imaginary "displacement" of the dipole, jz_0 say, and to write:

$$r = \left\{x^2 + y^2 + (z + jz_0)^2\right\}^{1/2} \tag{12}$$

Assume that z_0 is large, and expand (12) binomially thus:

$$r \doteq z + jz_0 + \frac{x^2 + y^2}{z + jz_0} \tag{13}$$

So long as $x^2 + y^2$ is much less than z_0^2 we can write:

$$\Pi_x = \frac{jz_0}{z + jz_0} \exp\left\{-jkz - \frac{jk(x^2 + y^2)}{2(z + jz_0)}\right\} \tag{14}$$

Though different in form, this equation is actually equivalent to (3) for the special case p=0, $\ell = 0$, as the following substitutions will demonstrate:

$$\left.\begin{array}{rcl} \dfrac{kw_o^2}{2} &=& z_o \\[2em] \dfrac{w^2}{w_o^2} &=& 1 + \dfrac{z^2}{z_o^2} \\[2em] R &=& z + \dfrac{z_o^2}{z} \\[2em] \Phi &=& arc\,tan\!\left(\dfrac{z}{z_o}\right) \end{array}\right\} \qquad (15)$$

When (15) is combined with (14) the result is:

$$\Pi_x = \frac{w}{w_o}\,exp\left\{-j(kz-\Phi) - \rho^2\!\left(\frac{1}{w^2}+j\frac{k}{R}\right)\right\} \qquad (16)$$

We recognise this as the fundamental so-called TEM_{oo} mode of which (3) is the general case. However, calculating the field components from (16) we find that finite axial components of E and H exist, so clearly the TEM description is misleading. There is another point we need to make before considering the higher-order modes. That is that the field derived from (16) does not have the required symmetry. This is easily corrected by adding the field associated with a magnetic dipole oriented along the y- axis, and similarly "displaced" by jz_o along the z-axis. When this is done in such a way that the two dipoles make equal contributions to the x-component of the electric field for large values of z, the resultant field has circularly-symmetrical equi-phase surfaces as required for the open resonator theory.

6.   HIGHER-ORDER MODES

The theory for the higher-order modes can be obtained in exactly the same way as that for the fundamental mode, though we now start with electric and magnetic multipoles rather than dipoles. In principle, this is straightforward, but it is actually very complicated in detail. For this reason we give the x-component of the electric field for the case p=0 only.

The following formula is for the Series B set of modes:

$$E_x = -j \frac{w_0}{w} exp\left(-\frac{\rho^2}{w^2}\right)\left\{\left(\frac{\rho}{w}\right)^\ell \cos\ell\vartheta\right\} \times$$

$$\left[\sin\left\{kz - (\ell+1)\Phi + \frac{k\rho^2}{2R}\right\}\right.$$

$$-\frac{\ell^2 + 3\ell + 2}{k^2 w w_0}\sin\left\{kz - (\ell+2)\Phi + \frac{k\rho^2}{2R}\right\}$$

$$+\frac{(2\ell+3)\rho^2}{k^2 w_0^2 w^2}\sin\left\{kz - (\ell+3)\Phi + \frac{k\rho^2}{2R}\right\}$$

$$\left.-\frac{\rho^4}{k^2 w_0^3 w^3}\sin\left\{kz - (\ell+4)\Phi + \frac{k\rho^2}{2R}\right\}\right]$$

When combined with a similar formula for Ey, the transverse field patterns are as shown in Fig 2. In the general case when p=0 does not hold, the equations for the field components are far more complicated, and will not be given here. They are given in Ref.1., together with a much more detailed discussion.

Yu and Luk (4) have shown that there is a close analogy between open resonator modes and optical fibre modes. Specifically, an appropriate notation is $EH_{\ell-1,p+1}$ for the $A_{p,\ell}$ modes, and $HE_{\ell+1,p+1}$ for the $B_{p,\ell}$ modes. Note that the radial and angular indices are in a different order in the EH/HE notation.

The next step is to construct standing wave solutions by adding or subtracting the conjugate of (17). We then find the shape of the nodal surfaces on which the tangential electric field vanishes; mirrors of this special shape could be used, and we should then have a theory of a particular kind of open resonator. For given values of p and ℓ the nodal surfaces for the A and B modes are identical. However, in practice spherical mirrors are used for reasons of manufacture, so the final step is to deform the mirrors into the required spherical shape using a perturbation theory to calculate the corresponding change in the resonant frequency. The resonant frequencies for an open resonator with spherical mirrors for the Series A modes are then found to be:

$$f_{p\ell q}^A = \frac{c}{2D}\left[q + 1 + \frac{2p+\ell+1}{\pi}arc\cos\left(1 - \frac{D}{R}\right)\right.$$

$$\left.+ \frac{2p^2 + 2p\ell - \ell^2 + 2p + 5\ell - 2}{4\pi kR}\right] \quad (18)$$

For the Series B modes the corresponding formula is:

$$f_{p\ell q}^{B} = \frac{c}{2D}\left[q + 1 + \frac{2p+\ell+1}{\pi}\arccos\left(1-\frac{D}{R}\right)\right.$$

$$\left. + \frac{2p^2 + 2p\ell - \ell^2 + 2p - 3\ell - 2}{4\pi kR}\right] \qquad (19)$$

We now notice a striking result: in general these two frequencies are not equal for the same mode numbers. From the simpler scalar theory outlined earlier, TEM modes having the same value of $2p+\ell+1$ have the same resonant frequency. In particular, modes having the same values of p and ℓ have the same resonant frequency. As mentioned before, this turns out not to be true using full vector theory, except for modes with $\ell=0$. In all other cases there is a small difference which depends on the mode number ℓ but is independent of p, at least to the order of accuracy to which we are working. The difference is given by:

$$f_{p\ell q}^{A} - f_{p\ell q}^{B} = \frac{2\ell}{\pi kR}\left(\frac{c}{2D}\right) \qquad (20)$$

We now note a rather fundamental fact. In general the linearly polarised "TEM" modes through which the A and B modes were originally obtained by algebraic addition cannot now be synthesised (as they could before) by combining A and B modes.

This is because the modes have different resonant frequencies; only in the case $\ell=0$ is the synthesis of linearly polarised modes possible. Even in this case the modes synthesised are only approximately linearly polarised, as shown by Cullen and Yu (5) in some detail for the special case of the p=0, $\ell=0$ mode which is the mode of greatest interest in applications. Equation (20) was verified by Yu and Luk (2) in an extensive series of experiments. By way of illustration we show the calculated and measured values of the difference in frequency between the A and B modes for the special case p=0 in Table 2 below. The resonant frequencies are in the range 9.5 to 11.8 GHz depending on the mode numbers q and ℓ. The mirror separation is 136.14 mm and the radius of curvature of the mirrors is 240 mm.

Table 2

Mode numbers		Frequency difference (GHz)	
q	ℓ	Theory	Experiment
5	1	0.0118	0.01209
7	1	0.0145	0.01517
7	2	0.0279	0.02889
7	3	0.0403	0.04206

The theory can therefore be regarded as verified for the p=0 modes, and the more extensive measurements referred to above provide convincing support for cases in which p is finite.

6.   APPLICATIONS

Two applications of the open resonator in microwave measurements, both involving the fundamental p=0, ℓ=0 mode, will now be given; the first is to the measurement of the permittivity of dielectric materials. It is a simple matter to derive formulae for the resonant frequency of an open resonator with a dielectric slab symmetrically located with its centre on the central plane of the resonator. The key equations are as follows:

$$\frac{1}{n} \cot\left(nkt - \Phi_t\right) = \tan\left(kd - \Phi_d\right) \qquad (21)$$

$$-\frac{1}{n} \tan\left(nkt - \Phi_t\right) = \tan\left(kd - \Phi_d\right) \qquad (22)$$

where n is the refractive index, t is the half-thickness of the dielectric slab, and Φ_t and Φ_d are phase angles specified below. For convenience we have written d=D/2. Equation (21) is for symmetric modes with an antinode of E_x in the central plane, whilst (22) is for the antisymmetric case. The phase angles are given by

$$
\begin{aligned}
\Phi_t &= \arctan\left(t/nz_0\right) - \arctan\left(1/nkR_1\right) \\
\Phi_d &= \arctan\left(d'/z_0\right) - \arctan\left(1/kR\right) \\
&\quad -\arctan\left(t/nz_0\right) - \arctan\left(1/kR_2\right)
\end{aligned}
\qquad (23)
$$

where

$$
\begin{aligned}
R_1 &= t + \frac{n^2 z_0^2}{t} \qquad R_2 = \frac{t}{n} + \frac{n^2 z_0^2}{t} \\
z_0 &= \sqrt{\{d'(R-d')\}} \qquad d' = d + \frac{t}{n}
\end{aligned}
\qquad (24)
$$

In the derivation of these formulae, the field matching at the air/dielectric interfaces has been carried out to first order only. A variational calculation is used to improve these results;details are given in reference (6). Formulae can also be obtained relating the Q factor of the open resonator to the loss tangent of the dielectric

material. Results of measurements using this technique have shown it to be capable of very high accuracy; indeed it is probably the most accurate method available for frequencies in the millimetre wave band. The literature contains much information now, including measurements on anisotropic materials. In any measurement method, the best test of reliable results is to measure the same quantity for a number of different values of the parameters of the measuring apparatus. The open-resonator method lends itself particularly well to this kind of cross-check, since it is possible by altering the separation of the mirrors to change the axial mode number. Moreover, if the radius of curvature of the mirrors is also altered, a further check is possible. Table 3 shows results due to Lynch (7) in which both parameters are varied.

Table 3

Wavelength (mm)	Axial mode number	Refractive index
R=914 mm		
8.862	5	1.50127
8.862	7	1.50128
R=152.4 mm		
9.192	5	1.50484
8.718	5	1.50598

These results are for polypropylene, and are extracted from a larger table in (6); they do, however, include the largest and smallest results from that table. Varying the thickness of the sample provides another check on the accuracy of the method. Table 4 below is extracted from a more extensive table given by Cook, Jones and Rosenberg (8).

Table 4

Sample thickness	Refractive index	Loss tangent (microradians)
5.83	1.5359	173.7
8.75	1.5360	167.2
10.16	1.5357	172.8
11.59	1.5360	171.9

These results are for polyethylene. Tables 3 and 4 provide good evidence for the accuracy of the open-resonator technique, which can now be regarded as well-established.

We now turn to the second application, the direct measurement of total scattering cross-sections at microwave frequencies. This application is interesting in demonstrating an important difference between open and closed resonators. A small loss-free perturbing object in a closed resonator will change its resonant frequency, but to first order will have no effect on its Q-factor. In an open resonator, however, the

dominant effect may well be on the Q-factor, since in this case energy can
be scattered out of the resonator and lost. This difference can be put to
good use in measuring total scattering cross-sections for a class of
objects having symmetry about the central plane. The theory is given by
Cullen (9), and space does not permit its inclusion here. The essence of
the matter is that a travelling wave can be synthesised from two standing
waves displaced a quarter wavelength apart in space and in time quadrature.
Then by measuring the change in the Q-factor of the open resonator when
the object is inserted in each of two positions on the axis one quarter-
wavelength apart, the scattering cross-section can be deduced. Fig.3
shows a comparison of theory and experiment for the scattering cross-
section of a sphere, as found by Cullen and Kumar (10). The results for
brass spheres agree well with the Mie theory, but for steel spheres the
cross-section is somewhat larger. This is due to the fact that the steel
spheres absorb energy as well as scattering it, so in this case it is the
'extinction cross-section' which is being measured.

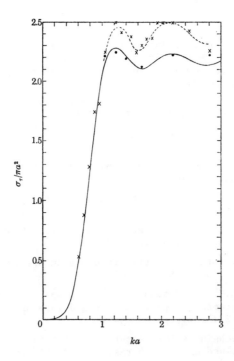

Fig.3. Variation of total scattering cross-section of a
conducting sphere with ka. ———, calculated values (Mie's)
for a perfectly conducting sphere; o, measured values for
brass spheres; x measured values for steel spheres.

7. CONCLUSIONS

In this paper we have set out some of the results of the complex source-point theory as applied to open resonators. We have also indicated why the open resonator is an important device in microwave and especially in millimetre-wave measurement techniques. Placing the theory on a sound foundation is therefore important practically, as well as being of considerable academic interest.

8. ACKNOWLEDGEMENTS

The author wishes to acknowledge the great assistance of Dr. Luk in providing the information about higher-order modes outlined rather briefly here. The work of Dr. Luk and the late Dr. Yu is a major contribution to our understanding of the behaviour of the open resonator. The author wishes also to pay his own tribute to Dr. Yu for a delightful and productive period of collaboration brought to an end by his untimely death. Finally, the author wishes to acknowledge the benefit of a long collaboration with Dr. A.C.Lynch, whose knowledge of how dielectric materials behave, and whose experimental skill and theoretical insights have been immensely valuable.

REFERENCES

1. Felsen L.B. "Complex-source-point Soltuions of the field equations and their relation to the propagation and diffraction of Gaussian beams". Proc. Symp. Mathematical theory of electro-magnetism, Rome Italy, Feb. 1974, pp 19-22.

2. Luk, K.M. and Yu, P.K. "Complex-source-point theory of Gaussian beams and resonators" Proc.IEE, Vol.132. Pt.J. 1985, pp 105-113.

3. Weinstein, L.A. "Open resonators and open waveguides". Golem Press 1969, pp 246-247.

4. Luk, K.M. and Yu, P.K. "Mode designations of laser beams". Proc. IEE, Vol.132, Pt. J. 1985, pp 191-194.

5. Cullen, A.L. and Yu, P.K. "Complex source-point theory of the electro-magnetic open resonator", Proc. Roy.Soc. A Vol. 366, 1979, pp 155-171.

6. Yu, P.K. and Cullen, A.L. "Measurement of permittivity using an open resonator, I", Proc. Roy.Soc. A. Vol 380, 1982, pp 49-71.

7. Lynch, A.C. "Measurement of permittivity using an open resonator II", Proc. Roy. Soc. A Vol. 380 1982 pp 73-76.

8. Cook, R.J., Jones, R.G. and Rosenberg, C.B. "Comparison of cavity and open-resonator measurement of permittivity and loss angle at 35 GHz" Trans. IEEE Vol. IM-23, 1974, pp 438-442.

9. Cullen, A.L. "New techniques for measurement of total scattering cross-section by the use of an open resonator." Electron. Lett. Vol.3, 1967, pp 557-558.

10. Cullen, A.L. and Kumar, A. "The absolute determination of extinction cross-sections by the use of an open resonator". Proc. Roy.Soc. A Vol. 315, 1970, pp 217-230.

SYNTHESIS AND ANALYSIS OF REFLECTOR ANTENNAS FOR SHAPED BEAMS

PROFESSOR P.J.B. CLARRICOATS

QUEEN MARY COLLEGE, UNIVERSITY OF LONDON.

1. INTRODUCTION

Reflector antennas which radiate with a shaped pattern are required in both civil and military applications. In radar and surveillance, shaped beams are demanded and control of sidelobe radiation is an important feature. However it is in satellite communication and satellite broadcasting that the pattern requirements are most exacting and the means for their achievment most challenging for the antenna designer. This is especially the case for spacecraft antennas where the objective is usually to provide optimum gain within the prescribed coverage region or regions and a rapid fall-off outside the coverage zone wherein sidelobes should not rise above prescribed levels. Furthermore, many systems call for dual-polarised operation and a constraint on the level of crosspolarisation is then customary.

Two means to achieve these objectives are now available and are discussed herein. The first utilises a paraboloidal reflector fed from an array whose elements are so positioned and excited that, ideally, the pattern specifications are met. The advantage of an array-fed antenna lies in the potential for precise pattern control and the scope for in-orbit beam reconfigurability, when active elements are included in the beam-forming-network. The principal disadvantage of an array-fed reflector lies in the complexity of the beam-forming-network if the array contains many elements, coupled with reduced efficiency due to losses in the network.

The second means to shape the antenna pattern uses just one feed element or exceptionally just a few, illuminating an appropriately shaped reflector. With a typical single offset reflector of size less than 100 wavelengths, the pattern control is not as precise as with a paraboloidal reflector and an array but if an additional reflector is added to form a dual-offset shaped reflector, the pattern control is then comparable. Obviously there is no scope for in-orbit reconfigurability, a disadvantage to be set against the elimination of a complex and heavy array and beam-forming network.

The design of both types of shaped beam antenna will now be described where extensive use is made of the results of Tun, Bergmann and Zhou Hai whose research at Queen Mary College has been carried out under the supervision of Dr. R.C. Brown and the Author. Following an account of the array method for shaped beam design, two techniques are described for the design of a shaped reflector antenna with a simple feed. In one, geometric optics (GO) is used to fashion the amplitude and phase in the aperture of the main reflector, corresponding to the desired far-field. In the other, an optimisation search routine is used to shape the reflector so that the far-field corresponds directly to the desired far-field. The latter method is simple in concept and execution but it requires considerable

B. de Neumann (ed.), Electromagnetic Modelling and Measurements for Analysis and Synthesis Problems, 103–123.
© 1991 *Kluwer Academic Publishers. Printed in the Netherlands.*

computing power if the design is to be obtained quickly. With improvements in computers in the foreseeable future this may be less of a constraint than at present.

2. THE SYNTHESIS OF ANTENNA PATTERNS USING AN ARRAY FEED
2.1. Introduction

This Section is concerned with the synthesis of a shaped beam by means of a single reflector fed from an array feed as illustrated in Figure 1.

Each element of the feed array produces a focussed beam in a direction approximately determined by its position in the focal plane and the equivalent focal length of the antenna system. These focussed component beams with correct weightings can be combined to produce a number of specific coverage patterns.

Next we describe an initial synthesis procedure that can be used to determine the approximate dimensions of the required antenna system quickly. We then consider a number of optimisation techniques which can be used to perform a final synthesis. Some results obtained for the synthesis of a specific coverage pattern employing a single offset reflector antenna are then presented.

2.2. Idealised Model for Initial Synthesis

In an idealised model, a focussing reflector antenna system is completely defined by its equivalent focal length f_o and the aperture diameter D. From the knowledge of the required angular coverage, one can obtain the corresponding focal-field area by applying

$$\sin(\Theta_p)e^{j\psi}{}_p = \frac{-(x_o + jy_o)}{f_o} \; ,$$

see Figure 2. Essentially, the equation relates the far field beam angle Θ_p in radians to the focal field position r as $\Theta_p = r/f_o$. Note that the size of the focal field area can be varied by choosing different values of f_o.

On the other hand, the aperture diameter D determines the directivity gain and beamwidth of the component beam for a given illumination taper which in turn is determined by the feed aperture diameter. The finite size of the feed horn also determines the closest feed-packing attainable and consequently the minimum spacing between the component beams.

Since a shaped beam antenna has an upper limit for the coverage gain, the principal merit for increasing its aperture diameter is to improve the pattern resolution by decreasing the beamwidths of individual component beams as shown in Figure 1. Note that for the larger antenna in the same Figure, a greater number of feed elements is needed to provide the same coverage, thus requiring a more complex feed-network. It must be remembered that the f_o/D ratio also influences the scanning performance and it is a design factor worthy of attention.

In most cases, the size of the antenna is dictated by the volumetric constraints of the spacecraft launcher. Once the antenna diameter and equivalent focal length have been determined, the angular coverage pattern can be directly translated into a focal field area, into which a number of feed elements can be placed depending on the required resolution.

In the case of generating multiple spot-beams, an increase in the aperture diameter gives rise to a higher directivity gain and narrower beamwidths. If the beams are widely spaced, each spot beam can be generated by a single feed horn placed at the correct position in the focal plane. The feed horn diameter must be such that it produces the desired

illumination taper at the edge of the reflector thus controlling the far field sidelobe levels. For far-out scanned beams, gain loss and pattern degradation caused by phase aberrations in the aperture can be compensated by replacing the single feed with a cluster of smaller feed elements excited appropriately.

Difficulties can arise if these individual spot-beams are to be spaced closely. It is generally impossible to achieve a low cross-over level between the beams as defined in Figure 3 without accepting a very strong illumination at the reflector edge which causes high sidelobes in the far field.

2.3. Shaped Beam Specifications

Before exploring the procedure for synthesising a shaped radiation pattern employing an array-fed offset reflector antenna, it is worthwhile to consider the specifications of a shaped beam pattern as its characteristics are quite different from those of the more common pencil-beam antennas.

Coverage Gain -

Ideally, a shaped beam antenna should provide a uniform gain over the entire coverage area. If one denotes the total angular coverage by a solid angle Ω, the maximum directivity gain attainable for a uniform coverage is given as follows:

$$G_{max} = 10 \ \log_{10}((4\pi)/\Omega) \quad \text{in dB}$$

In practice, meeting the gain requirements for the end of coverage region tends to be a criticial factor, as the far-out scanned beams suffer from considerable gain losses. Also, it must be ensured that the required coverage gain is maintained over the specified frequency band.

Sidelobe Suppression outside Coverage -

If frequency reuse is employed by means of spatial isolation between two adjacent shaped beams, it is imperative that the near-in sidelobes of the beams are strongly suppressed outside the coverage. One must also allow for a transition region in which the copolar gain is falling rapidly to the required suppression level. This rate of fall is represented by the so-called end of coverage gain slope.

Crosspolar Suppression -

If frequency reuse is employed by means of isolation between two orthogonally polarised field vectors, one hand of polarislation must be maintained at a power level well below the other both inside and outside the coverage area. The required suppression must be maintained between the two vectors in the same far-field direction inside the coverage zone and should 'not' be measured relative to the peak copolar level. A typical isolation requirement is between -27 to -33 dB and it must be maintained over the entire frequency band. Note that the crosspolarisation due to feed-array mutual coupling tends to be highly frequency dependent.

Pattern Resolution -

This parameter is best measured by the size of the component beams generated by individual elements in the feed array as shown in Figure 1. Unless the reflector surfaces are shaped, its pattern resolution can be directly related to the number of feed elements. It is an important parameter to be examined if the coverage patterns are to be reconfigured electroncially.

2.4. Optimisation Techniques

Once the far field patterns of the antenna for each feed element in the array-feed have been computed and stored, these patterns can be simply

summed with correct weightings to provide a shaped beam pattern. Thus it avoids the time consuming process of repeating the reflector analysis calculations. Guided by proceeding computations, the optimisation program can generate new sets of weightings until the closest shaped beam pattern to the specified target is achieved.

Each complex excitation coefficient of the feed is represented by two real variables corresponding to the real and imaginary parts. If 'N' is the number of feed elements, optimisation has to be performed in a '2N' variable space along the steepest descent path to a possible minimum value of a predefined error function. Steepest descent search is performed by a standard NAG (National Algorithmic Group) computer library routine. Different error function definitions can be used resulting in a number of optimisation techniques such as least squares, minimax, etc.

Definitions of Error Functions
Least Square Method -
Let the desired far-field pattern, described my M co- and cross-polar values in the far field be denoted by T^{co} and T^x. Each of these target values may be accompanied by weighting coefficients w^{co} and w^x so that some of the points may take priority over the others. If the far field radiation values are given by f^c_o and f^x, the error function is given by the following expression.

$$E = \sum_{i=1}^{M} w_i^{co}(f_i^{co} - T_i^{co})^2 + \sum_{j=1}^{M} w_j^{x} (f_j^{x} - T_j^{x})^2$$

Minimax Method -
Instead of summing up all the 'difference-squared' contributions from the far-field points as in the previous case, only the maximum value is taken as the error function. In this case, derivatives of the error function may possess discontinuities which can affect the steepest descent search process.

$$E = \max\left[w_1(f_1 - T_1)^2, \; w_2(f_2 - T_2)^2, \ldots\ldots\right]$$

Least pth Index Envelope Method -
This method has been first explored by Bandler for microwave filter network optimisation and later adopted by Bird for a shaped beam application. It is an extension of the previous minimax method by including all the far-field points in a simultaneous manner. The optimisation process attempts to force both the copolar and crosspolar patterns into upper and lower limit envelopes as shown in Figure 4 while maintaining continuity of the error function derivatives. p is the performance index value which controls the selection of 'difference values' $(f_i - T_j)$ in order of magnitude. If p is set to a very large value, the error function value is mainly given by the maximum 'difference value' at a single far-field point. Thus the optimisation method becomes a minimax procedure as p tends to infinity. p is usually set to a positive integer number in order to simplify the task of exponentiation.

Let us define 4 sets of 'difference functions' for i=1,m where m is the number of far field points.

$$^{up}F_i{}^{co} = {}^{up}w_i{}^{co} \left(f_i{}^{co} - {}^{up}T_i{}^{co}\right) \quad \text{for } i=1,M$$

$$^{lo}F_i{}^{co} = {}^{lo}w_u{}^{ci} \left(f_u{}^{co} - {}^{lo}T_i{}^{co}\right) \quad \text{for } i=1,M$$

$$^{up}F_i{}^{x} = {}^{up}w_i{}^{x} \left(f_i{}^{x} - {}^{up}T_i{}^{x}\right) \quad \text{for } i=1,M$$

$$^{lo}F_i{}^{x} = {}^{lo}w_i{}^{x} \left(f_i{}^{x} - {}^{lo}T_i{}^{x}\right) \quad \text{for } i=1,M$$

Let Rmax = Max (F_i) and q = sgn (Rmax). p. The following flow chart defines the error function E.

Rmax > 0	Rmax < 0
(Not all requirements met)	(All requirements met but still searching for best)

$$E = R_{max}\left[\sum_{i=1}^{M} \{H(F)/R_{max}\}^q \right]^{1/q} \qquad E = R_{max}\left[\sum_{i=1}^{M} \{F/R_{max}\}^q \right]^{1/q}$$

$$H(F) \quad \begin{aligned} &= F \text{ if } F>0 \\ &= 0 \text{ if } F<0 \end{aligned}$$

Since the crosspolar lower envelope is usually not required, either $^{lo}F_i{}^{x}$ is discarded or $^{lo}w_i{}^{x}$ is set to zero.

Initial Values for Optimisation

The optimisation program requires initial starting values for the complex feed excitation coefficients. Since the subsequent steepest-descent paths tend to be influenced by this starting point, care must be taken to provide a realistic set of excitation coefficients. Since there is usually a direct correspondence between the feed positions and far-field angular coverage zone, a set of uniform amplitude excitations with equal phase has been found satisfactory.

Directivity Gain

Since the far-field pattern is a superposition of a number of component beams with different weightings, the total pattern gain must be renormalised for each set of excitations.

If C_i denotes the complex feed excitation level for the ith feed, the total far field 'power' pattern must be divided by R where

$$R = \sum_{i=1}^{N} |C_i|^2$$

Approximate Component Beam Models

In some circumstances, a cluster of feed elements can be grouped together to form a single component beam such as a clean beam. In this way the number of feed excitation variables for the optimisation can be reduced. This approximate model can be used for an initial crude optimisation. Similarly, isotropic feed models producing $J_1(u)/u$ type component beams can also be employed.

2.5. Prototype Shaped Beam Antenna

Shaped beam antennas employing array feeds are used on board communication satellites e.g. Intelsat 5 and 6. Figure 5 shows a typical requirement for a European coverage. We shall study a synthesis procedure for a similar

coverage employing an array fed single offset parabolic antenna in circular polarisation. The reflector and feed elements are given in Figures 6 and 8.

In the synthesis procedure, mutual coupling and crosspolarisation effects are not taken into account since they cannot be controlled effectively by the copolarised feed excitations. Figure 7 shows the upper and lower copolar target envelopes which are then expressed in a 41 by 41 two-dimensional-cosine-grid. The corresponding feed-array positions are given in Figure 8 arranged on a planar focal surface. A high density 41 by 41 grid has been deliberately used to evaluate the feasibility of specifying very high resolution contours.

If circular polarisation is employed, inherent crosspolarisation due to the offset geometry of the reflector is not present and the feed crosspolarisation is directly projected into the far-field. The copolar contour is shown in Figure 9. Agreement with the original template is excellent. The crosspolar contour, shown in Figure 10, indicates a number of lobes with the peak being at -29 dB below the copolar maximum. These peak crosspolar levels are found to become quite high as the feed element aperture sizes are made quite small (1 wavelength diameter) and strong mutual coupling due to the excitation of TE21 modes is present. If the aperture size is increased, the peak level would fall to a smaller value due to a decrease in the coupling levels.

3. THE SYNTHESIS OF ANTENNA PATTERNS USING GEOMETRIC OPTICS
3.1. Introduction
This Section describes how the prescribed far-field of a dual offset reflector antenna can be synthesised using a method based on geometric optics. First however the amplitude and phase of the antenna aperture field must be obtained from the far-field pattern by means of a search procedure. Then using the GO synthesis technique the coordinates of the reflector surfaces are determined for a given feed pattern. To complete the design procedure surface-fitting followed by a physical optics analysis of the antenna is required.

3.2. Determination of the Aperture Field
Figure 11 shows a dual offset reflector while an example of a typical shaped beam contour is defined by the firm line in Figure 12. Within that contour the gain is prescribed as exceeding 27 dBi while within the region defined by a dotted line, it is desired that it should exceed 25 dBi. The axes of Figure 12 are U and V as in Section 2. Here the reflector diameter is 37 wavelengths compared to 50 wavelengths in the previous example.

The amplitude and phase distribution over the aperture are parameterised by the simple forms

$$AMP = a + b\,(1-r^2)p$$

$$PHA = Ax + By + Cx^2 + Dy^2 + Exy + \sum_{mn} C_{mn}f_m(x)f_n(y)$$

where the aperture coordinates x and y are normalised here to the range $[-\pi/2, \pi/2]$ to allow convenient expression of sinusioids $f_m(x)$ and $f_n(y)$:

$$f(x) = 1,\sin(1x),\cos(1x),\sin(2x),\cos(2x),\ldots.$$

for m,n = 1, 2, 3, 4, 5,....

Most phase distributions over an aperture for shaped beam applications

are basically of bowl shape and hence the quadratics in the phase expansion reduce the number of Fourier terms required. Here a 5 x 5 series is used. The parameters, a, b and p are chosen to give a 10 dB field taper over the aperture.

The coefficients of the phase model are varied in a quasi Newton search loop in which the far-field pattern is calculated and compared to the specification at Nyquist sample points within the coverage. The search minimises an objective function which is a one-sided sum of the fourth power of pattern differences, viz

$$F = \sum_i F_i{}^4$$

$$f_i = [(G_c-G_t)/\Delta Gs]$$

$$F_i = f_i \text{ if } f_i < 0; \text{ otherwise } F_i = 0$$

where G_c, G_t and ΔG_t are the calculated gain, target gain and tolerance respectively. In our example the target gain was set at 27 dBi and tolerance at 10%.

The resulting aperture field has the radiation pattern shown in Figure 12. Its peak gain is 32 dBi, and the pattern fully meets the specification. The phase pattern over the aperture (not shown) is very smooth.

3.3. Geometric Optics Synthesis of the Aperture Field

Once the amplitude and phase of the aperture field are known we may utilise the laws of geometric optics to determine the coordinates of the dual reflector.

We assume a one to one mapping of the feed rays (Θ,ϕ) to aperture points (x,y) as shown in Figure 11. The points (x,y) can be expressed in terms of (Θ,ϕ) by a transcendental equation which involves a real function $L(\Theta,\phi)$ and the aperture path length (x,y). Conservation of energy requires that $L(\Theta,\phi)$ satisfy a second-order, non-linear partial-differential equation of Monge-Ampere type involving $L(\Theta,\phi)$, (x,y), the aperture power $G(x,y)$ and the feed pattern $I(A,\phi)$

$$L = \frac{\overline{\omega}-2\eta e^L+(2e^L+d-1)\overline{\zeta}}{1+d-\overline{\omega}\zeta-\overline{\omega}\zeta+(1-d)\overline{\zeta}\eta}$$

defines a relation between η and ω through the functions $I(\omega)$ and $L(\eta)$. and

$$|L_{\eta\eta}-L_\eta{}^2|^2-(L_{\eta\overline{\eta}} - b)^2 = \pm \frac{I(\eta)}{G(\omega)} H(\omega,\eta,1,L)$$

where

$$b = -\frac{4e^L\gamma}{2\beta-h\delta}$$

$$H = B \{[hl_{\omega\overline{\omega}} - (1-2|1_\omega|^2)]^2-|hl_{\omega\omega}+21_\omega^2|^2)$$

$$B = \frac{8\gamma^2}{(1-\delta)(2\beta-h\delta)(1+|\eta|^2)^2}$$

$$\gamma = |1 + \eta L_\eta|^2$$

$$\delta = 1 - \sqrt{1 + 4|1_\omega|^2} \quad \text{and} \quad h = D/K$$

The Equation is the complex form of the Monge-Ampere equation. The choice of sign corresponds to elliptic and hyperbolic forms.

For the elliptic form there are two types of solution depending on the sign of $(L_{\eta\bar{\eta}} - b)$. By noting

$$c_1 = \text{Real}[L_{\eta\bar{\eta}} + L_{\eta\eta} - L_\eta^2 - 2b]$$

and

$$c_2 = \text{Real}[L_{\eta\bar{\eta}} - L_{\eta\eta} + L_\eta^2 + 2b]$$

one type of solution requires c_1 or c_2 positive and the other both positive.

The synthesised antenna geometry is shown in Figure 11. The offset height is $x_0 = 26.3\lambda$, and the distance from feed to subreflector is $r_0 = 26.3\lambda$. The total path length at the centre of the aperture is 63.7λ. The feed pointing angle β_0 is $12.6°$, and the cone semi-angle β_c is $14°$. The feed model has a power pattern $I = \cos^{125}\Theta'$, where Θ' is the angle from the feed axis.

3.4. Physical Optics Analysis of the Dual Reflector

The output of the GO synthesis leads to a set of points typically between 600 and 1000 on the surfaces of the subreflector and main reflector together with the phase and amplitude of the aperture field. The latter can be compared directly with the target values or, following aperture-integration, with the desired far-field. However, this strategy overlooks subreflector diffraction and furthermore a description of the surfaces is needed in order to manufacture an antenna. In studies reported here the discrete reflector points have been fitted with a quadratic and a Fourier series containing either 3 x 3 or 5 x 5 terms. Typical rms residual errors are less than 2°. The model has continuous second derivatives, as are required in the PO diffractive analysis and this is performed using typically 1000 points on the subreflector and 4000 on the main reflector.

For the example of Section 3.2 Figures 13 and 14 show the final co-polar and crosspolar radiation patterns. For the copolar pattern, the difference between the PO result and the target far-field of Figure 12 lies in diffraction by the subreflector which was about 17 wavelengths in size. Nonetheless the desired coverage has been achieved with a margin with a crosspolar peak at -5.4 dBi which lies 38 dB below the copolar maximum.

4. SYNTHESIS BY OPTIMISATION USING DIFFRACTION

In the synthesis of a shaped beam by the GO technique of Section 3 there are four steps as in Method 1 of Table 1. A more direct approach is that of Method 2 which is described as synthesis by direct optimisation. In this approach the reflector surface is characterised by a polynomial and Fourier series as follows:

$$Z = a\,x + b\,y + c\,x^2 + d\,x\,y + e\,y^2$$
$$+ \sum_{k}^{N_x} \sum_{l}^{N_y} C_{kl}\,f_k(x)\,f_l(y)$$

where $f_k(x) = 1$, $\sin x$, $\cos x$, $\sin 2x$, etc.

for $k = 0, 1, 2, 3$,

For convenience x and y were normalised here to the range $[-\pi/2, [\pi/2]$.

Thus the reflector shape is controlled by the coefficients in the above equation and these can be varied in order that the gain of the antenna within the field of view conforms to that specified at the sample points, to within a prescribed tolerance. The gain of the antenna is calculated using physical optics currents on the main reflector and either GO ray tracing or PO on the subreflector (both techniques have been tried). Then as the block schematic of Figure 15 shows, the desired and specified gains $G(\alpha, \beta)$ are "compared" through a one-sided sum of squared differences as below.

$$F - \sum_{i}^{Nu} \sum_{j}^{Nv} \Delta_{ij}^2 \quad \text{for} \quad \Delta_{ij} = \frac{G_{ij}^c - G_{ij}^s}{\delta G_{ij}} < 0$$

This summation covers Nyquist cells on a regular grid of direction cosines (α, β).
For convenience here all figures use (u, v) coordinates defining a grid by

$$N_u \text{ integral values of } u_i = D\alpha_i$$

and similarly for N_v, v_j and β_j

A number of examples have been successfully attempted. Figure 16 shows the template in uv space of a typical spacecraft antenna pattern (the so-called Contoured Beam Reflector Antenna=COBRA pattern suggested by the European Space Agency) see Figure 5. The points are sampled at Nyquist rate with a reflector of 55 wavelengths diameter. Within continental coverage the desired gain is 29.5 dBi and, as Figures 17 and 18 show, this has been achieved with both single and dual reflector antennas. Specifications for the antennas are in Table 2. The dual-reflector contains more detail in the coverage region, as a consequence of the greater freedom in phase and amplitude over the main reflector.
The method has also been applied to a reflector intended to illuminate two geographical regions using two feeds. The template is shown in Figure 19 together with the final copolar contour. Details of the reflector and feed are in Table 2. Both the gain in the coverage zones and the maximum crosspolarisation was achieved.
Evidently the synthesis method using diffractive optimisation works satisfactorily but compared to the GO synthesis method it requires a high computing rate. Figure 20 shows how the objective function depends on iteration number for the last example. When using a Cray 1s computer with vectoristion each iteration occupies about 20 seconds of CPU time. However, in a total of 15 minutes a complete design is available with minimal operator intervention.

5. CONCLUSIONS

The paper has considered two types of reflector antenna for the generation of shaped beam radiation patterns namely, the array-fed reflector and the shaped reflector with a simple feed. In the latter, two methods of design have been considered. The array design has the advantage of precise pattern control subject to constraints imposed by the reflector size and

TABLE 1.	
METHOD 1	METHOD 2
GO SYNTHESIS	SYNTHESIS BY DIRECT OPTIMISATION
(A) OBTAIN APERTURE FIELD FROM FAR-FIELD	(A) INSERT PO PREDICTION IN SEARCH PROGRAM WITH REFLECTOR SURFACE(S) AS VARIABLE(S)
(B) USE GO SYNTHESIS TO CREATE POINTS ON REFLECTOR(S)	
(C) FIT SURFACES OF REFLECTOR(S)	(B) SET BOUNDS ON COVERAGE ZONE AND SIDELOBES
(D) USE PO TO PREDICT FAR-FIELD PATTERN	(C) CONTINUE SEARCH UNTIL ACTUAL AND DESIRED PATTERNS AGREE TO WITHIN SPECIFIED TOLERANCES

TABLE 2

COBRA beam with a single reflector antenna. The geometrical parameters are :

$Hx=31\lambda$, $Ax=27.5\lambda$, $By=27.5\lambda$, $f=50\lambda$, $\Theta*=28.08$ deg.

The feed is placed at $(0.,0.,0.)$, $\Theta_0=32.02$ deg., the feed taper is 17.4 dB which corresponds to $n=16$.

COBRA beam with a dual reflector antenna. The geometrical parameters are :

$Hxmr=31.5\lambda$, $Axmr=27.5\lambda$, $Bymr=27.5\lambda$, $f=45$, $Hxsr=-14.284\lambda$, $Axsr=11.0\lambda$, $Bysr=12.5\lambda$, $F=23.606\lambda$, $e=0.38367$, $\alpha=-10.306$ deg.

The feed taper is 21.81 dB, which corresponds to $n=40.464$ with $\Theta*=19.9$ deg. $\Theta_0=23.099$ deg.

AFSAT beam with a single reflector in conjunction with two feeds. The geometrical parameters are :

$Hx=25\lambda$, $Ax=20\lambda$, $By=20\lambda$, $f=40\lambda$, $\Theta_0=33.8$ deg.

The feed is placed at $(0.,2\lambda,0.)$ and $(0.,-2\lambda,0.)$ respectively. $\Theta*=27.3$ deg. The excitation coefficients are $(0.4289,-0.4686)$ and $(0.5318, 0.5598)$ respectively. The taper is 12.15 dB with $n=11.812$.

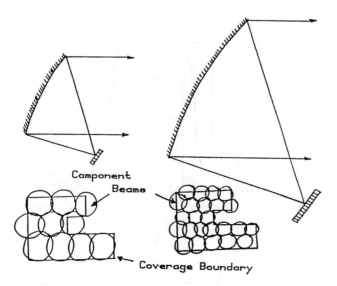

FIGURE 1. ANTENNA SIZE & PATTERN RESOLUTION.

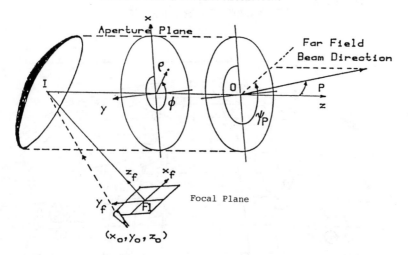

FIGURE 2. FOCAL & APERTURE PLANE COORDINATE SYSTEMS.

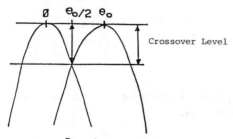

FIGURE 3 BEAM CROSSOVER LEVEL.

114

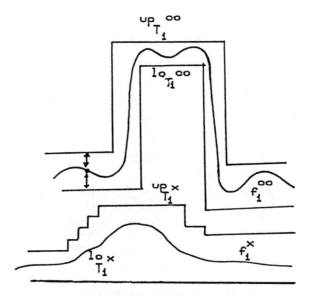

FIGURE 4. Co- & Cross-polar Envelopes

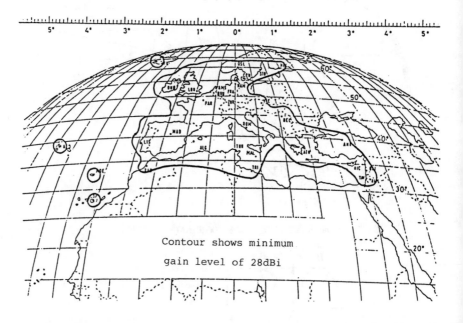

Contour shows minimum
gain level of 28dBi

FIGURE 5. A Typical Requirement for a European
Shaped Beam Coverage.

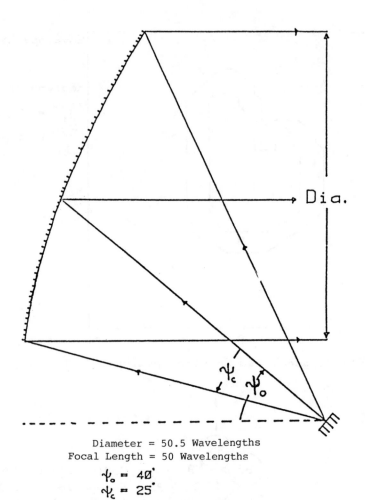

Diameter = 50.5 Wavelengths
Focal Length = 50 Wavelengths

$\psi_o = 40°$
$\psi_c = 25°$

FIGURE 6. SINGLE OFFSET PARABOLIC REFLECTOR ANTENNA
WITH ARRAY FEED.

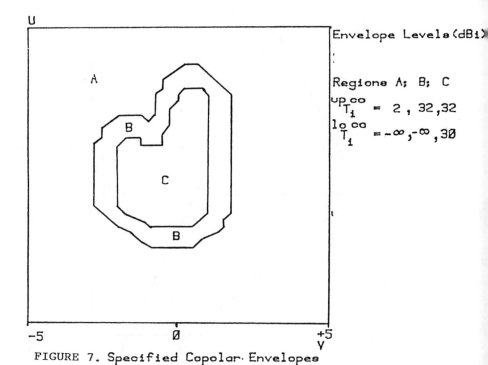

FIGURE 7. Specified Copolar Envelopes

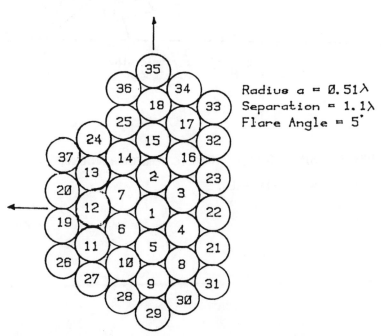

FIGURE 8. 37-Element Dominant Mode Conical Horn Array.

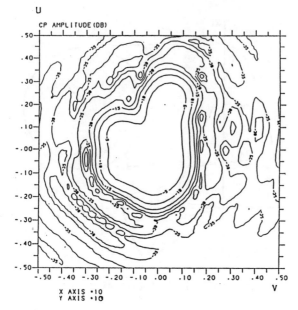

FIGURE 9. Circularly Polarised Far Field Contour of Prototype Antenna fed by a 37 Element Feed Array. Mutual Coupling included.

Peak Level = -29 dB below Copolar

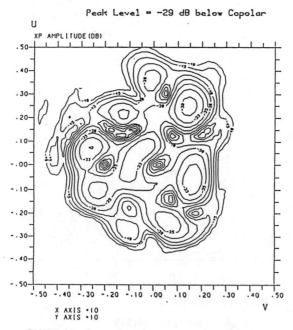

FIGURE 10. Corresponding to Figure 11 Crosspolar Contour

118

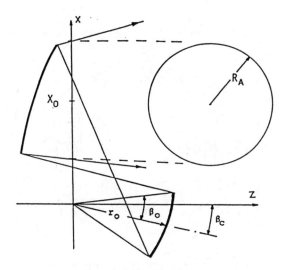

FIGURE 11 GREGORIAN. CONFIGURATION

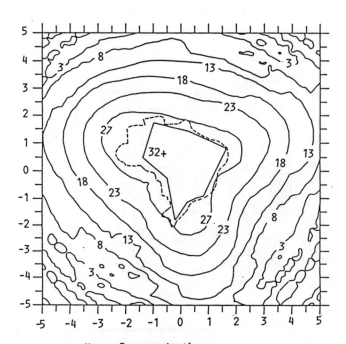

FIGURE 12 - MODEL PATTERN (dBi)

AXES ARE DIRECTION COSINES X37

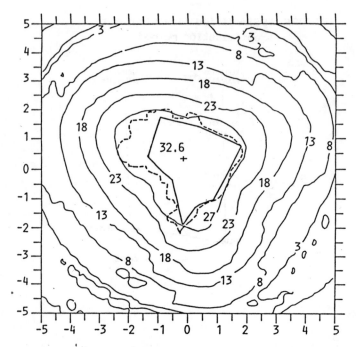

FIGURE 13 COPOLAR PATTERN FROM PO ANALYSIS (dBi)

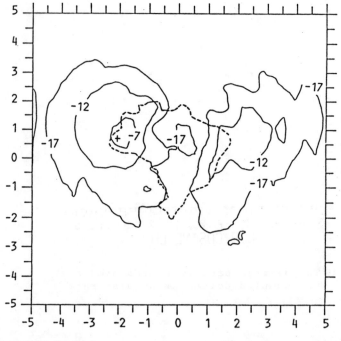

FIGURE 14 CROSSPOLAR PATTERN FROM PO ANALYSIS (dBi)

120

FIGURE 15Simplified flow diagram for
optimisation method

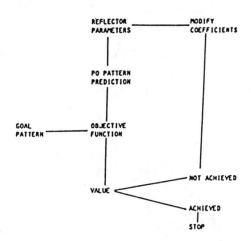

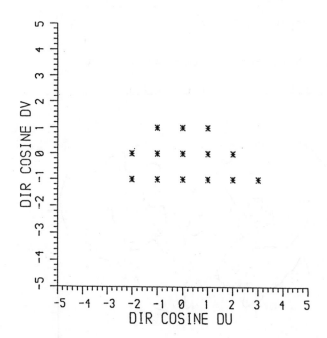

FIGURE 16 The desired pattern represented by
the sampled points at Nyquist rate

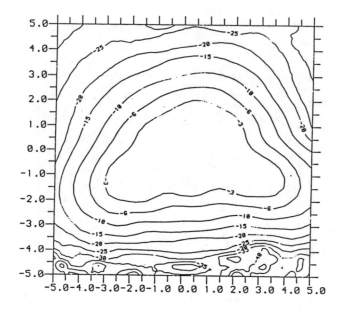

FIGURE 17 The radiation pattern for the single
reflector synthesis.   G = 32.328dBi

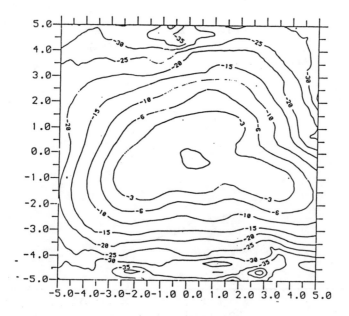

FIGURE 18 The synthesised coplar radiation
pattern for a dual reflector.
G = 33.145dBi.

FIGURE 19. Copolar Contours and Sample Points.

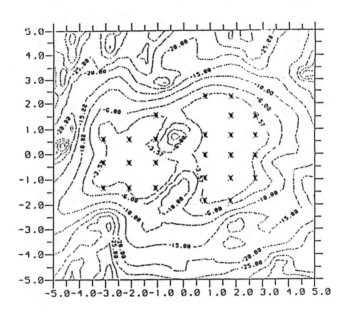

FIGURE 20.

Optimisation Procedure of AFSAT.
Array.

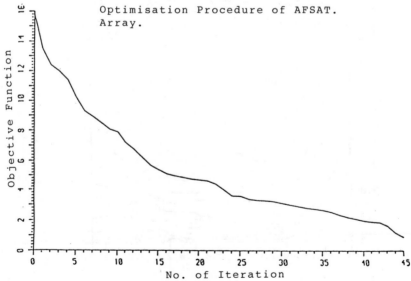

the minimum spacing of array elements. It also offers the potential for in-orbit reconfigurability. By contrast the shaped reflector offers the advantage of a simple feed system at the sacrifice of reconfigurability. Two methods of design have been examined; in one the aperture field is first determined from the far-field, then a GO synthesis procedure used to create the aperture field. Additional steps involve fitting the reflector surfaces and PO analysis of the final antenna. Alternatively we have shown that the antenna can be designed by an optimisation procedure in which the reflector is shaped to create the desired far-field pattern. Since PO is employed in the search, the final reflector performance is known and data is stored which could be used directly to manufacture the reflectors. The disadvantage of the method lies in the high demand for computing rate which could usually be met only with a large and presently expensive computer. As computer costs fall the method is likely to gain favour with the antenna community. It has the additional advantage of giving sensitivity information directly, which will be valuable when manu-facturing tolerances are set. Designers having access to both of the methods may find advantage in using GO synthesis to give starting values for their later use in the Optimisation Method.

6. ACKNOWLEDGMENT

The Author gratefully acknowledges the help of Drs. Roy Brown, Soe Min Tun, Jose Bergmann and Mr. Zhu Hai whose results have been used in the preparation of the Paper.

7. REFERENCES

1. CLARRICOATS, P.J.B., TUN, S.M. and PARINI, C.G.: 'Effects of mutual coupling in conical horn arrays', Proc. IEE, 131, Pt. H, 1984, pp.165-171.
2. CLARRICOATS, P.J.B., TUN, S.M. and BROWN, R.C.: 'The performance of offset reflector antennas with array feeds', Proc. IEE, 131, Pt. H, 1984, pp.172-178.
3. SCHRUBEN, J.S.: 'Formulation of reflector-design problem for lighting fixture', J. Opt. Soc. Am62, December 1972,
4. WESTCOTT, B.S., 'Shaped reflector antenna design', Research Studies Press, 1983.
5. BERGMANN, J.R., BROWN, R.C. and CLARRICOATS, P.J.B.: 'Dual reflector synthesis for specified aperture power and phase', Electronics Letters 85, 21, pp.820-821.
6. BROWN, R.C., CLARRICOATS, P.J.B. and ZHOU, H.: 'Optimum shaping of reflector antennas for specified radiation patterns", Elec. Letts., Vol. 21, No. 24, 21st November, 1985, pp.1164-1165.
7. BROWN, R.C., CLARRICOATS, P.J.B. and ZHOU, H.:'Shaped-beam reflector antennas design using optimisation', Military Microwaves' 86, 24th-26th June 1986, Brighton, England. pp.146-151.
8. BROWN, R.C., CLARRICOATS, P.J.B. and ZHOU H.: 'Synthesis of satellite beams using reflector antennas', Procs. International Conference on Antennas and Propagation, 1987, Session 13A, Paper 4.

A scheme to analyze problems of resonant size using the Conjugate–Gradient method and the Fast Fourier Transform.

M. F. Cátedra, Jesús G. Cuevas, E. Gago.
E.T.S.I. de Telecomunicación. Universidad Politécnica de Madrid.
28040 Madrid. SPAIN.

1. INTRODUCTION.

Electromagnetic problems can be clasified according to the size of the objects involved:

−Electrically small : the size of the bodies is much smaller than a wavelength.
−Resonant: the size of the bodies is of the order of a wavelength.
−Electrically large: the size of the bodies is much greater than a wavelength.

The first kind of problems can be solved by the Moment Method (MM) but in most cases the treatment gets troublesome. For resonant problems MM becomes inefficient in terms of CPU time and computer memory and GTD is still inapplicable. Hybrid methods are valid only for a few special geometries. Finally for the third kind of problems GTD is applicable when the bodies have smooth surfaces.

The technique presented here, CG+FFT[1], has its application in the first two problems, specially in the second one, where CPU time and memory requirements grow up when matrix methods are employed.

The memory difficulties pointed above can be partially avoided by applying iterative methods to the resolution of the chosen equation, [2],[3], [6],[7]. In spite of its memory–saving capacity, iterative methods have the disadvantage of being slow, and some of them do not always converge. In the case of the Conjugate–Gradient method, (CG), its convergence is intrinsic and its speed is improved by using the Fast Fourier Transform algorithm (FFT). It is not the aim of this paper to go into mathematical details about the CG method, but just to give some ideas about its performance that will be neccesary to understand how CG+FFT works (for further details see [1],[2],[3],[6],[7],[8],[9]).

The CG algorithm mentioned below was taken from [2] and it is applied to solve, in an iterative way, operator equations of the form:

$$LI = Y \qquad\qquad\qquad (1)$$

where L is the operator, Y is the excitation function and I is the unknown function. At each iteration an approximation $\hat{I}$ to I is computed and to obtain $\hat{I}$ it is necessary to apply operator L and its adjoint L^a over known

---

[1] CG+FFT stands for 'Conjugate–Gradient plus Fast Fourier Transform' technique

B. de Neumann (ed.), Electromagnetic Modelling and Measurements for Analysis and Synthesis Problems, 125–143.
© 1991 Kluwer Academic Publishers. Printed in the Netherlands.

functions. The iterations end when the residual error is smaller than a certain pre-established value.

The fundamental idea lying in the CG+FFT technique is as follows: when the two operators L(.) L^a(.) appearing in the GC method have convolution nature (which is a very common situation) they can be easily evaluated, in an approximate way, by using the FFT algorithm, once the operator has been discretized.

By using the FFT, [4],[5], together with the CG method a technique is obtained (CG+FFT) that avoids the memory requirement problems of matrix methods as well as overcomes the CPU time problems inherent to iterative methods.

To summarize, the most salient features of MM and CG+FFT are shown next. N being the number of unknown in the discretizing process, we have:

(i)   CG+FFT needs to store a few N-sized vectors in computer memory in opposite to the N^2 storage requirement in the case of MM.

(ii)   Time neccesary to solve (1) by MM can be put as $t_{MM} = KN^3 + K'N^2$ while for CG+FFT we have $t_{CG+FFT} = K''N^2 \ln N$, which reduces in practice to $t_{CG+FFT} = K''N \ln N$

(iii) Another advantage of the CG+FFT technique is that being an iterative approach it does not propagate roundoff errors.

Once the performance of CG+FFT has been outlined the rest of the paper describes operations L(.) and $L^a(.)$ in several electromagnetic situations.

2. ELECTROSTATIC PROBLEMS.

Electrostatic problems are presented here since being simple in its formulation they could be helpful in understanding the operator discretizing process and the operator expression in terms of FFT.

In the electrostatic problems we try to find the charge distribution on a flat plate of arbitrary shape at a constant potential V_0 (Fig. 1). The operator equation is found by making constant the electrostatic potential for points r on the surface plate S_c.

$$\int_{S_c} G(r - r') \sigma(r') dr' = V_0 \qquad ; \; r \in S_c \qquad (2)$$

where $G(r)=1/|r|$ is the electrostatic Green's function and $\sigma(r)$ is the surface charge density. Then the operator L is the electrostatic potential at any point (L is an integral operator). Because no complex magnitude appears in (2) the operator is selfadjoint. Next step is the discretization of the operator. To do this the plate geometry is inscribed in a grid

of $N_x \times N_y$ small rectangles, each one of area $\Delta x \, \Delta y$ as in figure 1. Next, the charge density is expanded as a pulse series:

$$\sigma(r) = \sum_{i=o}^{N_x} \sum_{j=o}^{N_y} \sigma^D(i,j) \, P_{ij}(r) \tag{3}$$

where $P_{ij}(r)$ is the pulse function with value 1 on the ij–rectangle and zero on the rest. $\sigma^D(i,j)$ is a two-dimensional sequence of unknown coefficients.

Dirac's delta functions, δ, centered on each rectangle are chosen as testing functions. By substituting (3) in (2) and testing with δ–functions the following discretized operator is obtained

$$[L(\sigma)]_{ij} = \sum_{i'j'} \sigma^D(i',j') \int_{r_{i'j'}^-}^{r_{i'j'}^+} G(r_{ij}-r')ds' =$$

$$\tag{4}$$

$$= \Delta x \Delta y \sum_{i'j'} \sigma^D(i'-i, j'-j) = \Delta x \, \Delta y \, \sigma^D \otimes G^D$$

where the discrete Green's sequence $G^D(i,j)$ is approximated as follows:

$$G^D(i,j) = \frac{1}{4\pi\epsilon} \begin{cases} [(i\Delta x)^2+(j\Delta y)^2]^{-\frac{1}{2}} & i,j\neq0 \\ \dfrac{1}{\Delta x \, \Delta y} \displaystyle\int_{-\Delta x/2}^{\Delta x/2} \int_{-\Delta y/2}^{\Delta y/2} (x^2+y^2)^{-\frac{1}{2}} \, dxdy & i=j=0 \end{cases}$$

Expression (4) is a discrete convolution and the sign $\otimes$ stands for it. This convolution will be computed by the FFT algorithm and so it is necessary to do the following considerations:

(i)   In order to take into account all charges, the length of non–zero values of sequence G^D must be at least twice σ^D. (fig. 2)

(ii) The convolution will be computed as a circular convolution. Because the size of G^D is twice as much as that of σ^D, then the period of circular convolution must be twice the size of σ^D in order to avoid overlapping in the zone of charge (fig. 3).

The final form of the discrete operator (and its adjoint) is:

$$[L(\sigma)]_{ij} = \Delta x \, \Delta y \, FFT^{-1} \{ \bar{\sigma}^D(n,m) \, \bar{G}^D(n,m) \} \tag{5}$$

where $\bar{\sigma}^D(n,m) = FFT\{\sigma^D(i,j)\}$ and $\bar{G}^D(n,m) = FFT\{G^D(i,j)\}$, (last equation is valid only for samples that belong to the conductor).

Several electrostatic cases have been solved using this CG+FFT approach. As an example the charge distribution on a circular disk at 1 Volt is shown in figure 4. The agreement between the CG+FFT result and the analytical solution is very good even in the asymptotic rim zone.

Finally it is important to point out that because of the simplicity of the integral operator (for electrostatic problems), the selection of basis or testing functions is not very critical. This is the reason for which pulse functions and point matching for testing give good results.

3. ELECTRODYNAMIC PROBLEMS. BIDIMENSIONAL CONDUCTOR PLATES

In the case of electrodynamic problems, the procedure is similar to the one of the previous section, but now taking into account that the operator is no longer selfadjoint and an expression for the adjoint operator needs to be obtained. Also the integro—differential nature of operators involved now needs a careful choice of basis and testing functions. In fact these functions play a fundamental part in the accuracy and in the rate of convergence of this technique.

The operator equation is derived from EFIE,

$$-\hat{n} \times E^i = \hat{n} \times E^s =$$

$$= \hat{n} \times \{k^I \int_{S_c} G(r-r') \, J(r')ds' + k^C \, \nabla \int_{S_c} G(r-r')\nabla'J(r')ds' \, \} \tag{6}$$

where $G(r-r')=\exp(-jk|r-r'|)/|r-r'|$; $k=2\pi/\lambda$; $J=(J_x,J_y)$ is the unknown induced surface density current; E^i and E^s are the incident and scattered fields respectively, $K^I=-j\omega\mu/4\pi$, $K^C=1/(4\pi\epsilon\omega)$, and $\hat{n}$ is the normal to S_c. Splitting (6) in its components, the following expression for the operator is left:

$$\begin{bmatrix} L(J_x) \\ L(J_y) \end{bmatrix} = \hat{n} \times \{ \int_{S_c} ds' \begin{bmatrix} k^I G(r-r')+k^C\dfrac{\partial}{\partial x}G(r-r')\dfrac{\partial}{\partial x'} & k^C\dfrac{\partial}{\partial x}G(r-r')\dfrac{\partial}{\partial y'} \\ k^C\dfrac{\partial}{\partial y}G(r-r')\dfrac{\partial}{\partial x'} & k^I G(r-r')+k^C\dfrac{\partial}{\partial y}G(r-r')\dfrac{\partial}{\partial y'} \end{bmatrix} \begin{bmatrix} J_x(r') \\ J_y(r') \end{bmatrix} \} \tag{7}$$

The basis functions must be chosen to meet boundary conditions at the plate edge, (e.g. any current component must vanish when it is normal to an edge of the plate). A possible set of basis functions are rooftop functions T^x_{i1} T^y_{i1}, fig. 5, wich are Cartesian products of triangles and pulses. These functions have been widely used in MM applications . Testing functions are one-dimensional pulses R^x_{i1} R^y_{i1} (fig. 5). These types of basis and testing functions are differently oriented depending on which component we are expanding, J_x or J_y. (see fig. 6)

Other functions that will be necessary are the partial derivatives of the rooftop functions. These are defined as a pair of two-dimensional pulses, and appear in fig. 5 (only for x-component).Also in figure 5, P_{i1}^x is defined. The definition of P_{i1}^y is analog to the one for the x-component. Finally other functions being used are the two -dimensional pulses, P_{i1}^{qx} P_{i1}^{qy}, equally centered and with equal volume that T_{i1}^x and T_{i1}^y respectively (fig. 6).

These definitions lead to the following expressions:

$$J_x(x,y) = \sum_{i=1}^{N_x} \sum_{l=1}^{N_y+1} J_x^D(i,l)T_{i1}^x \quad ; \quad J_y(x,y) = \sum_{i=1}^{N_x+1} \sum_{l=1}^{N_y} J_y^D(i,l)T_{i1}^y$$

$$\frac{\partial J_x}{\partial x} = \sum_{i=1}^{N_x+1} \sum_{l=1}^{N_y+1} \frac{J_x^D(i,l)-J_y^D(i-1,l)}{\Delta x} P_{i1}^q$$

$$\frac{\partial J_y}{\partial y} = \sum_{i=1}^{N_x+1} \sum_{l=1}^{N_y+1} \frac{J_y^D(i,l)-J_y^D((i,l-1)}{\Delta y} P_{i1}^q$$

(8)

Next step is to substitute (8) in (7) and test with R_{i1}^x and R_{i1}^y functions. The procedure to do these operations depends on whether the term has derivatives or not.

a) For terms of the form $\int G(r-r')J_x(r')ds'$, the procedure is to substitute J_x of (8) but changing $T_{i',1}^x$, by its approximation $P_{i',1}^{q,x}$. (the pulse equally centered and with equal volume that $T_{i',1}^x$). After that the testing with R_{i1}^x leads to an integral wich is solved by the midpoint rule. The final result is:

$$\Delta x \sum_{i'} \sum_{1'} J_x^D(i',1') \int_{S_c} G(x_i-x', y_1^--y') P_{i',1}^x, \ ds' = J_x^D(i,1) \otimes G^D(i,1)$$

(9)

where:
$$G^D(i,1) = \int_{-\Delta x/2}^{\Delta x/2} \int_{-\Delta y/2}^{\Delta y/2} G(x_i-x', y_1^--y')dx'dy'$$

$$x_i^{\pm} = (i\pm\tfrac{1}{2})\Delta x; \ y_1^{\pm} = (1\pm\tfrac{1}{2})\Delta y$$

b) For the other terms (which have derivatives) it follows:

$$\int_{x_i^-}^{x_i^+} dx \frac{\partial}{\partial x} \int_{S_c} G(x-x',y_1^--y') \sum_{i'} \sum_{1'} \frac{(J_x^D(i',1')-J_x^D(i',1'-1)}{\Delta y} P_{i',1}^q, \ ds' =$$

$$\frac{1}{\Delta y} \sum_{i'} \sum_{1'} (J_x^D(i',1')-J_x^D(i',1'-1)) \int_{S_c} [G(x_i^+-x',y_1^--y')-G(x_i^--x',y_1^--y')]P_{i',1}^q, ds'$$

$$- \frac{1}{\Delta y} \; [J_x^D(i,1) - J_x^D(i,1-1)] \otimes [G^D(i+1,1) - G^D(i,1)] \tag{10}$$

The discrete function G^D in the last expression is the same as the one which appears in a) due to the spatial invariance of Green's function. According to a) and b) the final expression for the discretized operator $L(J)$ is:

$$
\begin{bmatrix} [L(J_x)]_{i1} \\ [L(J_y)]_{i1} \end{bmatrix} = \begin{bmatrix} g_x(i,1) & 0 \\ 0 & g^y(i,1) \end{bmatrix} . \text{FFT}^{-1} \;\{
$$

$$
\begin{bmatrix} C_x^I + C_x^q (1-F_x^*(n))(F_x(n)-1) & C_y^q (1-F_y^*(m))(F_x(n)-1) \\ C_x^q (1-F_x^*(n))(F_y(m)-1) & C_y^I + C_y^q (1-F_y^*(m))(F_y(m)-1) \end{bmatrix} . \begin{bmatrix} \overline{G}^D(n,m) \overline{J}_x^D(n,m) \\ \overline{G}^D(n,m) \overline{J}_y^D(n,m) \end{bmatrix} \}
$$

$$\tag{11}$$

where;

$$C_x^I = \frac{-j\omega\mu A_x}{4\mu} \;\; ; \;\; C_x^q = \frac{-j}{4\mu\epsilon\omega\Delta x} \;\; ; \;\; F_x(n) = \exp(j2\pi n/2(N_x+1))$$

$$C_y^I = \frac{-j\omega\mu A_x}{4\mu} \;\; ; \;\; C_y^q = \frac{-j}{4\mu\epsilon\omega\Delta y} \;\; ; \;\; F_y(m) = \exp(j2\pi m/2(N_y+1))$$

$$\overline{G}^D = \text{FFT}(G^D) \qquad \overline{J}_x^D = \text{FFT}(J_x^D)$$

and $g^x(i,1)$ and $g^y(i,1)$ take into account the existence of a conductor in the rectangle. Factors $(1-F_\alpha^x(.))$ come from application of the shifting property of FFT to sequences G^D and σ^D. Bar over sequences means FFT.

The discrete adjoint operator is derived directly from (11). This expression can be written in a symbolic way as:

$$LJ = \begin{bmatrix} L_{xy} & L_{xy} \\ L_{yx} & L_{yy} \end{bmatrix} \tag{12}$$

(where e.g. L_{xy} is the part of the operator that applies over J_y to give the x component of E field).

The inner product is defined as:

$$<A,LJ> \; = \; <A_x, L_{xx} J_x> + <A_y, L_{xy} J_y> + <A_y, L_{yx} J_x> + <A_y, L_{yy} J_y>$$

where each of these terms is of the form:

$$<B_\alpha, D_\alpha> = \int_{S_c} B_\alpha \; D_\alpha^* \; ds \approx \Delta x \Delta y \sum_i \sum_1 B_\alpha^D(i,1) \; D_\alpha^{D*}(i,1) \tag{13}$$

Now applying the definition of the adjoint operator we have

$$<A,LJ> \; = \; <L^a A, J> \; \Rightarrow \; \begin{cases} <A_x, L_{xx}J_x> \; = \; <L^a_{xx}A_x, J_x> \\ <A_x, L_{xy}J_y> \; = \; <L^a_{yx}A_x, J_y> \\ <A_y, L_{yx}J_x> \; = \; <L^a_{xy}A_y, J_x> \\ <A_y, L_{yy}J_y> \; = \; <L^a_{yy}A_y, J_y> \end{cases} \tag{14}$$

that finally leads to:

$$\begin{bmatrix} [L(J_x)]_{i1} \\ [L(J_y)]_{i1} \end{bmatrix} = \begin{bmatrix} g_x(i,1) & 0 \\ 0 & g^y(i,1) \end{bmatrix} . FFT^{-1} \{$$

$$\begin{bmatrix} c_x^{I\,*} + c_x^{q\,*}(1 - F_x(n))(F_x^*(n)-1) & c_x^{q\,*}(1 - F_x(n))(F_y^*(m)-1) \\ c_y^{q\,*}(1 - F_y(m))(F_x^*(n)-1) & c_y^{I\,*} + c_y^{q\,*}(1 - F_y(m))(F_y^*(m)-1) \end{bmatrix} . \begin{bmatrix} \overline{\overline{G}}^D(n,m)\overline{J}^D_x(n,m) \\ \overline{\overline{G}}^D(n,m)\overline{J}^D_y(n,m) \end{bmatrix} \}$$

$$\tag{15}$$

As can be seen the last expression is the transpose and complex conjugate of (11)

Some results obtained using the CG+FFT technique, outlined here, are depicted in figures 7 - 11. In general, agreement is very good between CG+FFT results and measurements or results obtained by other techniques.

Special care has been given to analyze slot antennas as an application of CG+FFT to planar structures. Slot antennas, [16], are widely used in broadband applications. These antennas are made by widening a slot line with a certain profile, giving rise to several variants, LTSA, ETSA, etc (see figure 12). Here slot antennas are modelled by a rectangular grid of 64 by 64 samples. The analysis is made considering the antenna as a radiator and a theoretical model of the feed is needed. This model is highly critical for the H-plane pattern. The choice has been a short dipole lying on the antenna plane at the begining of the transmission line, perpendicularly to the longitudinal axis of the antenna, as shown in fig. 10.

Using CG+FFT the radiation pattern (magnitude and phase) can be obtained, and also the center of phase (which is very important for array applications). Figures 13,14 show radiation patterns for both E-plane and H-plane for the geometry of figure 12. Agreement between CG+FFT-predicted results and analytical or measured results is good in general. It is necessary to bear in mind that the experimental results were obtained, in receiving, by placing a diode between the conductors of the antenna, and the dynamic range of diode was not high enough, [16].

4. THREE-DIMENSIONAL PROBLEMS

For three-dimensional bodies the formulation is based on volumetric rather than surface currents. This formulation has been chosen because both

dielectric and conductor bodies will be studied. For conducting bodies the operator to be discretized is:

$$E^s(J) = \frac{-j\omega\mu_o}{4\pi} \int_V J(r')G(r-r')dv' + \frac{1}{j\omega\epsilon_o 4\pi} \nabla \int_V G(r,r')\nabla'J(r')dv' \qquad (16)$$

where v is the volume of body, J is the volumetric current density, E^s is the scattered field at point r, and G(r) is the three-dimensional free--space Green's function.

The operator equation is obtained forcing the total electric field t vanish inside the body. This leads to:

$$E^i = - E^s(J) \qquad ; r \epsilon v \qquad (17)$$

Where E^i and E^s are the incident and scattered fields respectively. For dielectric bodies and taking into account the equivalence theorem, the operator equation is left:

$$E^i = \frac{J^{eq}}{j\omega(\epsilon-\epsilon_o)+\sigma} - E^s(J^{eq}) \qquad (18)$$

where E^s is the operator of (16).

The discretizing process for these operators is similar to the one outlined in previous sections. First, the body geometry is inscribed in a cubic mesh of $(N_x+1).(N_y+1).(N_z+1)$ small cells. In these cells three--dimensional rooftop functions are defined as basis. Three-dimensional rooftops functions are cartesian products of a triangle and two pulses in the orthogonal directions. For instance to expand the x component of current the rooftop function is:

$$T^x_{ijk} = \Lambda(x-x_i)\Pi(y-y_1^-)\Pi(z-z_k^-) \qquad (19)$$

Similarly, testing functions are pulses along a direction and δ's along the others, so for the x component we have:

$$R^x_{ijk} = \Pi(x-x_i)\delta(y-y_1^-)\delta(z-z_k^-) \qquad (20)$$

The rest of the procedure is similar to the two-dimensional case.

This formulation has the additional advantage of being profitably applied in problems with several conductivities and/or different electric permittivities.

Results for RCS of several dielectric and conductor bodies are presented in figures 15,17. The agreement is very good in general. In the cases where it is not so good, we believe that it is due to the small number of samples employed in modelling the body.

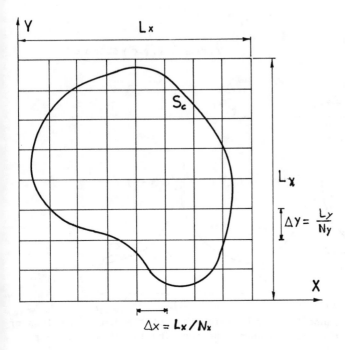

$$\sigma(r') = \sum_{i=0}^{N_x} \sum_{j=0}^{N_y} \sigma^{-D}(i,j) \, P_{ij}(r')$$

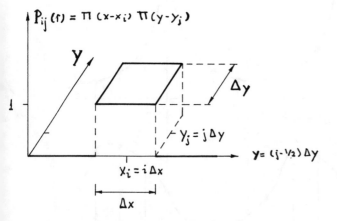

Fig. 1. Geometry and basis functions for the electrostatic case.

134

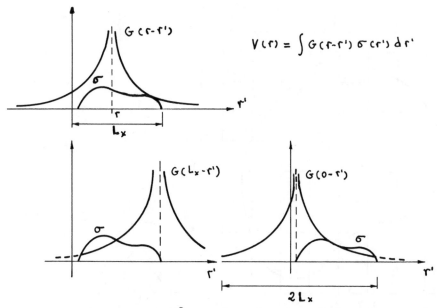

$$V(r) = \int G(r-r')\, \sigma(r')\, dr'$$

Fig. 2. Length of secuence G^D must be twice that of the non zero values of σ^D in order of take into account the contribution of all charges.

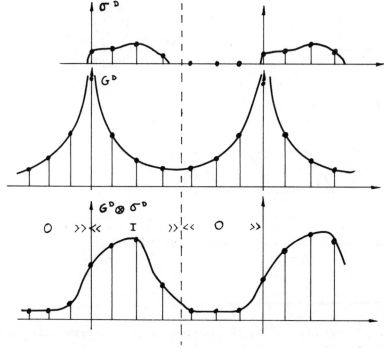

Fig. 3. Length of the period for circular convolution must be twice that of non zero values of σ in order to avoid overlapping (0) in the zone of charge (I).

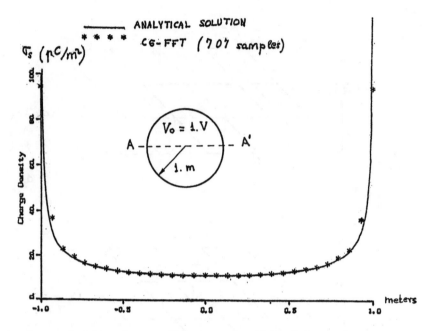

Fig. 4. Charge density along a diameter on a disk of 1 m. of radius, at
1 V of potential.

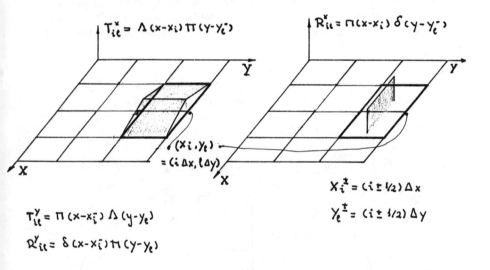

$$T_{i\ell}^x = \Lambda(x-x_i)\,\Pi(y-y_\ell^-)$$

$$R_{i\ell}^x = \Pi(x-x_i)\,\delta(y-y_\ell^-)$$

$$(x_i, y_\ell) = (i\Delta x, \ell\Delta y)$$

$$X_i^{\pm} = (i\pm\tfrac{1}{2})\Delta x$$

$$Y_\ell^{\pm} = (i\pm\tfrac{1}{2})\Delta y$$

$$T_{i\ell}^y = \Pi(x-x_i^-)\,\Lambda(y-y_\ell)$$

$$R_{i\ell}^y = \delta(x-x_i)\,\Pi(y-y_\ell)$$

Fig. 5. The Basis and testing functions for the x components are depicted
and defined. For y-components they are only defined.

136

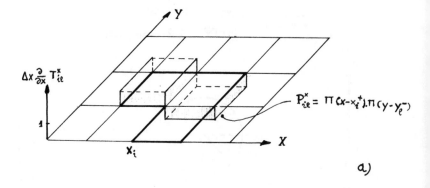

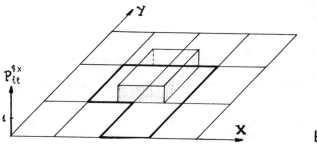

Fig. 6. Other functions being necessary to discretize the operator. Pulses in a) appears in terms having derivatives. Pulses in b), which approximate $T^x_{i1}(T^y_{i1})$, appear in terms where no derivatives exist.

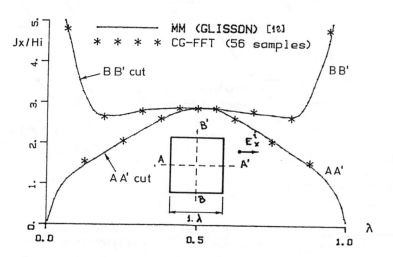

Fig. 7. Amplitude of the copolar surface current on a square plate of side $1.\lambda$ illuminated by a plane wave as shown.

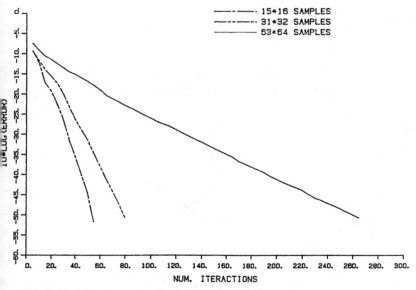

ERROR FOR RECTANGULAR PLATE (DIMENSIONS : 3.3 , 3.3)

Fig. 8. Convergence study for the copolar surface current on a square plate of side 3.3λ illuminated by normally incident plane wave.

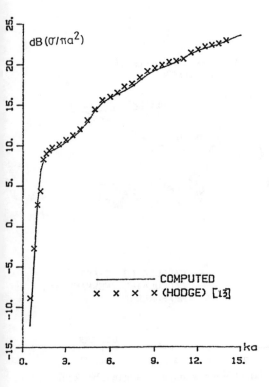

Fig. 9. RCS versus size of a disk. Normal incidence. Backscatter.

138

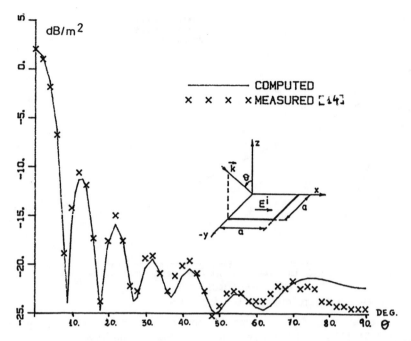

Fig. 10. RCS of square plate versus aspect angle (a=3.3 λ, f=10 GHz), backscatter.

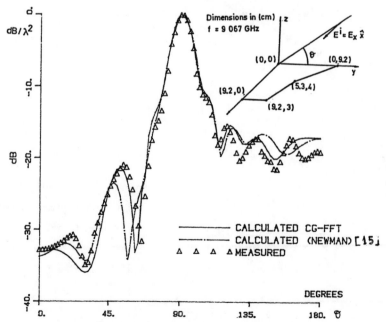

Fig. 11. RCS of the polygon indicated versus aspect angle, backscatter.

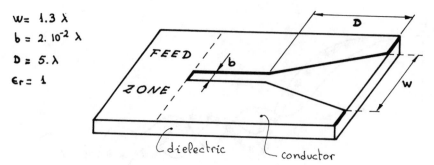

$W = 1.3 \lambda$

$b = 2.10^{-2} \lambda$

$D = 5.\lambda$

$\epsilon_r = 1$

Fig. 12. Geometry and dimension of a LTSA.

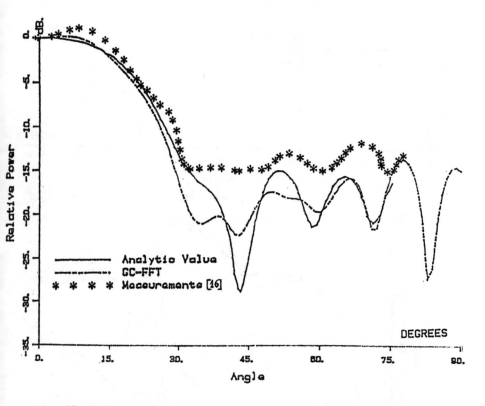

Fig. 13. E—Plane relative power pattern for LTSA of figure 12.

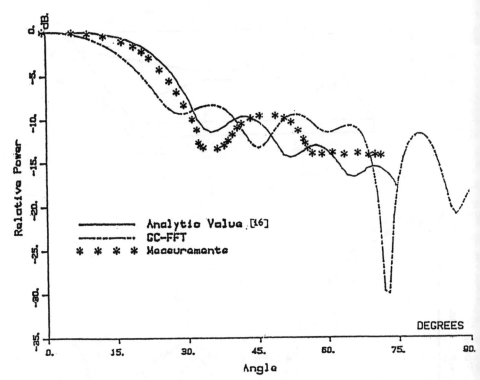

Fig. 14. H-plane relative power pattern for LTSA of figure 12.

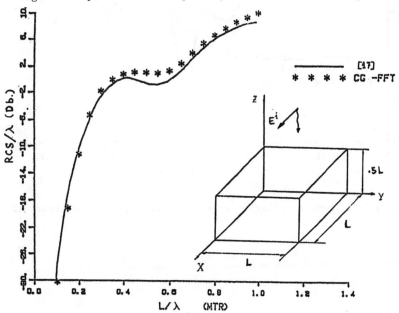

Fig. 15. Backscatter RCS for a perfect conducting parallelepiped with dimensions (L, L, 0.5L).

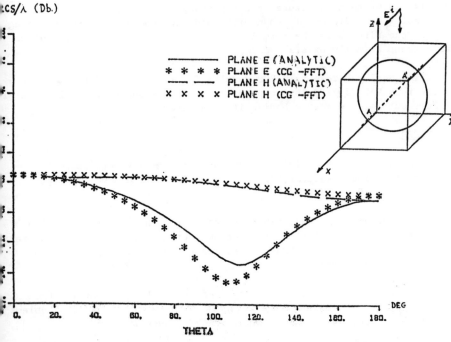

Fig. 16. RCS of a metalic sphere versus aspect angle (Ka=1.). Bistatic scattering.

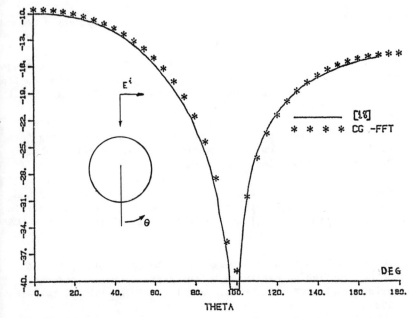

Fig. 17. Bistatic radar cross section of homogeneous dielectric sphere Ka =1., ϵ_r = 4.

142

As can be seen in figures corresponding to 3–D cases the formulation works better with dielectric than with conductor bodies, the reason being that for the first ones the formulation is more realistic because volumetric current actually exists.

Time employed to obtain RCS results for conductor bodies is about 5 minutes in a VAX 750/FPS 164 computer system (maximun theoretical speed: 1 Mflops), but for dielectric bodies it is shorter.

REFERENCES

[1]   M.R. Hestenes, E. Stiefel , "Methods of conjugate gradients for solving linear systems," J. Res. Nat. Bur. Stand., vol. 49, pp. 409–436, 1952.

[2]   T.K. Sarkar, S.M. Rao, "The application of the conjugate gradient method for the solution of electromagnetic scattering from arbitrarily oriented wire antennas." IEEE Trans. on A. P., vol. AP–32, pp. 398–403, Apr. 1984.

[3]   A.F. Peterson, R. Mittra, "Method of conjugate gradient for the numerical solution of large–body electromagnetic scattering problems" J. Opt. Soc. Am. A. vol. 2, pp. 971–977, June 1985.

[4]   D.F. Elliot, K.R. Rao, "Fast transforms, algorithms, analyses, applications," Academic Press, 1982.

[5]   R.N. Bracewell, "The Fourier Transform and its applications," Second Edition, McGraw–Hill, 1978.

[6]   T.K. Sarkar, E. Arvas, S.M. Rao, "Application of FFT and the conjugate gradient method for the solution of electromagnetic radiation from electrically large and small conducting bodies," IEEE Trans. on A.P. vol. AP–33, pp. 635–640, May 1986.

[7]   M.F. Catedra, "Solution to some electromagnetic problems using fast Fourier Transform with conjugate gradient method," Electr. Lett., vol. 22, pp. 1049–1051, Sept. 25, 1986.

[8]   T.K. Sarkar, E. Arvas, "On a class of finite–step iterative methods (conjugate directions) for the solution of operator equation arising in electromagnetics," IEEE Trans. on A.P., vol. AP–33, pp. 1058–1066, October 1985.

[9]   A.F. Peterson, R. Mittra, "Convergence of the conjugate gradient method when applied to matrix equations representing electromagnetic scattering problems," IEEE Trans. on A.P., vol. AP–34, pp. 1477–1454, Dec. 1986.

[10]   A. W. Glisson, D. R. Wilton, " Single and efficient numerical methods

for problem of electromagnetic radiation and scattering from surfaces ", IEEE Trans. on A. P., vol. AP–28, pp 593 – 603, Sept. 1980.

[11] L. W. Pearson, " A technique for organizing large moment calculations for use with iterative solution methods ", IEEE Trans. on A. P., vol. AP–33, pp 1031 – 1033, Sept.

[12] S.M. Rao, D. Wilton, A.W. Glisson, "Electromagnetic scattering by surfaces of arbitrary shape," IEEE Trans. on A.P., vol. AP–30, pp. 409–418.

[13] D.B. Hodge, "Scattering by circular metallic disk," IEEE Trans. on A.P., vol. AP–28, pp. 707–712, Sept. 1980.

[14] W.C. Anderson, "RCS prediction techniques: A review and comparison with experimental measurements," Military Microwaves, MM (86), pp. 475–486, Brighton, England, June 1986.

[15] E.H. Newman, P. Tulyathan, "A surface patch model for polygonal plates," IEEE Trans. on A.P., vol. AP–30, pp. 588–593, Dec. 1982.

[16] R. Janaswamy, D.H. Schaubert, D.M. Pozar, "Analysis of the transverse electromagnetic mode linearly tapered slot antenna," Radio Sci., vol. 21, no. 5, pp. 797–804, Sept 1986.

[17] L.L. Tsai, "Radar Cross Section of a Simple Target: a Three–Dimension Conducting Rectangular Box", IEEE Transaction on AP, vol. AP–25, no. 6, pp. 882–884, November 1977.

[18] K. Umashankar, A. Taflove, Sadasiva M. Rao, "Electromagnetic Scattering by Arbitrary Shaped Three–Dimensional Homogeneous Lossy Dielectric Objects", IEEE Transactions on AP, vol. AP–34, no. 6, pp.402, January 1970.

A method to analyze scattering from general periodic screens using
Fast Fourier Transform and Conjugate Gradient method.

M. F. Cátedra, Rafael P. Torres, Jesús G. Cuevas.
E.T.S.I. de Telecomunicación. Universidad Politécnica de Madrid.
28040 Madrid. SPAIN.

INTRODUCTION.

Periodic structures employed as scatterers exibit some interesting
properties that give them a wide range of application. Among these
properties we have the discrete nature of its scattered spectrum and the
polarization properties of the scattered field.

The periodic structures that will be analyzed in this paper are
planar periodic structures also refered to as frecuency selective surfaces
or periodic screens. They consist of a patches of a certain shape, S_0,
inscribed in a rectangular cell of area $T_x T_y$ and periodically arrayed with
periods T_x and T_y, as can be seen in figure 1.
Periodic surfaces are used as, [1]:

- Beam splitter (the surface works as a spatial separator for
  different frecuencies).
- Polarizers (converting lineal polarized fields in circularly
  polarized).
- Radar Cross-section reducers.
- Filters.

With an inmediate extension of the technique developed in this paper
other geometries, such as that of figure 2, can be analyzed. The geometry
in figure 2-a is easily analyzed using three-dimensional basis and testing
functions. That in figure 2-b can also be analyzed with our technique in
combination with general scattering-matrix formulation.

Periodic structures as scatterers have been studied by several
authors [1]-[8]. Among others, the most popular approaches are
variational techniques, equivalent circuit representation, harmonic
matching and scattering formulation. This last one is the most powerful. It
allows to obtain the currents on the scatterers and in turn all the
magnitudes of interest such as reflection and transmission coefficients,
cross-polarization levels in the scattered fields, etc.

The scattering formulation appears in several versions that differ
in the numerical procedure that is used to solve the equations. The Moment
method, (MM), with entire-domain basis functions, [4], is limited to
standard geometries, MM with subsectional basis functions, [2], [6], [7],
is more flexible but has the disadvantages inherent in matrix methods where
the size of the problems that can be handled is limited by the large
computer memory and long CPU-time required.

An iterative numerical procedure such Conjugate-Gradient Method,
(CG), [10], permits an increase of the size (relative to wavelength) of

R. de Neumann (ed.), Electromagnetic Modelling and Measurements for Analysis and Synthesis Problems, 145–160.
© 1991 Kluwer Academic Publishers. Printed in the Netherlands.

involved geometries but the time it consumes is stil too long due to iterative character of the solution. Nevertheless the Electric-Field Integral Equation (EFIE) of the problem can be represented by an operator involving convolutions that can be performed in the K-domain after calculating approximately the Fourier transform with the Fast-Fourier Transform (FFT) algorithm. Our resulting combined procedure CG+FFT, [6], [11], [12] has reduced memory and CPU-time requeriments. It is similar to the one described in [8], the main differences being in the discretizing procedure. Our formulation, however, is computationally more effective as will be shown in section 4.

In the Introduction of [11] it is outlined how the CG+FFT works for finite structures. Except for the particularities due to periodicity the developpement of the method for periodic screens follows similar lines. The differences due to periodicity show themselves in the expression of the operator: For finite structures the Green's function is infinite in extension and the use of FFT produces undesired aliasing that has to be avoided. On the other hand the Green's function for periodic structures is built from the infinite aliased individual Green's functions, one for each unity current element placed on each cell at the same local position. As will be seen in next sections, this aliasing is obtained in a natural way by using FFT.

In section 2 the nature of scattered field is treated conceptually rather than in a formal way. Section 3 is devoted to establish the operator equation, also is devoted to obtain the periodic Green function and finally the expression of the operator as a product of transforms. Finally, in section 4 some results obtained with this technique are given and compared with results obtained by other procedures.

2. SPECTRUM SCATTERED BY A PLANAR PERIODIC SURFACE.

To know the scattered field it is necessary to solve the operator equation taking into account all the parameters involved such as geometry, periods, incident direction, etc. Nevertheless a qualitative description of the situation can be obtained considering only the periods, (T_x, T_y), and the incident direction (θ, ϕ).

When a plane wave,

$$E^i = E_0 \exp(-jk^i.r) \tag{1}$$

where $r=(x,y,z)$, $k^i=2\pi/\lambda(\mathrm{sen}\theta\cos\phi,\mathrm{sen}\theta\,\mathrm{sen}\phi,\cos\theta)$ impinges on a planar periodic surface (fig. 1), a discrete refflected and/or transmitted plane wave spectrum appears, (see fig. 3-a). The number and nature of the propagating waves deppend upon periodicity (through T_x and T_y), and can be wieved by a representation in k-space (Fig. 3-b). The spatial or Floquet's modes that can propagate lie inside the visible region. Those outside this region are evanescents. For $T_{max} = \max\{T_x,T_y\}< \lambda/2$ only the fundamental reflected, r(0,0), or transmitted, t(0,0), waves appears.

3. FORMULATION OF THE OPERATOR EQUATION.

When a field is incident on a periodic surface all fields and currents must comply with Floquet's theorem. So the current on the $(i,1)$-cell is:

$$J_{i1}(x,y) = J(x,y) \exp(-jk_{0x}(x+iT_x)).\exp(-jk_{0y}(y+1T_y)). \qquad (2)$$

where $J(x,y)$ is a common function for all cells, and $0 \le x \le T_x$, $0 \le y \le T_y$.

The tangencial component of the scattered field is given by:

$$E^s(r) = \int_S \mathcal{G}(r-r') \, J^S(r')ds' \qquad (3)$$

where $r=(x,y)$, $r'=(x',y')$, S is the total surface where any current exist, $J^S=(J_x^S,J_y^S)$ is the surface current density, and $\mathcal{G}$ is given as follow:

$$\mathcal{G} = \begin{bmatrix} k^I G((r-r')+k^C \partial/\partial x g(r-r')\partial/\partial x' & k^C \partial/\partial x G(r-r')\partial/\partial y' \\ k^C \partial/\partial y G(r-r')\partial/\partial x' & k^I G(r-r')+k^C \partial/\partial y G(r-r')\partial/\partial y' \end{bmatrix}$$

being $K^I=-j\omega\mu/4\pi$, $K^C=-j/4\pi\epsilon\omega$, $G(r)=\exp(-j2\pi|r|/\lambda)/|r|$.

Equation (3) states that $E^s(r)$ is the sum of the integrations of the current over each latice cells weighed with a diadic $\mathcal{G}$ as shown in figure 4-a for a one dimensional case). But except for the Floquet factor[1], the same result can be obtained by making periodic the function G and making the integration over a single cell as in figure 4-b. Now if the Floquet's factor is included in the Green's function, the scattered field can be put as:

$$E^s(r) = \int_{S_0} \mathcal{G}_F(r-r') \, J(x',y') \, \exp(-jk_{0x}x')\exp(-jk_{0y}y')ds' \qquad (4)$$

where S_0 is the surface of a reference cell and $\mathcal{G}_F$ is defined as:

$$\mathcal{G}_F(r-r') = \sum_i \sum_1 \exp(jk_{0x}iT_x)\exp(jK_{0y}1T_y) \, \mathcal{G}((x+iT_x)-x',(y+1T_y)-y') \qquad (5)$$

An operator equation can now be written through the EFIE:

$$-E^i = -E_0 \exp(-jk_{0x}x) \exp(-jk_{0y}y) =$$
$$\int_{S_0} \mathcal{G}_F(r-r') \, J(r')\exp(-jk_{0x}x')\exp(-jk_{0y}y')ds' \qquad (6)$$

or in a more complete form:

---

[1] Exponential factors in (2) are called Floquet's factors

$$-E^i(x,y)=\begin{bmatrix} k^I\int_S ds'G_F(r-r')\Phi(r')+k^C\partial/\partial x \int_S ds'G_F(r-r')\Phi(r')[\partial/\partial x'-jk_{0x}] & ; \\ k^C\partial/\partial y \int_S ds'G_F(r-r')\Phi(r')[\partial/\partial y'-jk_{0x}] & \end{bmatrix} ;$$

$$\begin{bmatrix} k^C\partial/\partial y \int_S ds'G_F(r-r')\Phi(r')[\partial/\partial y'-jk_{0x}] \\ k^I\int_S ds'G_F(r-r')\Phi(r')+k^C\partial/\partial x \int_S ds'G_F(r-r')\Phi(r')[\partial/\partial x'-jk_{0x}] \end{bmatrix} \cdot \begin{bmatrix} J_x(x',y') \\ J_y(x',y') \end{bmatrix}$$

$$(7)$$

where $G_F(r-r') = \Sigma\ \Sigma\ \exp(-jk_{0x}iT_x)\exp(-jK_{0y}1T_y)\ G((x+iT_x)-x',(y+1T_y)-y');$
$\Phi(r)=\exp(-jk_{0x}x)\exp(-jk_{0y}y)$, and $0\leq x,x'\leq T_x$, and $0\leq y,y'\leq T_y$.

3.1 EXPRESSION OF THE OPERATOR AS A PRODUCT OF TRANSFORMS.

All the integrals In equation (6) can be put in the form of one of two classes of convolutions deppending on whether or not they include derivatives. These two kinds of convolutions are expressed generically as:

$$V^I(r)=k^I\int_S ds'G_F(r-r')\Phi(r')J_\alpha(r') \tag{8-a}$$

$$V^C(r)=k^C\int_S ds'G_F(r-r')\Phi(r')M_\alpha(r') \tag{8-b}$$

where $M_\alpha(r)=[\partial/\partial\alpha-jk_{0\alpha}]J_\alpha(r)$

Both expressions in (8) can be computed with the help of the Fourier Transform (f) as follows:

$$V^{I,C}(r)=k^{I,C}f^{-1}\{\overline{G}_F(k)\cdot\frac{\overline{J}_\alpha(k+k_0)}{\overline{M}_\alpha(k+k_0)}\} \quad ; \qquad\qquad k_0=(k_{0x},k_{0y}) \tag{9}$$

the bar over functions means Fourier Transformed. Now defining $\overline{G}_F^d=\overline{G}_F(k-k_0)$ the equation (9) is writen as:

$$V^{I,C}(r)=k^{I,C}\ \Phi(r)f^{-1}\{\overline{G}_F^d(k)\cdot\frac{\overline{J}_\alpha(k)}{\overline{M}_\alpha(k)}\} \tag{10}$$

(The nature and properties of $\overline{G}_F^d(k)$ will be given a few lines below). Taking (10) into account the equation (7) takes the form:

$$-E_x^i(r)=k^I\Phi(r)f^{-1}\{\overline{G}_F^d(k)\overline{J}_x(k)\}+k^C\Phi(r)[\partial/\partial x-jk_{0x}]f^{-1}\{\overline{G}_F^d(k)\overline{M}_x(k)\}+$$

$$k^C\Phi(r)[\partial/\partial x-jk_{0x}]f^{-1}\{\overline{G}_F^d(k)\overline{M}_y(k)\} \tag{11-a}$$

$$-E_y^i(r)=k^I\Phi(r)f^{-1}\{\overline{G}_F^d(k)\overline{J}_y(k)\}+k^C\Phi(r)[\partial/\partial y-jk_{0y}]f^{-1}\{\overline{G}_F^d(k)\overline{M}_y(k)\}+$$

$$k^C\Phi(r)[\partial/\partial y-jk_{0y}]f^{-1}\{\overline{G}_F^d(k)\overline{M}_x(k)\} \tag{11-b}$$

Terms in 11 wich are multiplied by k^I are called inductive terms and the ones multiplied by k^C are the capacitive terms.

The last equations involves continuous functions as $\bar{J}_\alpha$, $\bar{M}_\alpha$, but it will be shown that $\bar{G}_F^d$ has discrete nature, so we do not need to know $\bar{J}_\alpha$ or $\bar{M}_\alpha$ at every point since they are always multiplied by $\bar{G}_F^d$. To show this it is convenient to make use of Shah function $Ш_{Tx}(x)$ and its properties [9]. Shah function is defined as:

$$Ш_{Tx}(x) = \sum_{-\alpha}^{\alpha} \delta(x-nT_x) \tag{12}$$

According to this definition :

$$G_F(r) = G(r) \otimes [(\exp(jk_{0x}x \, Ш_{Tx}(x))(\exp(jk_{0y}y \, Ш_{Ty}(y))] \tag{13}$$

and by Fourier transforming both sides of this equation it is left:

$$\bar{G}_F(k) = (2\pi)^2/(T_xT_y)\bar{G}(k_x,k_y)[Ш_{2\pi/Tx}(k_x-k_{0x}) \, Ш_{2\pi/Ty}(k_y-k_{0y})] =$$

$$(2\pi)^2/(T_xT_y) \sum_i \sum_l \bar{G}(k_x,k_y)\delta(k_x-(2\pi i/T_x+k_{0x}))\delta(k_y-(2\pi l/T_y+k_{0y})) \tag{14}$$

being $\bar{G}(k_x,k_y) = 2\pi/\sqrt{((2\pi/\lambda)^2-k_x^2-k_y^2)}$
Finally, taking into account the definition of $\bar{G}_F^d$ we have:

$$\bar{G}_F^d(k_x,k_y) = (2\pi)^2/(T_xT_y)\sum_i\sum_l \bar{G}(k_x+k_{0x},k_y+k_{0y})\delta(k_x-(2\pi i/T_x))\delta(k_y-(2\pi l/T_y)) \tag{15}$$

The above analytical process can be pictorially viewed in figure 5 (for a one-dimensional case).

On the other hand, the tangential E-field has continuous (and periodic) nature, so that in order to work on a computer the equations have to be discretized. This discretization is made by testing the fields and expanding the current in a suitable basis. Subsectional basis and testing functions will be employed. Both set of functions are defined in the subdomains resulting from discretizing the geometry of the reference cell as shown in figure 6. The discretizing process which is similar that of planar finite structures, [11], is only outlined here and proceeds as follows.

The basis functions must be chosen to meet boundary conditions at the plate edge, (e.g. any current component must vanish when it is normal to an edge of the plate). A possible set of basis functions are rooftop functions T_{il}^x T_{il}^y, wich are Cartesian products of triangles and pulses. These functions have been widely used in MM applications. Testing functions are one-dimensional pulses R_{il}^x R_{il}^y.

Other functions that will be necessary are the partial derivatives of the rooftop functions. These are defined as a pair of two-dimensional pulses. Finally other functions being used are the two-dimensional pulses, P_{il}^{qx} P_{il}^{qy}, equally centered and with equal volume that T_{il}^x and T_{il}^y respectively.

These definitions lead to the following expressions:

$$J_x(x,y) = \sum_{i=1}^{N_x} \sum_{1=1}^{N_y+1} J_x^D(i,1)T_{i1}^x \quad ; \quad J_y(x,y) = \sum_{i=1}^{N_x+1} \sum_{1=1}^{N_y} J_y^D(i,1)T_{i1}^y$$

$$\frac{\partial J_x}{\partial x} = \sum_{i=1}^{N_x+1} \sum_{1=1}^{N_y+1} \frac{J_x^D(i,1)-J_y^D(i-1,1)}{\Delta x} P_{i1}$$

$$\frac{\partial J_y}{\partial y} = \sum_{i=1}^{N_x+1} \sum_{1=1}^{N_y+1} \frac{J_y^D(i,1)-J_y^D(i,1-1)}{\Delta y} P_{i1} \tag{16}$$

where $P_{i1} = P_{i-\frac{1}{2},1}^x = P_{i,1-\frac{1}{2}}^y$

Next step are: a) expansion of current and b) testing the fields.

a)

We have to compute $\bar{J}_\alpha(r)$ and $\overline{\partial/\partial x\, J_\alpha(r)}$ at points $k_{pq} = (2\pi p/T_x, 2\pi q/T_y)$ where $\bar{G}_F^d$ does not vanish.

– For inductive terms we have:

$$J_x(r) = \sum_i \sum_1 J_x^D(i,1)P_{i1}^x(r)$$

and therefore $\bar{J}_x(r)$ is as follows

$$\bar{J}_x(r) = \begin{cases} \tilde{J}_x^D(p,q)\ \bar{P}_{00}^x(k_{pq}) & ; \quad \text{for } k=k_{pq} \\ ?\ ?\ ?\ ? & ; \quad \text{for } k\neq k_{pq} \end{cases} \tag{17}$$

where $\tilde{\ }$ stands for FFT and $\bar{\ }$ stands for Fourier Transform. Expression (17) means that $\bar{J}_x$ has been approximated by the FFT of the coefficients in the expansion of current.

– For capacitive term we have:

$$\overline{\partial/\partial x\, J_x(r)} = (\Delta x)^{-1}[1-\exp(-j2\pi p/N_x)]\ \tilde{J}_x^D(p,q)\ \bar{P}_{00}^x(k_{pq}) \tag{18}$$

b)

– For the inductive term it holds:

$$\int_{x_i^-}^{x_i^+} dx\ \Phi(x,y_1^-)V^I(x,y_1^-) \approx \Delta x \Phi(x_i,y_1^-)V^I(x_i,y_1^-) \tag{19}$$

where $V(x,y)$ was defined in (10). By using the same approximation as in (18) we have:

$$\int_{x_i^-}^{x_i^+} dx\ \Phi(x,y_1^-)V^I(x,y_1^-) \approx k^I \Delta x \Phi(x_i,y_1^-)\text{FFT}^{-1}\{\bar{G}_F^d(k_{pq})\ \tilde{J}_x^D(p,q)\ \} \tag{20}$$

– For the part with derivatives it holds

$$\int_{x_i^-}^{x_i^+} dx\ \Phi(x,y_1^-)\partial/\partial x\ V^C(x,y_1^-) \approx \Phi(x_i,y_1^-)[V(x_i^+,y_1^-)-V(x_i^-,y_1^-)] \qquad (21)$$

and with the same approximations mentioned above is left:

$$\int_{x_i^-}^{x_i^+} dx\ \Phi(x,y_1^-)\partial/\partial x\ V^C(x,y_1^-)$$

$$\approx \Phi(x_i,y_1^-)FFT^{-1}\{\overline{G}_F^d(k_{pq})\ J_x^D(p,q)[\exp(j2\pi p/N_x)-1]\} \qquad (22)$$

exponential factor takes into account the displazament of secuence $V(x_i^+,y_1^-)$ with respect to $V(x_i^-,y_1^-)$.

With expressions (17), (18), (20), and (22) and the ones that arises for y–components, the final form for the operator equation, in term of product in the transform domain, is:

$$\begin{bmatrix} E_{0x}^i(i,1) \\ E_{0y}^i(i,1) \end{bmatrix} = \begin{bmatrix} g_x(i,1)\ ;\ 0 \\ 0\ ;\ g^y(i,1) \end{bmatrix}.FFT^{-1}\ \{$$

$$\begin{bmatrix} C_x^I+C_x^q(1-F_x^*(p))(F_x(p)-1)\ ; & C_y^q(1-F_y^*(q))(F_x(p)-1) \\ C_x^q(1-F_x^*(p))(F_y(q)-1)\ ; & C_y^I+C_y^q(1-F_y^*(q)(F_y(q)-1)) \end{bmatrix} . \begin{bmatrix} \overline{G}^D(k_{pq})J_x^D(p,q)\overline{P}_{00}^x(k_{pq}) \\ \overline{G}^D(k_{pq})J_y^D(p,q)\overline{P}_{00}^y(k_{pq}) \end{bmatrix} \}$$

$$(23)$$

where; $F_\alpha(n) = \exp(j2\pi n/N_\alpha)-jk_{0\alpha}\Delta\alpha$; $C_\alpha^I = -j\omega\mu\Delta\alpha/4\pi$; $C_\alpha^q = -j/4\pi\epsilon\omega\Delta\alpha$; $\alpha:x,y,$ and the other magnitudes and symbols has been defined before.

Equation (23) is very similar to [11]–(11) for planar finite structures, the only diference being in the deffinitions of factors F_α. The periodicity of the structure is taking into account by adding $-jk_{0\alpha}\Delta\alpha$ to the factor F_α for finite structures. So, the computer effort is similar in solving a isolated finite cell or the complete periodic structure.

Note that in (23) the bar over function denotes Fourier Transform that has an analytic expression, while symbol ~ denotes FFT that has to be evaluated numerically.

The adjoint operator obtained by simply trasposing and complex conjugating the dyad in (23).

152

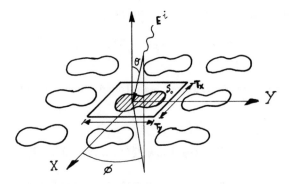

Figure 1. Geometry for a planar periodic structure. Shadowed parts are metal patches or holes in a metal plane.

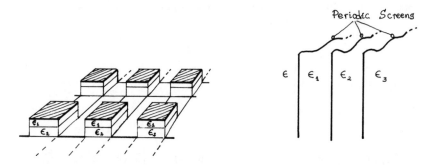

Figure 2. Geometries and arrangements that can be analyzed with the CG+FFT technique with modifications in the formulation employed for the geometry of fig. 1.

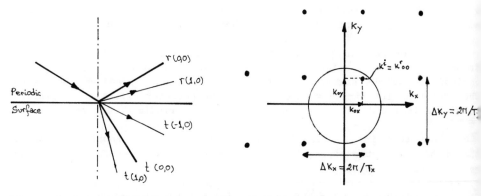

Figure 3. Scattered plane-wave spectrum for a plane periodic surface. a) Ray representation (modes r(0,0) and t(0,0) correspond to the modes that follows the Reflection and Transmission Snell laws respectively). b) K-space representation. Propagating modes lie inside the visible region bounded by the circle of radius $2\pi/\lambda$.

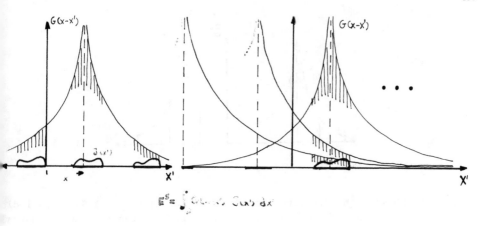

$$E^g = \int G(x-x') \, J(x') \, dx'$$

Figure 4. a) Scattered E-field obtained by weighing J with G and integrating. b) The same result as in a) obtained by making periodic G and integrating over a single cell.

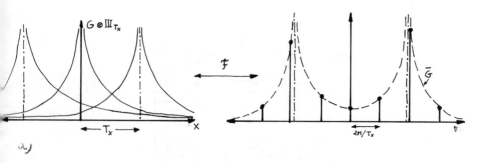

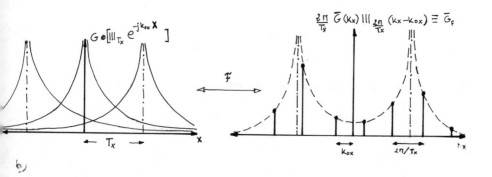

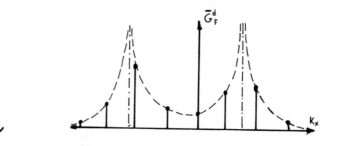

Figure 5. Showing the discrete nature of the function $\overline{G}_F^d$. a) Shows periodic Green functions (left) and its Fourier Transform under normal plane wave incidence. b) Same as in a) but with oblique incidence. c) Final form for $\overline{G}_F^d$ obtained by making periodic $\overline{G}_F$ in b).

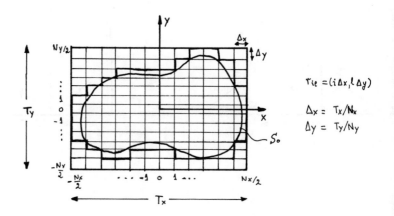

Figure 6. Discretization of the geometry.

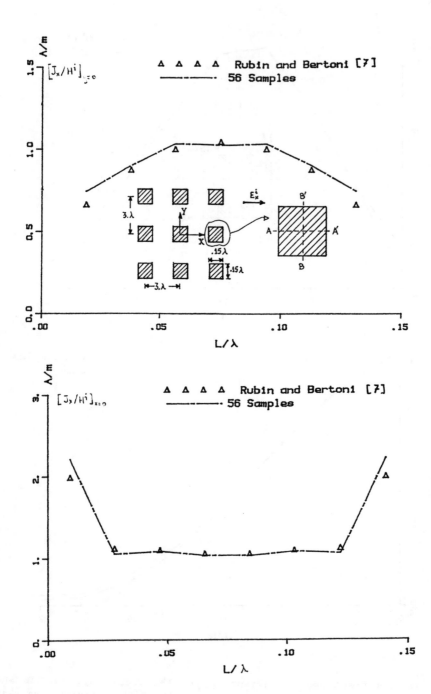

Figure 7. Amplitude of the co-polar surface current in a plate of the periodic arrangement shown. Incidence is normal.

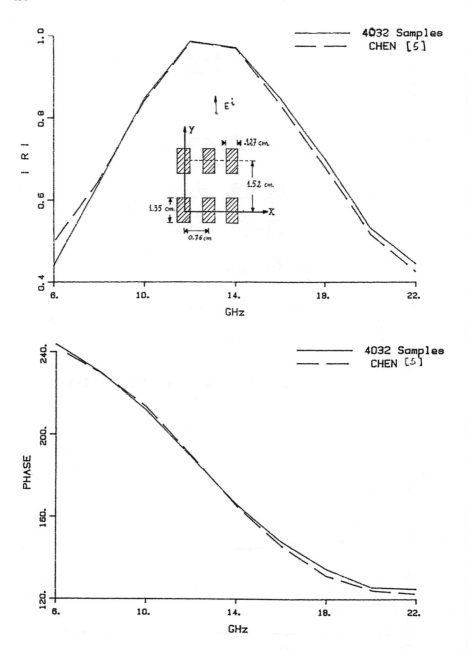

Figure 8. Reflection coefficient, R, of the rectangular lattice shown
Normal TE plane-wave incidence with E-field parallel to y-axis
a) Magnitude of R, b) Phase of R.

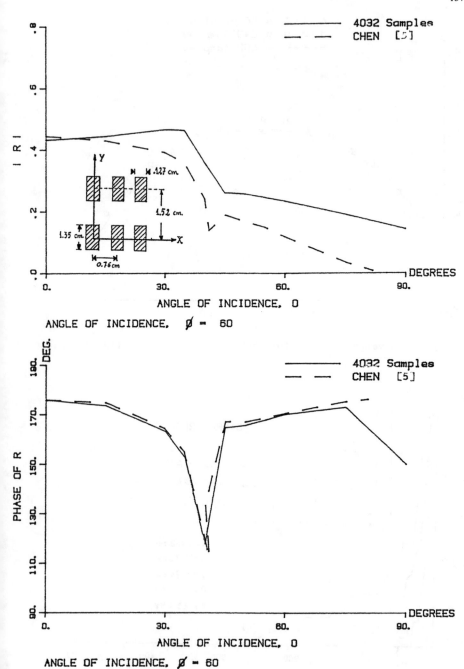

Figure 9. Reflection coefficient for TM wave due to TE incidence versus aspect angle. a) Magnitude. b) Phase.

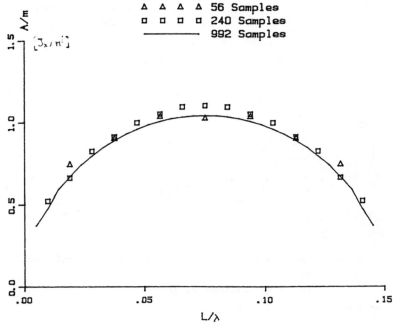

Figure 10. Convergence study for the surface current of fig. 7.

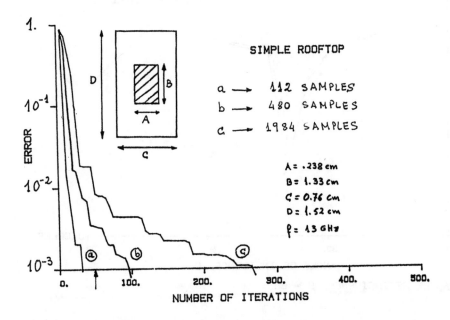

Figure 11. Relative residual error versus number of iterations. The arrows
mark the number of iterations used by a similar algorithm [8]

4. RESULTS.

Having obtained the expressions for the operator and its adjoint, the CG+FFT technique solves the operator equation with a CG algorithm, [10], as explained in [11]. Several types of results are presented in figures 7–12. Figure 7 shows the form of the induced current in the conducting part of an array cell under TE–plane wave incidence. Figure 8 gives the reflection coefficient versus frecuency for another array. Figure 9 shows reflection coefficient versus aspect angle for a cross–polarized situation (a reflected TM wave due to an incident TE) situation. Convergence studies are shown in figures 10 and 11. In figure 11 relative residual error versus number of iteration is depicted. The arrow mark the number of iteration employed by the algorithm in [8] and for a similar number of unknowns (121) of curve (a).

The agreement of our result with those published previously is good in the majority of the cases given in the figures. All results were obtained in a VAX 750/FPS 164, and the time employed to solve a typical structure (1984 samples), was of the order of a few minutes.

AKNOWLEDGEMENTS.

This work has been suported in part by the Spanish Navy and by the Spanish Advisory Comission for Scientific and Technological Research (CAICYT), Proyect No 2261/83.

REFERENCES.

[1] T. Cwik, R. Mittra, K. C. Lang, T. K. Wu. "Frecuency selective screens." IEEE A. P. Soc. Newsletter, vol. 29, No. 2, April 1987.

[2] B. J. Rubin. "Scattering from a periodic array of apertures or plates where the conductors have arbitrary shape, thickness, and resistivity", IEEE Trans. on AP., vol. AP–34, pp 1356–1365, Nov.1986

[3] T. Cwik, R. Mittra. "Scattering from general periodic screens", Electromagnetics, vol. 5, No 4, pp 263–283, 1985.

[4] Ch. Tsao, R. Mittra., "Spectral–domain analysis of frecuency selective surfaces comprised of periodic arrays of cross dipoles and Jerusalem crosses." IEEE Trans. on AP., vol. AP–32, pp 478–486, May 1984.

[5] Ch. Chen. "Scattering by a two–dimensional periodic array of conducting plates." IEEE Trans. on AP., vol AP–18, No 5, Sept. 1979.

[6] M. F. Cátedra, Jesús G. Cuevas. "A method to analyze scattering from general periodic screens using Fast Fourier Transform and Conjugate Gradient method." 1987 IEEE AP–S Int. Symp. Dig. Blacksburg, VA, pp. 91–94.

[7] B. J. Rubin, H. L. Bertoni. "Reflection from a periodically perforated plane using a subsectional current approximation", IEEE Trans. on AP., vol. AP–31, pp 829–836, Nov. 1983.

[8] C. H. Chan, R. Mittra. "Convergence studies of Conjugate Gradient method applied to frecuency selective surface problems using subdomain basis functions." 1987 IEEE AP–S Int. Symp. Dig.

Blacksburg, VA, pp. 87–90.

[9]   R. N. Bracewell. "The Fourier Transform and its applications." 2nd. edition, McGraw-Hill, 1978.

[10] T. K. Sarkar. "The application of the conjugate gradient method for the solution of operator equation arising in electromagnetic scattering from wire antennas.", Radio Science, Vol. 19, No 5, pp. 1156–1172, Sep.-Oct. 1984.

[11] M. F. Cátedra, Jesús G. Cuevas, E. Gago. "A scheme to analyze problems of resonant size using the Conjugate-Gradient method and the Fast Fourier Transform." In this Proceedings.

[12] M. F. Cátedra. "Solution to some electromagnetic problems using Fast Fourier Transform with Conjugate Gradient Method." Electronics Letters, vol. 22, No. 20, pp 1049–1051, Sep. 1986.

THE APPLICATION OF THE CONJUGATE GRADIENT METHOD FOR THE SOLUTION OF OPER-
ATOR EQUATIONS IN ELECTROMAGNETICS

TAPAN K. SARKAR

DEPARTMENT OF ELECTRICAL AND COMPUTER ENGINEERING
SYRACUSE UNIVERSITY
SYRACUSE, NY 13244-1240

ABSTRACT
    The objective of this paper is to survey many of the popular methods
utilized in solving numerical problems arising in electromagnetics. Historically, the matrix methods have been quite popular. One of the primary
objectives of this paper is to introduce a new class of iterative methods,
which have advantages over the classical matrix methods in the sense that a
given problem may be solved to a prespecified degree of accuracy. Also,
these iterative methods (particularly conjugate gradient methods) converge
to the solution in a finite number of steps irrespective of the initial
starting guess. Numerical examples have been presented to illustrate the
principles. The conjugate gradient method solves a Toplitz system in $\theta(N)$
steps when applied in conjunction with Fast Fourier Transform. The conjugate gradient method essentially performs a singular value decomposition for
ill-conditioned systems. This is illustrated by the solution of the deconvolution problem.

I. INTRODUCTION
    The basic objective of this paper is to survey the various numerical
methods utilized in the solution of operator equations arising in electromagnetics. This survey first starts with the concept of an ill-posed problem and well-posed problems. This is presented in section two. If a numerical method is to be presented in detail, including its rate of convergence,
it is essential that the problem be considered well-posed. In the third
section the classical eigenfunction expansion is presented. In section four
the well known method of moments is described. Some criteria on the choice
of expansion and weighting functions are presented. This is particularly
critical if one is interested in studying carefully the rate of convergence
of a numerical method. However, for formal (blind) application of the
method (as is presently done) the choices are not too critical. This is
because in a numerical method one is looking for an approximate solution,
and it is not too critical whether the expansion functions chosen for a particular problem are complete or not. In section five, the philosophy of the
iterative methods is presented. The primary reason for going to a new generation of methods over the method of moments is to seek a solution with a
certain prescribed degree of accuracy. Also, the iterative methods require
less core storage than matrix methods. Moreover, with the advent of a new
generation of computers which are geared towards vector processing, iterative methods may prove to be quite efficient. Two versions of the conjugate gradient methods are presented in section six, along with numerical
examples.
    In the discussion of the various methods, only references which are dir-
ectly relevant are noted. No attempt has been made to cite the earliest

B. de Neumann (ed.), Electromagnetic Modelling and Measurements for Analysis and Synthesis Problems, 161–188.
© 1991 Kluwer Academic Publishers. Printed in the Netherlands.

sources. In many cases, additional references may be found in the papers mentioned.

II. CONCEPT OF ILL-POSED AND WELL-POSED PROBLEMS

Most problems of mathematical physics can be formulated in terms of an operator equation:

$$Ax = Y, \tag{1}$$

where A, in general, is given linear integrodifferential operator and x is the unknown to be solved for a particular given excitation Y. If it is assumed that the solution to (1) exists for the given excitation y, then symbolically, the solultion to (1) can be written as:

$$x = A^{-1}Y. \tag{2}$$

Observe if the operator A is singular (i.e., the solutions of the equation Ax = 0 are nontrivial, $x \neq 0$) then (1) has multiple solutions. Such a case occurs, when one is interested in the analysis of electromagnetic scattering from a dielectric or conducting closed body, at a particular frequency which corresponds to one of the internal resonances of the body. We shall assume, at the present moment that such a situation does not arise. If one is really interested in the analysis for the case when Ax = 0 has a nontrivial solution, then the present techniques can be modified to treat those special cases. Finally, we assume A^{-1} is a bounded operator. This implies that small perturbations of the excitation only produce small perturbations of the solution as:

$$||x|| \leq ||A^{-1}|| \ ||Y||, \tag{3}$$

where $||x||$ defines a norm as:

$$||x||_2 = \int_{e_1}^{e_2} x(z)\overline{x(z)}dz \tag{4}$$

where the overbar denotes complex conjugate.

Observe in this case we have introduced the mean square norm, even though other norms could be defined:

$$||x||_1 = \int_{e_1}^{e_2} |x(z)|dz, \tag{5}$$

$$||x||_\infty = \max_{e_1 \leq z \leq e_2} |x(z)|. \tag{6}$$

The reason we chose the mean square norm is because it is easy to prove convergence of numerical techniques under norm (4) rather than (5) or (6).

If either of the following three conditions:

(1) the solution x exists for a given set of excitation Y,
(2) the solution x is unique,
(3) small perturbations in Y result is small perturbations of x,

are violated then we call the problem ill-posed. The treatment of ill-posed problems is described elsewhere. For our presentation, we will assume that the problem is well-posed [1].

So far we have discussed only the mathematical nature of the problem; now we will discuss how we obtain the solution x.

III. CLASSICAL EIGENFUNCTION APPROACH

Historically, the first technique developed is the eigenvector expansion. In this case, we choose a set of eigenvectors ϕ_i , for the operator A, i.e.,

$$A\phi_i = \lambda_i \phi_i, \tag{7}$$

where λ_i is the eigenvalue corresponding to the eigenvector ϕ_i. So we make a representation of the solution x by:

$$x = \sum_{i=1}^{\infty} a_i \phi_i, \tag{8}$$

where a_i are the unknowns to be solved for. We substitute (8) in (1) and obtain:

$$\sum_{i=1}^{\infty} a_i A_i \phi_i = \sum_{i=1}^{\infty} a_i \lambda_k \phi_k = Y. \tag{9}$$

Since eigenvectors corresponding to different distinct eigenvalues are orthogonal, we can multiply both sides of (9) by the eigenvector ϕ_j and integrate the product over the domain of the operator A (let us say from e_1 to e_2), i.e.,

$$<c,d> \triangleq \int_{e_1}^{e_2} c(z)d(z)dz, \tag{10}$$

$$a_j = \frac{1}{\lambda_j} \frac{<y,\phi_j>}{<\phi_j,\phi_j>}$$

Once the unknown coefficients are given by (10) the solution is obtained as:

$$x = \sum_{i=1}^{\infty} \frac{<y,\phi_i> \phi_i}{<\phi_i,\phi_i>\lambda_i} \tag{11}$$

Observe the solution is quite straightforward provided the series converges. The series will converge if λ_i does not approach zero. If λ_i approaches zero, then this is equivalent to the statement that $||A^{-1}||$ is not bounded and the problem is ill-posed. This particular situation occurs when the operator A is an integral operator with a kernal that is square integrable, i.e.,

$$\int_{e_1}^{e_2} K(p,q)x(q)dq = y(p), \tag{12}$$

where:

$$\int_{e_1}^{e_2} \int_{e_1}^{e_2} |K(p,q)|^2 dpdq < \infty . \tag{13}$$

Equation (12) is popularly known as Fredholm equation of the first kind. It is well known that all Fredholm equations of the first kind with square integrable kernels are ill-posed. However, if the kernel of a Fredholm equation of the first kind has a singularity, then the smallest eigenvalues approach zero at a much slower rate than for a kernel which is quite smooth. In most electromagnetic problems, we encounter a Fredholm equation of the first kind (as for example Hallen's integral equation); however, the kernel has a <<log>> singularity. This may explain why the solution procedure for such problems remains relatively stable as opposed to an antenna pattern synthesis problem where the kernel is very smooth, and as the order of

approximation is increased, spurious oscillations appear in the solution. The major problem with (11) is the determination of the eigenvectors. For simple geometries and other geometries which conform to one of the separable coordinate systems, the eigenfunctions are relatively easy to find. However, for arbitrary geometries, the determination of the eigenvectors itself is a formidable task. For this reason, we take recourse to the computer for computing certain integrals and so on. Once we set the problem up on a computer, we can no longer talk about an exact solution as the computer has only finite precision. The next generation of numerical techniques under the generic name of <<method of moments>> revolutionized the field of electromagnetic theory. Problems which were quite difficult to solve by eigenvector expansion and problems with arbitrary geometries can now be attacked with ease.

IV. METHOD OF MOMENTS

In the new technique <<the method of moments>> [2, 3] we choose a set of expansion functions ψ_i, which need not be the eigenfunctions of A. Moreover, the functions ψ_i need not even be orthogonal. The only requirement on ψ_i is that we know them. Now the solution x can be approximated as:

$$x = x_N = \sum_{i=1}^{N} a_i \psi_i. \tag{14}$$

Observe the upper limit in (14) is N and not ∞, because it is not practically feasible to do an infinite sum on the computer. As an example, consider the partial sum S_n of the following series.

$$S_n = 1 + \frac{1}{2} + \frac{1}{3} + \ldots + \frac{1}{n}.$$

Suppose we set the problem up on the computer and tell the computer to stop summing when the relative absolute difference of two neighbouring partial sums is less than a very small value, i.e.,

$$|S_{n+1} - S_n| \leq 10^{-20}.$$

The computer in all cases will give a convergent result. However, in the actual case the solution of (11) is:

$$\lim_{n \to \infty} S_n = \infty.$$

This is the reason whenever we solve the problem on a computer, we always talk in terms of an approximate solution, i.e., we are now solving the problem in a finite dimensional space N instead of the original problem which has the setting in an infinite dimensional space. As is clear from the above example, the approximate solution yielded by the computer may be nowhere near the exact solution. The natural question that now arises(like the example of the summation of the above series)is what guarantee do we have for the approximate solution x_N to converge to x as $N \to \infty$. This is still an open question for the mathematicians. But suffice it to say that the convergence of $x_N \to x$ depends on the boundedness of the inverse operator A and on the choice of the expansion functions ψ_i. In our presentation we will avoid answering the question as to what happens when $N \to \infty$.

We now substitute (14) in (1) and obtain:

$$Ax = Ax_N = \sum_{i=1}^{N} a_i A\psi_i = Y.$$

This results in an equation error:

$$E = Ax_N - Y = \sum_{i=1}^{N} a_i A\psi_i - Y. \tag{16}$$

We now weigh the error E to zero with respect to certain weighting functions, ω_j, i.e.,

$$\langle E, \omega_j \rangle = \int_{e_1}^{e_2} E(z)\omega_j(z)dz = 0. \tag{17}$$

The result is

$$\sum_{i=1}^{N} a_i \langle A\psi_i, \omega_j \rangle = \langle Y, \omega_j \rangle \tag{18}$$

Equation (18) is the key equation of the method of moments. The method gets its name from (17) where we are taking the functional moment of the error E with respect to the weighting functions, and equating them to zero.

Observe that there is essentially an infinite choice of weighting functions. However the choices of the weighting functions are not arbitrary. The weighting functions must satisfy certain specific criteria which are discussed in section IV.2.

By utilizing a formal approach, if we choose:

(a) the weighting functions to be delta functions, we get the method of point matching;

(b) the weighting functions to be the same as expansion functions, we get Galerkin's method, and finally;

(c) the weighting functions to be $A\psi_i$, we obtain the method of least squares.

The properties of the various methods are available in [4-5].

IV.1. On the choice of expansion functions.

The expansion functions chosen for a particular problem have to satisfy the following criteria:

(i) The expansion functions should be in the domain of the operator in some sense, i.e., they should satisfy the differentiability criterion and must satisfy the boundary conditions for an integro-differential operator [6, 7]. It is not at all necessary for each expansion function to exactly satisfy the boundary conditions. What is required is that the total solution must satisfy the boundary conditions at least in some distributional sense. The same holds for the differentiability conditions. When the boundary and the differentiability conditions are satisfied exactly, we have a classical solution, and when the above conditions are satisfied in a distributional sense, we have a distributional solution.

(ii) The expansion functions must be such that $A\psi_i$ forms a complete set for i = 1,2,3...for the range space of the operator. It really does not matter whether the expansion functions are complete in the domain of the operator, what is important is that ψ_i's must be chosen in such a way that $A\psi_i$ is complete, as is going to be shown later on. It is interesting to note that when A is a differential operator, the ψ_i's have to be linearly dependent for $A\psi_i$ to form a complete set. This is an absolute necessity as illustrated by the following example.

Consider the solution to the following differential equation:

$$-\frac{d^2x}{dz^2} = 2 + \sin z, \text{ for } 0 \leq z \leq 2\pi, \tag{19}$$

with the boundary conditions:

$$x(z = 0) = 0 = x(z = 2\pi). \tag{20}$$

A normal choice for the expansion functions would be to take sin iz for all i, i.e.,

$$x = \sum_{i=1}^{\infty} a_i \sin iz. \tag{21}$$

These functions satisfy both the differentiability conditions and the boundary conditions. Also since the operator in (19) is self adjoint both Galerkin's method and the method of least squares would yield identical solutions. We do not consider point matching for this example.

The above choice of expansion functions leads to the solution:

$$x = \sin z. \tag{22}$$

It is quite clear that (22) does not satisfy (19), and hence is not the solution. Where is the problem? Perhaps the problem may be that sin iz does not form a complete set in (21) even though they are orthogonal on the interval $[0, 2\pi]$. Therefore, in addition to the <<sin>> terms we add the constant and the <<cos>> terms in (21). This results in:

$$x = a_0 + \sum_{i=1}^{\infty} [a_i \cos iz + b_i \sin iz], \tag{23}$$

where a_i and b_i are the unknown constants to be solved for. Now the total solution of (23) has to satisfy the boundary conditions (20). Observe (23) is the classical Fourier series solution and hence it is complete in the interval $[0, 2\pi]$. Now if we solve the problem again by Galerkin's method and the method of least squares, we find the solution to be still:

$$x = \sin z,$$

the same as (22) and we know that it is not the correct solution. What exactly is still incorrect? The problem is that even though a_0, $a_i \cos iz$; $b_i \sin iz$ form a complete set for x, $A\psi_i$ — the operator A operating on the expansion function ψ_i — do not form a complete set. This is because $A\psi_i$, for i = 1,2,3,...are merely $c_i \cos iz$, $d_i \sin iz$, where c_i and d_i are certain constants. Note that from the set $A\psi_i$, the constant term is missing. Therefore the representation of x by (23) is not correct.

From the above discussion it becomes clear that we ought to have the constant term in $A\psi_i$, which implies that the representation of x must be of the form:

$$x = \sum_{i=1}^{\infty} [a_i \sin iz + b_i \cos iz] + a_0 + cz + dz^2 \tag{24}$$

and the boundary conditions (20) have to be enforced on the total solution. Observe that the expansion functions (1, z, z^2, sin iz, cos iz) in the interval $[0, 2\pi]$ and in the limit $N \to \infty$ form a linearly dependent set. This is because the set 1, (sin iz), (cos iz) can represent any function such as z and z^2 in the interval $[0, 2\pi]$. The final solution is obtained as:

$$x = z(2\pi - z) + \sin z, \tag{25}$$

which turns out to be the exact solution.

This simple example illustrates the mathematical subtleties that exist with the choice of expansion functions in the method of moments.

IV.2. On the choice of weighting functions

It is important to point out that the weighting functions in the method of moments have to satisfy certain conditions also. Since the weighting function weights the residual to zero, we have:

$$<E,w_j> \triangleq < \sum_{i=1}^{N} a_i A\psi_i - Y,w_j> \triangleq <Y_N -Y,w_j> = 0. \qquad (26)$$

As the error $Y_N - Y$ is orthogonal to the weighting functions, it is clear that the weighting functions should be able to reproduce Y_N and to some degree Y. From this discussion it is clear that Figure 1.5 on p.18 of [2] is not proper. This is because Y_N and the space of weighting functions are different and hence the error is not orthogonal to Y_N ($= S(Ax_N)$) by the terminology of [2]. Therefore, the approximation yielded in [2] is neither minimum norm nor unique.

Since the weighting functions approximate Y_N, they must be in the range of the operator. In other words, the weighting functions must form a basis in the range of the operator, and in the limit $M \to \infty$ must be able to represent any excitation Y. So if the weighting functions w_i cannot approximate Y to a high degree of accuracy, the solution x_n obtained by this technique of minimizing the residuals may not produce a solution which resembles the true solution. Any method which enforces the inner product to zero in (5) must satisfy the above criterion. For Galerkin's method $w_i = \psi_i$ and for the method of least squares $w_i = A\psi_i$. Since both Galerkin's and the method of least squares are special cases of the more general method of minimizing residuals, the two methods must satisfy the requirements for the weighting functions [7-9].

In short, it is seen by relating $<Y_N - Y,w_j> = 0$ in (26) to the theory of functional approximations that:

a) The weighting function w_i must be in the range of the operator or, more generally, in the domain of the adjoint operator.

b) Since the weighting functions are orthogonal to the error of the approximation (w_j) should span Y_N.

c) As $N \to \infty$, $Y_N \to Y$. Therefore, the weighting function should be able to represent the excitation Y in the limit.

If the weighting functions do not satisfy the above criterion, then a meaningful solution may not be obtained by the method of moments. This was first observed by Harrington [3,p.139], when he wrote, <<... In general the method of moments does not minimize the distance from the approximate I (I_N in this case) to the exact I, although it may in some special cases.>>

V. PHILOSOPHY OF ITERATIVE METHODS

For the method of moments approach, the expansion functions are selected a priori and the numerical procedure has to do with the solution for the unknown coefficients which multiply the expansion functions. For the iterative methods, the form of the solution is not selected a priori but evolves as the iteration progresses. This is very important from a philosophical point of view. In the method of moments it is essential to choose the expansion and the weighting functions in a particular fashion. If this is not done carefully, the method of moments may not work. This was pointed out by Harrington in his classic paper [3].

For the iterative methods we do not have to worry about the proper selection of the expansion and weighting functions. Since the solution evolves as the iteration progresses, one need not worry about the proper choice of expansion functions. Also for the iterative methods the weighting functions are dictated by the expansion functions and hence, if the

iterative method is handled properly, convergence is guaranteed [11-17]. In addition to this philosophical difference, there are two additional practical differences.

(a) In the method of moments, the maximum number of unknowns is limited by the largest size of the matrix that one can handle on the computer. Therefore, we would like to propose a solution procedure where we arrive at the solution without ever having to explicitly form the square matrix in (18). One could also achieve the above goal by computing each element of the square matrix in (18) as needed, but this procedure may be inefficient.

(b) Once we have solved for the unknowns in (18) by the conventional techniques, we do not have any idea about the quality of the solution, i.e., we do not know the degree of accuracy with which the computed solution satisfies (1) or how close it is to the exact solution. Moreover, with the conventional techniques [5, 13] when we increase the number of unknowns from N to 2N we are not even guaranteed to have a <<better>> solution.

It is primarily for the above two reasons that we propose to use a class of iterative methods where some <<error criterion>> is minimized at each iteration and so we are guaranteed to have an improved result at each iteration. Moreover, with the proposed methods, one is assured to have the exact solution in a finite number of steps. In summary, we prefer to use the new class of finite step iterative methods over the conventional matrix methods because the new techniques address the three very fundamental questions of any numerical method, namely,

a) the establishment of convergence of the solution,
b) the investigation of the rapidity of convergence, and
c) an effective estimate of the error to our satisfaction.

The first attempt to answer the above three questions was made by Sarkar et al. in [10], where they tried to answer the questions in terms of the solution of a matrix equation. They utilized (18), the discretized form of the original operator equations (1). Sarkar and Rao [11] extended the approach to the solution of an operator equation arising in electrostatic problems. Their approach is similar to the earlier approaches of Hestenes [16], Hayes [17]. In [11], Sarkar and Rao utilized the method of steepest descent which in general, is not a finite step iterative method. They later extended the solution procedure to the computation of electromagnetic radiation and scattering [12-15] from wire antennas, by utilizing the conjugate gradient method. The conjugate gradient method is a finite step iterative method and converges to the solution for any initial guess. In [11-15], the authors solved the operator equation directly by passing the formation of the matrix as is done in the method of moments. Recently, Van den Berg [18] had presented a unified error minimization scheme which is an extension of [10] in the sense that he deals with functionals instead of vectors.

In this paper, we broaden the approaches of [10-18] along the lines of [10] and present a class of iterative methods which always converge to the solution in a finite number of steps, irrespective of what the initial guess is. These classes of iterative methods are known as the method of conjugate directions. Both the conjugate gradient method and the well known Gaussian elimination techniques belong to this class. In this paper, we will confine our attention only to the class of conjugate gradient methods.

Development of conjugate direction methods.

The method of conjugate directions is a class of iterative methods that

always converges to the solution at the rate of a geometric progression, in a finite number of steps, irrespective of the initial starting guess [19]. The fundamental principle of the method of conjugate directions is to select a set of vectors P_i such that they are A-conjugate or A-orthogonal, i.e.,

$$<AP_i, P_j> = 0 \text{ for } i \neq j. \tag{27}$$

The practical significance of (27) is explained in the following way. Suppose we want to approximate the solution X by X_N so that:

$$X = X_N = \sum_{i=1}^{N} \alpha_i P_i, \tag{28}$$

then:

$$AX = AX_N = \sum_{i=1}^{N} \alpha_i AP_i = Y. \tag{29}$$

The vectors P_i span the finite dimensional space in which the approximate solution is sought. By taking the inner product of both sides with respect to P_j and enforcing (27) we obtain:

$$\alpha_i <AP_i, P_i> = <Y,P_i>,$$

or:

$$\alpha_i = \frac{<Y,P_i>}{<AP_i,P_i>}, \tag{30}$$

and therefore:

$$X = \sum_{i=1}^{N} \frac{<Y, P_i>}{<AP_i,P_i>} P_i. \tag{31}$$

It is clear from (31) that if we could select a set of A-orthogonal vectors P_i, then the construction of the numerical solution would be quite straightforward. The computational complexity of this algorithm then reduces to computing some inner products and scalar ratios.

Equation (31) actually represents the fundamental basis of the conjugate direction methods. Now so far we have not said how to choose the functions P_i. When the functions P_i are selected a priori, we get the direct methods like Gaussian elimination and when the vectors P_i are determined iteratively, we get an iterative method. Before we present the iterative version of the conjugate directions, we discuss a particular direct form of the conjugate direction method, namely Gaussian elimination. In the next section we present the generalized version (applicable to arbitrary operator equations) of Gaussian elimination and demonstrate where efficiency/accuracy in the solution procedure can be achieved. We also point out the weak point of Gaussian elimination and the reason why we would like to pursue an iterative method.

It is important to point out that when A is a matrix (31) actually is the computational form of Gaussian elimination. For this case, the vectors P_i are selected as follows. We start with a set of coordinate column vectors F_i, such that F_i has a 1 in location i and zero elsewhere. From the F_i vectors we get the A-orthogonal P_i vectors by utilizing the Gram-Schmidt decomposition procedure. So we let:

$$P_i = F_i = \begin{bmatrix} 1 \\ 0 \\ 0 \\ \cdot \\ \cdot \\ \cdot \\ 0 \end{bmatrix}, \tag{32}$$

and obtain:

$$P_2 = F_2 - \frac{\langle AF_2, F_1 \rangle}{\langle AF_1, F_1 \rangle} F_1, \tag{33}$$

and so on. In general:

$$P_{i+1} = F_{i+1} - \sum_{j=1}^{i} \frac{\langle AF_{i+1}, P_j \rangle}{\langle AP_j, P_j \rangle}. \tag{34}$$

The vectors P_i thus generated are substituted in (31) and we obtain the solution yielded by Gaussian elimination for matrices. This development can be easily extended to the general operator equation by treating F_i's as functions rather than vectors. Even though the above computation process is straightforward it has the following advantages/disadvantages.

(1) One can obtain any meaningful solution only after one has gone through the compuation of the N terms of the series prescribed by (31). This can be very time consuming particularly if N is large. However, this method would be quite advantageous if N is small (e.g. $N \le 50$).

(2) Second, even after all the terms have been computed to yield the solution, we are not sure about the accuracy of the solution. When N is large, the round-off error in the compuation of P_i in (34) can build up to cause erroneous results. There is no way to rectify these round-off errors and they propagate as N gets large. However, one can partially rectify the round-off errors if the vectors P_i are computed in an iterative fashion.

In short, for small N the direct method is efficient (e.g. for $N \le 50$ and for well conditioned matrix A, it is well known that Gaussian elimination is a very efficient way to solve equation (1)). However, if N is large, we would prefer a method where we know the error in the computed solution. That is why we lean towards an iterative method, where a suitable error criterion is minimized at each iteration and so if any truncation or round-off error is generated in the computation of P_i, it is corrected.

Almost an infinite number of ways to select P_i exists. However, in our approaches we select P_i from the residuals computed after each iteration. Since the residual:

$$R = Y - AX, \tag{35}$$

is proportional to the gradient of the quadratic functional $F(X)$:

$$\begin{aligned} F(X) &= \langle R, R \rangle = \langle Y - AX, Y - AX \rangle \\ &= ||Y - AX||^2, \end{aligned} \tag{36}$$

to be minimized, this particular choice of P_i from the residuals R is called the method of conjugate gradient. In the next section we present the basic philosophy of the generalized conjugate gradient method and develop two specialized cases, which are compuationally straightforward.

VI.   A CLASS OF CONJUGATE GRADIENT METHODS

For the class of conjugate gradient methods that we are going to present in this paper, we propose to deal with the following functional:

$$F(X) = \langle R, SR \rangle, \tag{37}$$

where:

S = a hermitian positive definite matrix (assumed known but has not been defined yet).

So at the i + 1 iteration, we minimize the functional:

$$F(X_i) = \langle R_i, SR_i \rangle \tag{38}$$

where

$$R_i = Y - AX_i. \tag{39}$$

We propose to minimize $F(X_{i+1})$ by incrementing X_i in the following way:

$$X_{i+1} = X_i + A_i P_i, \tag{40}$$

where a_i is chosen to minimize $F(X_{i+1})$. This is achieved by equating:

$$\frac{\partial F(X_{i+1})}{\partial a_i} = 0 = \frac{\partial}{\partial a_i} \times \langle Y - AX_i - a_i AP_i, S(Y - AX_i - a_i AP_i) \rangle \tag{41}$$

This results in:

$$\frac{\partial}{\partial a_i} [\langle R_i, SR_i \rangle - a_i \langle AP_i, SR_i \rangle - \bar{a}_i \langle R_i, SAP_i \rangle + |a_i|^2 \langle AP_i, SAP_i \rangle] = 0. \tag{42}$$

The functional $F(X_{i+1})$ is minimized if:

$$a_i = \frac{\langle R_i, SAP_i \rangle}{\langle AP_i, SAP_i \rangle}. \tag{43}$$

We also observe that the residuals can be generated recursively by:

$$R_{i+1} = Y - AX_{i+1} = R_i - a_i AP_i. \tag{44}$$

By comparing (43) and (44) it is clear that:

$$\langle R_j, SAP_i \rangle = 0 \text{ for all } i \neq j \text{ and } j > i, \tag{45}$$

or :

$$\langle SR_j, AP_i \rangle = 0.$$

Equation (45) implies that certain orthogonality conditions must be satisfied if the residuals are to be computed recursively by utilizing (43).

So far we have not specified how to generate the vectors P_1. We obtain the search direction vectors P_i recursively as:

$$P_{i+1} = KG_{i+1} + b_i P_i \qquad (46)$$

where G_{i+1} are certain functions yet undetermined. K is an arbitrary Hermitian positive definite operator yet undefined. We now select the parameter b_i such that the vectors P_i are A–orthogonal with respect to the following inner product:

$$\langle AP_i, SAP_j \rangle = 0. \qquad (47)$$

Equation (47) guarantees that the method will converge in a finite number of steps. We prefer deriving the parameter b_i from the concept of A–orthogonalization rather than from the principle of the second minimization step [18]. This is because the principle of A–orthogonalization guarantees that the method will converge in a finite number of steps, whereas the second minimization described in [18] usually does not.

By enforcing (47) in (46) we get:

$$b_i = -\frac{\langle A*SAP_i, KG_{i+1} \rangle}{\langle AP_i \; SAP_i \rangle} \qquad (48)$$

where A* is the adjoint operator for A and is defined by:

$$\langle Au, v \rangle = \langle u, A^* u \rangle. \qquad (49)$$

Equations (39), (40), (43), (44), (46) and (48) describe the generalized conjugate direction method.

We have yet to specify the operators S and K and the vectors G_i. When we select the vectors G_i as the coordinate vectors, i.e. G_i has an 1 in the ith position and zero elsewhere, then we get the generalized Gaussian elimination method. When we choose:

$$G_i = A*SR_i, \qquad (50)$$

then we get the class of conjugate gradient methods.

The conjugate gradient methods described by (39), (40), (43), (44), (46), and (50) all converge to the solution in a finite number of steps starting with any initial guess [16, 17, 19].

We see that for the class of conjugate gradient methods the functional F(X) is minimized at each iteration. This is because from (18):

$$F(X_i) - F(X_{i+1}) = \frac{\langle R_i, SAP_i \rangle^2}{\langle AP_i, SAP_i \rangle} > 0. \qquad (51)$$

In (51) the right hand side of the equation is always positive as S is a positive definite matrix. So that the error is minimized at each iteration and the sequence X_i converges.

Of course, it is very well and good to know that the sequence X_i converges, but it is much better to know how it converges. If we introduce an operator T such that:

$$T = KA*SA, \qquad (52)$$

and let:

$$q = \text{infimum [spectrum (T)]} =$$
smallest eigenvalue of operator T, when $\qquad$ (53)
T is a positive definite matrix,

and

$$Q = \text{supremum [spectrum (T)]} =$$
largest eigenvalue of operator T, when $\qquad$ (54)
T is a positive definite matrix,

then it can be shown that the functional F(X) is minimized at each iteration and that the rate of convergence at ith iteration is given by [14]:

$$F(X_i) \leq \left\{ \frac{2(1-\alpha)}{(1+\sqrt{\alpha})^{2i} + (1-\sqrt{\alpha})^{2i}} \right\}^2 F(X_0) \qquad (55)$$

and that the sequence of solution X_i converges to the exact solution h, so
that:

$$||X_n - h|| \text{ tends to zero faster than } \left(\frac{1-\sqrt{\alpha}}{1+\sqrt{\alpha}} \right)^n$$

where $\alpha = q/Q$.

In the next section we present two versions (special cases) of the conjugate gradient method. Each of the special cases has certain advantages.
The reason we chose these two special cases is because these particular two
lead to the simplest algorithm. We also derive the rate of convergence of
each of these techniques. Other special cases are also possible and they
are presented elsewhere [14, 20, 21].

VI.1. Minimization of residuals (conjugate gradient method A).

For this special case we set S = K = I = Identity operator, and so
from (52):

$$T = A*A, \qquad (57)$$

and:

$$F(X_i) = ||R_i||^2. \qquad (58)$$

The conjugate gradient method for this case starts with an initial guess X_0
and define:

$$R_0 = Y - AX_0, \qquad (59)$$

$$P_0 = G_0 = A*R_0. \qquad (60)$$

For i = 0, 1,... let:

$$a_i = \frac{||G_i||^2}{||AP_i||^2}, \qquad (61)$$

$$X_{i+1} = X_i + a_i P_i, \qquad (62)$$

$$R_{i+1} = R_i - a_i AP_i, \qquad (63)$$

$$G_{i+1} = A*R_{i+1} \tag{64}$$

$$P_{i+1} = G_{i+1} + b_i P_i, \tag{65}$$

$$b_i = \frac{||G_{i+1}||^2}{||G_i||^2} \tag{66}$$

By letting q and Q be the spectral bounds for T = A*A and setting:

$$\alpha = q/Q,$$

the characteristics of this method can be summarized as follows:

(i) The iteration (53)–(66) is such that X_i converges to the exact solution h for all initial guesses.

(ii) $||R_i||^2$ is the least possible in i steps for utilizing any iterative scheme of the form (62).

(iii) $||h - X_i||$ decreases at each iteration even though we do not know the exact solution h.

(iv) The following bound on the functional holds:

$$F(X_i) \leq \left\{ \frac{2(1-\alpha)^i}{(1+\sqrt{\alpha})^{2i} + (1-\sqrt{\alpha})^{2i}} \right\}^2 F(X_0). \tag{67}$$

Because $||R_i||^2$ is minimized this particular method essentially gives a sequence of least square solutions to Ax = Y, such that at each iteration the estimate X_n gets closer to h. This is the most widely used technique.

VI.2 Minimization of solution error (conjugate gradient method B)

For this case, we let:

$$S = (AA*)^{-1} \tag{68}$$

$$K = A*A, \tag{69}$$

$$T = A*A, \tag{70}$$

so that:

$$F(X_i) = ||h - X_i||^2 \tag{71}$$

Observe in (71) we are minimizing the error between the exact and the approximate solution even though we do not know the exact solution. Substitution of (68)–(70) into the generalized conjugate gradient algorithm yields this special version. Give X_0, let:

$$R_0 = Y - AX_0, \tag{72}$$

$$P_0 = A*R_0, \tag{73}$$

For i = 0, 1, ..., let:

$$a_i = \frac{||R_i||^2}{||P_i||^2}, \tag{74}$$

$$X_{i+1} = X_i + a_i P_i \qquad (75)$$

$$R_{i+1} = R_i - a_i A P_i, \qquad (76)$$

$$P_{i+1} = A^* R_{i+1} + b_i P_i, \qquad (77)$$

$$b_i = \frac{||R_{i+1}||^2}{||R_i||^2}. \qquad (78)$$

By defining q and Q to be the spectral bounds of $T = A^*A$ and $\alpha = q/Q$, we get the following results:

(i) The iteration is such that X_n converges to h for all i.

(ii) $||h - X_i||^2$ is the least possible in i steps for all possible iterative methods of the form (72), even though we do not know the exact solution.

(iii) $||R_i||^2$ may not decrease, in fact it may even increase. Often it oscillates.

(iv) The functional at each iteration is minimized and the following bound holds:

$$F(X_i) \leq \left\{ \frac{2(1 - \alpha)^i}{(1 - \sqrt{\alpha})^{2i} + (1 - \sqrt{\alpha})^{2i}} \right\}^2 F(X_0), \qquad (79)$$

where α is the ratio of the smallest to the largest spectral bounds for T.

Other special cases are available elsewhere [14].

VII. SOLUTION OF POCKLINGTON'S EQUATION FOR A STRAIGHT WIRE [22]

Consider a z-directed straight wire antenna of length L and radius a, irradiated by an incident field of intensity E^i. By enforcing the total tangential electric field on the wire to zero, the well-known Pocklington's equation is obtained:

$$\left(k^2 + \frac{d^2}{dz^2} \right) \int_0^L J(z')G(z,z')dz' = j\omega 4\pi\varepsilon_0 E^i_{tan}(z) \qquad (80)$$

where E^i_{tan} denotes the tangential component of the incident field on the wire, J is the unknown current distribution to be solved for, $G(z,z')$ is the Green's function given by

$$G(z, z') = \frac{1}{2\pi} \int_0^{2\pi} \frac{\exp(-jkR)}{R} d\phi. \qquad (81)$$

Here

$$R^2 = (z-z')^2 + 4a^2 \sin^2 \frac{\phi}{2}, \qquad (82)$$

and

$$k = 2\pi/\lambda \qquad (83)$$

with λ denoting the wavelength.

The integrodifferential equation (80) can be rewritten in a compact form as

$$AJ = Y \qquad (84)$$

where J is the unknown to be solved for the known excitation

$Y = -j4\omega\pi\epsilon_0 E_{tan}^i$, and A denotes the integrodifferential operator of (80) and we apply the Conjugate gradient method to solve (84).

The explicit expression for the adjoint operator A* is given by

$$A*V = (k^2 + \frac{d^2}{dz^2}) \int_0^L V(z)\overline{G(z, z')}dz \tag{85}$$

In the usual CG method (80) is solved by the iteration defined in (57)-(66). The computations are performed as followed. The wire is divided into N segments and the current is assumed to be a constant over each segment. The values of these constants obtained at the nth iteration are stored in the column vector J_n. Hence the ith element of J_0 is the initial guess for the current over the ith segment. The computation of the term AJ_0, can be performed in different ways. In the conventional CG method utilized in [12], AJ_0 was computed as follows.

First the convolution integral

$$\int_0^L J_0(\dot{z}')G(z_m,z')dz' = \sum_{i=1}^N J_0^i \int_{\ell_i}^{\ell_{i+1}} G(z_m, z')dz' = B(z_m) \tag{86}$$

was computed at the middle of each segment, calling the resulting function $B(z_m)$. Here Z_m denotes the midpoint of the mth segment, and $\ell_{i+1} - \ell_i$ is the length of the ith segment and J_0^i is the initial guess for the current over the ith segment. Then the operator $(k^2 + d^2/dz^2)$ was applied to $B(z_m)$ in a finite difference form to obtain AJ_0. Similar steps were used in the computations of the terms AP_n, $A*R_n$, etc. So even though at each iteration the computation of the above terms is equivalent to computing the entire impedance matrix of the method of moments, the iterative method appears to be faster for the large problems.

It was observed that in the conventional CG method the bulk of the computational time is used up in computing the terms like AP_n or $A*R_n$, both of which require the spatial derivative of a convolution integral. It is known that a convolution is transformed into a simple multiplication in the transformed domain. Similarly, derivatives transform into simple multiplications in the transformed domain. Hence in computing AJ_0 (for example), it would be computationally more efficient first to take the Fourier transform of both J_0 and G (to obtain $\tilde{J}_0$ and $\tilde{G}$, respectively, where the tide represents the transformed quantities), multiply them in the transform domain (which results in the transform of the convolution integral given in (86), and then multiply this result by $K^2 - k^2$. The resulting expression is the transform of AJ_0 and will be denoted by $\tilde{A}J_0$. AJ_0 can be obtained from $\tilde{A}J_0$ by simple inverse Fourier transformation.

The Fourier transform of the function $J_0(z)$ and the Green's function G are defined as follows:

$$\tilde{J}_0(k_z) = \int_{-\infty}^{\infty} J_0(z) \exp(-jk_z z)dz \tag{87}$$

$$\tilde{G}(k_z) = \frac{1}{2\pi} \int_{-\infty}^{\infty} \exp(-jk_z z)d\bar{z} \cdot$$

$$\cdot \int_0^{2\pi} \frac{\exp(-jk\sqrt{z^2+4a^2 \sin^2 \frac{\phi}{2}})}{(z^2 + 4a^2 \sin^2 \frac{\phi}{2})^{\frac{1}{2}}} d\phi = I_0[a\sqrt{k^2 - k_z^2}] K_0[a\sqrt{k^2 - k_z^2}]. \tag{88}$$

bserve that the expression for the transform of the Green's function is btained in a closed form in terms of the zeroth order modified Bessel unctions of the first and second kind denoted by I_0 and K_0, respectively. t is important to point out that the transform of the reduced kernel is iven by

$$G_{red}(k_z) = \int_{-\infty}^{\infty} \frac{\exp(-jk\sqrt{z^2 + a^2})}{\sqrt{z^2 + a^2}} \cdot \exp(-jk_z z)dz = K_0[a\sqrt{k^2-k_z^2}]. \quad (89)$$

ence, the utilization of the exact kernel in the computations introduces he additional factor $I_0[a\sqrt{k^2 - k_z^2}]$. Also note that both the exact and educed kernel transforms have a logarithmic (log) singularity as witnessed y the term $K_0[a\sqrt{k^2 - k_z^2}]$.

It is obvious then that, with the exact kernel used, the transform of he entire term AJ_0 becomes

$$\tilde{A}J_0 = (k^2 - k_z^2)\tilde{J}_0(k_z)I_0(a\sqrt{k^2 - k_z^2}) \, K_0(a\sqrt{k^2 - k_z^2}) \quad (90)$$

nd AJ_0 simply is the inverse transform of $\tilde{A}J_0$. At this point it is mportant to point out two interesting properties (90) namely the following.

1) The log singularity no longer exists with (90), because $(k^2 - k^2)$ ancels the log singularity of $K_0[a\sqrt{k^2 - k^2}]$, (as x log x → 0 for x → 0).

2) No finite difference derivative calculation is necessary. We feel his is a great advantage, particularly for the solution of electromagnetic cattering from electrically small structures. This is because the round-ff error in the finite difference approximation for the double derivative o longer exists! The numerical error comes from the fact that one sub-racts a large number from another large number to yield a small number,

$$\frac{d^2x}{dz^2} = \frac{X_{n+1} - 2X_n + X_{n-1}}{(\Delta z)^2}$$

The next question that arises is how to take the Fourier transform. This an efficiently be accomplished by utilizing a fast Fourier transform al-orithm. So both the forward and the inverse transforms are computed util-zing the FFT. However, unless careful attention is paid in the applica-ion of FFT, serious aliasing could void the computation. The evaluation f AJ_0 is now summarized as follows.

Step 1   Divide the wire into N segments with N unknowns for the current. Then the array J_0 is of dimension N.

Step 2   Pad the array J_0, with zeros so that the modified composite number of two (i.e., $M = 2^m$ and $M > 2*N$).

Step 3   Take the Fourier transform of the modified array by the FFT algorithm, yielding $J_0(k_z)$.

Step 4   Multiply this transform $\tilde{J}_0$ by $(k^2-k_z^2) \cdot I_0[a\sqrt{k^2-k_z^2}]K_0[a\sqrt{k^2-k_z^2}]$. The Bessel functions are computed numerically [5].

Step 5   Take the inverse transform of the product utilizing the FFT algorithm and divide each element of the array by M. The division by M is necessary to get the correct result.

Step 6   Select only the first N elements of the above array.

Step 7   The first N elements yield an estimate for AJ_0, without much aliasing. (The zero padding in step 2 is necessary to reduce the aliasing error.)

The computation for $A*R_0$ is done in a similar fashion, except that in tep (4) we multiply by the complex conjugate of the Green's function

(observe the Green's function is complex for $k_z > k$).

Equations (57)-(78) are applied in a routine fashion till the desired error criterion is satisfied. In our case, the error criterion is defined as

$$\frac{||AJ_n - Y||}{||Y||} = 10^{-4}. \tag{91}$$

We stop the iterations when (91) is satisfied.

As a first example consider the solution of the current distribution on a straight wire of 100λ long and 0.007 λ in radius illuminated by a broadside incident field of amplitude 1 V/m. With ten expansion functions per wavelength, we would be solving for 1000 unknowns. The CPU time required to solve this problem with an accuracy of 10^{-4} in the normalized residuals $||AJ_n - Y||/||Y||$ was 6 h (utilizing the exact kernel). This is in contrast to the original version of computation by the conjugate gradient method of 30 CPU h (utilizing the reduced kernel). The method of moments with Gaussian elimination (utilizing reduced kernel) took 45 CPU h. All computations are performed on a VAX 11/782.

As a second example consider the electromagnetic scattering from a 1600λ wire antenna with a 0.007λ radius illuminated by a broadside incident field of amplitude 1 V/m. With ten expansion functions per wavelength, the total number of unknowns would be 16 000. The CPU time taken by this method to solve the problem with an accuracy of 10^{-2} in the normalized residuals was eight hours, and for an accuracy of 10^{-3} in the normalized residuals, it was 10 h. We have also been able to solve for current distribution on a 3400λ antenna with a 0.007λ radius. By utilizing ten unknowns per wavelength, this would amount to solving for 34000 unknowns.

As a final example consider the solution of the electromagnetic scattering from a 0.007λ long antenna with a radius 0.0001λ irradiated by a plane wave of amplitude 1 V/m incident parallel to the wire antenna. We choose ten expansion functions for this problem and the solution is given in Table I (utilizing the exact kernel).

The accuracy in the solution is with $||AJ_n - Y||/||Y|| \leq 10^{-4}$. With conventional methods it is very difficult, if not impossible, to solve for the current with ten unknowns on a 0.007λ antenna.

TABLE I

k (spatial point)	Re[J]	Im [J]
1	0.132×10^{-12}	0.939×10^{-7}
2	0.973×10^{-12}	0.164×10^{-6}
3	0.924×10^{-12}	0.210×10^{-6}
4	0.627×10^{-12}	0.240×10^{-6}
5	0.153×10^{-12}	0.255×10^{-6}
6	0.174×10^{-12}	0.255×10^{-6}
7	0.588×10^{-12}	0.240×10^{-6}
8	0.797×10^{-12}	0.210×10^{-6}
9	0.123×10^{-12}	0.164×10^{-6}
10	0.161×10^{-12}	0.939×10^{-7}

As a final example, we consider the analysis of electromagnetic scattering from a 0.5λ antenna of radius 0.001λ, irradiated by a plane wave of amplitude 1 V/m. The antenna is subdivided into 30 segments. Methods A and B are now applied directly to the solution of the Pocklington's

equation for the thin wire antenna. So, in this formulation, we bypass the matrix formulation stage. Both methods A and B have been applied to solve the problem. The numerical computations obtained from the two techniques is presented in Table II. Observe that for method A, the magnitude of the residuals decrease monotonically.

TABLE II - Errors of the two versions of the conjugate gradient method (A & B)

Number of iterations	Normalized error = $\dfrac{\lvert\lvert AX_n - Y \rvert\rvert^2}{\lvert\lvert Y \rvert\rvert^2}$	
n	Method A	Method b
1	1,000	1,000
2	0.984	61.558
3	0.972	77.731
4	0.961	88.222
5	0.951	94.835
6	0.942	98.433
7	0.933	99.625
8	0.925	98.909
9	0.916	96.716
10	0.907	93.419
11	0.898	89.361
12	0.889	84.002
13	0.881	100.192
14	0.849	24.200
15	0.590	1.9308
16	1.045×10^{-17}	1.013×10^{-17}

For method B, even though the absolute error between the true solution and the solution at the end of each iteration decrease, the residuals actually increase and do not go down until the very end. Even though from a theoretical point of view method B provides a better approximation to the solution than method A, from a numerical standpoint it appears both methods perform equally well. The CPU time taken by both techniques is almost the same. Observe both the methods even converged to a good solution in fifteen iterations, starting from a null initial guess. The reason the methods concerged in 15 iterations is because the current distribution on the structure is symmetrical about the midpoint of the antenna. Hence, the operator would have tow sets of eigenvalues which are equal. Since the conjugate gradient method yields the solution in at most M steps, where M is the number of independent eigenvalues of the operator (15 in this case) the method converged in 15 iterations.

Also it is interesting to point out that the CPU time against the number of unknowns vary as N rather than N^2 for conventional methods. This is shown in Figure 1 .

VIII. ELECTROMAGNETIC SCATTERING FROM A FLAT RECTANGULAR PLATE

Consider a flat rectnagular plate of size L_1 x L_2 located in the x - y plane and irradiated by an incident field of intensity E^i. By forcing the total tangential electric field on the surface of the plate to be zero, we obtain the following electric field integral equation:

$$k^2 \hat{x} \int_0^{L_1} dx' \int_0^{L_2} dy' J_x(x',y') G(x,y)$$

$$+ \hat{y} \int_0^{L_1} dx' \int_0^{L_2} dy' J_y(x',y') G(x,y)$$

$$+ \left(\hat{x} \frac{\partial}{\partial x} + \hat{y} \frac{\partial}{\partial x} \right) \int_0^{L_1} dx' \int_0^{L_2} dy'$$

$$\cdot \left(\frac{\partial J_x(x',y')}{\partial x'} + \frac{\partial J_y(x',y')}{\partial y'} \right) G(x,y)$$

$$= -j\omega 4\pi\epsilon_0 [\hat{x} E_x^{inc}(x,y) + \hat{y} E_y^{inc}(x,y)] \tag{92}$$

where E_x^{inc} and E_y^{inc}, respectively, denote the x and y component of the incident electric field, and $G(x,y)$ denotes the Green's function defined as

$$G(x, y) = \frac{\exp(-jkR)}{R}. \tag{93}$$

Here

$$R = \sqrt{(x-x')^2 + (y - y')^2}. \tag{94}$$

We now solve (92) by the conjugate gradient method. The FFT is utilized in the computation of the Green's function with J_x, J_y, $\partial J_x/\partial x$ and $\partial J_y/\partial y$.

The following procedure is used. We first discretize the plate, assume certain number of unknowns for both $J_x(x,y)$ and $J_y(x,y)$. We then pad the array of J_x and J_y with zeros so that we have a two-dimensional array, either dimension of which is composite number of two. We then take the two-dimensional FFT for both J_x and J_y yielding $\mathfrak{J}_x$ and $\mathfrak{J}_y$. The transform for the Green's function is found to be

$$\tilde{G}(k_x, k_y) = \int_{-\infty}^{\infty} \exp(-jk_x x)\, dx \int_{-\infty}^{\infty}$$

$$\cdot \exp(-jk_y y)\, dy G(x,y)$$

$$\tag{95}$$

$$= \frac{2\pi}{\sqrt{k_x^2 + k_y^2 - k^2}}$$

so the first step in computing AJ_0 in the conjugate gradient method is performed in the transform domain as follows:

$$\tilde{AJ}_0 = k^2 [\hat{x}\mathfrak{J}_{x0}\tilde{G} + \hat{y}\mathfrak{J}_{y0}\tilde{G}] + [\hat{x}(jk_x)$$

$$+\hat{y}(jk_y)][jk_x\mathfrak{J}_{x0}\tilde{G}+jk_y\mathfrak{J}_{y0}\tilde{G}]. \tag{96}$$

We then obtain the function by taking the inverse transform of each component of $\tilde{AJ}_0$ separately. The two-dimensional inverse transform is defined by

$$AJ_0 = \frac{1}{4\pi^2} \int_{-\infty}^{\infty} \exp\ (+jk_x x)\ dk_x \int_{-\infty}^{\infty} \exp\ (+jk_y y) dk_y\ \tilde{A}J_0. \tag{97}$$

The rest of the computations in (8)-(12) are carried out accordingly and each vector component is treated separately. The iterations are continued till the following error criterion is satisfied

$$\frac{||AJ_n - Y||}{||Y||} \leq 10^{-4}.$$

As a first example consider a 1.0λ square plate irradiated by a normally incident field of intensity 377 V/m. We assume nine unknowns per wavelength so that the total number of unknowns we are solving for is 162. The results for the two cuts for J_x are shown in Fig. 2, when the incident field is x-oriented. The agreement between Rao's triangular patch modeling [10] and the conjugate gradient method appears to be quite reasonable.

As a second example, we consider electromagnetic scattering from a 7λ by 7λ rectangular plate. By utilizing ten unknowns per wavelength, this would amount to $2 \times 10 \times 7 \times 10 \times 7 = 9800$ unknowns. In this case, we chose M' = 256 and N' = 256 (i.e. 256 x 256 point FFT).

In Table III we summarize the results obtained for different degrees of accuracy and number of iterations. It is quite gratifying to see that a 0.1 percent accuracy can be obtained in only 410 iterations when we are solving for 9800 unknowns.

Also, we have solved the problem of electromagnetic scattering from a 0.005λ square plate, with seven expansion functions for each x and y direction variation (therefore, we have $2 \times 7 \times 7 = 98$ unknowns) by utilizing the same computer program and just changing the dimensions of the plate. In this case, we chose M' = 64 and N' = 64 (i.e. 64 x 64 point FFT).

So, from these limited numerical computations, it is clear that by utilizing the FFT and CG method, it is possible to solve for the currents on the surfaces of conducting plates which are electrically large and small without running into computational instabilities.

IX. EPILOGUE

The claim made in this paper is that the application of the FFT and the conjugate gradient method is far more superior in performance than even applying the method of conjugate gradient to the solution of the matrix equations. This is because the total time taken to solve the problem by matrix methods in p iterations is equal to

$$T_{matrix} \simeq T_{set\ up} + p \cdot N^2 \cdot T_{multiply} \tag{98}$$

where

$T_{set\ up}$ set up time required to compute N^2 elements of the matrix.

p   number of iterations required to solve the problem to a desired degree of accuracy.

N   number of rows (or columns) of the matrix equation to be solved for

$T_{multiply}$ time taken by the computer to perform a multiplication

In contrast, the time taken to solve the operator equation directly by the FFT and the conjugage gradient method is approximately equal to

$$T_{CGFFT} \simeq P \cdot (8N)[\log_2 (2N)+0.25] \cdot T_{multiply} \qquad (99)$$

This estimate comes from performing four FFT's per iteration in the computation of $A*R_n$ and AP_n on a 2N array. In addition one needs to compute two inner products for $|A*R_{n+1}|$ and $|AP_n|$, which require N multiplications.

Leaving behind the set up time required to compute all the matrix elements, which itself is quite big, we see that for N > 64, the FFT procedure becomes faster than the application of the conjugate gradient method to the solution of matrix equations.

X. DECONVOLUTION OF THE IMPULSE RESPONSE OF A CONDUCTING SPHERE

In this section, we obtain an impulse response h(t) from x(t) and y(t) in the time domain, without ever explicitly performing any computation in the frequency domain. We first measure directly the same response y(t) (called the output signal) when the target is present and then measures the response x(t) (called the input signal) when the target is replaced by a plane conducting sheet. The impulse response h(t) is then obtained iteratively from x(t) and y(t) in the time domain [22].

In this method, we minimize the functional F(h) given by

$$F(h) = (R,R) = (Ah-y, Ah-y) \qquad (100)$$

where A is the convolution operator

$$A = \int_0^\infty x(t-\tau)(\cdot) \, dr, \qquad (101)$$

and the inner product and the norm is given by (4). The conjugate-gradient method starts with an initial guess h_0 and generates P_0 and $A*R_0$ where A* represents the adjoint operator. With reference to the adjoint operator A* is the advanced convolution operator, which is defined as

$$A* \cdot z = \int_0^\infty x(t-\tau)z(t)dt = \int_0^\infty x(t)z(t+\tau)dt \qquad (102)$$

In defining (102), it has been assumed that x(t) is causal.

The numerical computation is done in the following way. We start with a array of elements for h_0, which is our initial guess. We next convolve h_0 with x, to yield another linear array, say Q. We then subract y from the array Q, resulting in R_0. The computation of the convolution can be speeded up utilizing the fast Fourier transform (FFT). When we plan to utilize the FFT, we pad the arrays h_0 and x with an approximately equal number of zeros. We then take the FFT of both h_0 and x, and multiply them. We then take the inverse FFT to yield the array B. We truncate the array B to form the array Q, so that Q has a dimension less than x, as we padded up the original arrays with zeros before taking FFT.

Next, we perform the advance convlution of x with R_0, resulting in the array $A*R_0$. Again the FFT can be utilized to speed up the computation. The remaining computations are done as described in (58)-(66).

The conjugate-gradient method always converges for any initial guess and for any functional equation with a bounded operator as long as it is implemented correctly. Even when y is not the range of the operator A (this may happen when y is contaminated by noise), the conjugate-gradient method still yields the minimum norm solution. Another important question in this numerical technqiue is how to terminate the iteration. Since the conjugate gradient method terminates in M steps (in absence of round-off error), where M is the number of independent eigenvalues of the operator in the space in which the problem is being solved. The conjugage gradient method seeks the solution, at each iteration, along the direction of the

eigenvectors corresponding to first large and then small eigenvalues in
descending order. For an ill-posed problem there are some small eigenvalues due to noise. Hence we would like to stop the iterative process before the method seeks a solution along the direction of the eigenvector
corresponding to small eigenvalues. Hence too few iterations may give a
solution that is not good enough, and for ill-posed problems, too many iterations might produce spurious oscillations in the solution as the method
seeks the solution along the eigenvectors corresponding to the small noise
eigenvalues. Hence the conjugate gradient method essentially performs a
singular value decomposition without actually computing the eigenvalues and
the criteria that we utilize to stop the iterative process is that the magnitude of the largest element in the residual must be below a certain value,
that is.

$$|R_n(t)| \leq W \text{ (a prefixed constant)} \tag{103}$$

We now discuss the method of choosing W. For most problems of interest, the
elements of the input and output are accurage up to the quantization errors
associated with each sample of $x(t)$ and $y(t)$. Assume that the quantization
errors $\Delta x(t_n)$ and $\Delta y(t_n)$ are independent, t_n being the sampling time, We
further assume that

$$|\Delta x(t_n)| \leq C \tag{104}$$

and

$$|\Delta y(t_n)| \leq C \tag{105}$$

where C is a constant dependent on the number of bits used in the quantization process. In our esperiments, eight-bit quantization has been used in
recording the digital form of the signal. Hence in our computation, we have
chosen the effective number of bits ε_p to be eight, which then yields the
constant C given by

$$C = 2 = \varepsilon_p = 2^{-8} = 0.00390625. \tag{106}$$

Now observe that the error in the solution after n-iteation would be

$$R_n(t) = \int_0^\infty x(t-\tau)h_n(\tau) \, d\tau - y(t).$$

as $n \to \infty$, we would have a different solution other than the exact solution
due to discretization error. Let us term the true impulse response by
$h_e(t)$. We now define a discretization error R_D due to sampling and express
it as

$$R_D(t) = \int_0^\infty [x(t-\tau) + \Delta x(t-\tau)]h_e(\tau) \, d\tau - y(t) - \Delta y(t)$$

$$\cong \int_0^\infty \Delta x(t-\tau)h_e(\tau) d\tau - \Delta y(t)$$

$$\leq C \int_0^\infty |h(\tau)| \, d\tau + 1 \quad \cdot \tag{107}$$

We now replace $h(\tau)$ in $h_n(\tau)$ in (107) and check at each iteration whether

$$R_n(t) \leq C \int_0^T |h_n(\tau)| d\tau + 1 \quad , \text{ for all } t, \tag{108}$$

i.e. the residuals $R_n(t)$ are of the same order of magnitude as the discretization errors.
We then check at each iteration whether

$$C \int_0^T \left| h_{\bar{n}}(t) \right| \, dt + 1 \leqq \left| R(t_n) \right| . \tag{109}$$

It was shown by Oettli and Prager [28] that the above inequality is a necessary and sufficient condition for h(t) to be a solution of (100). Note that we have replaced the upper limit in (108) by T. This is because in practice the data is recorded up to a finite time T.

The first target considered was a conducting sphere of 1.5 in diameter. The input signal x(t), the scattering from a plane conducting sheet, after 100 waveform averaging and the subtraction of the background noise, is shown in Fig. 3.. The same process was also applied to the target, and the resulting output signal is presented in Fig. 4. It is seen that both signals are relatively noise-free. We then applied the conjugate-gradient method to the x(t) and y(t) to obtain the impulse response h(t) of the target. For the 256 data points, it took eight iterations to obtain the impulse response shown in Fig. 5. It is noticed that the deconvolved response has a rather noisy tail. This is because the signal-to-noise ratio deteriorates as we progress in time. In order to minimize the effect of the noisy tail we changed the definition of the inner product by introducing an exponential weighting. Since the impulse response h(t) exists mainly in the first 300 ps, we modify the minimization criterion by defining a weighted inner product. Hence the modified error criterion is

$$(R_n, R_n) = \int_0^T R_n(t) R_n(t) \, \exp \, [-t/t_0] dt; \quad T = 2 \text{ ns}$$

where t_0 = 500 ps. The exponential weighting function tends to supress the effect of the noisy tail on the deconvolution process. With this modified functional, we applied the conjugate gradient method to the x(t) and y(t) and obtained the impulse response for an effective number of bits ε_p = 8.0, which required nine iterations for a 256 data point signal. The impulse response is shown in Fig.6 and its spectra in Fig. 7. The resonant frequencies are obtained as 6.6 and 9.0 GHz which are close to the theoretical values of 6.9 and 9.3 GHz. The same deconvolution technique with ε_p = 8.0 was then applied to these transient waveforms and yields fairly good impulse responses. The data processing then produced the spectra as summarized in Table IV. The measured and theoretical resonant frequencies are presented in Table IV. The closeness of the results show that the experimental technique and the modified conjugate gradient method is feasible for determining the impulse response of a target.

XI. CONCLUSIONS

This paper surveys the various numerical methods utilized in the solution of operator equations arising in electromagnetics. It is shown that the conjugate gradient method can be used very effectively to solve very large systems. Also the generalized biconjugate gradient method can be utilized as an alternative to preconditioning [25].

XII. REFERENCES

[1] Sarkar, T. K., Weiner, D. D., Jain, V. K. Some mathematical considerations in dealing with the inverse problem. IEEE Trans AP, USA, March 1981, pp.373-379.

[2] Harrington, R. F., Field computation by moment methods. The Macmillan, London (1968).

TABLE III
ACCURACY IN THE SOLUTION VERSUS NUMBER OF ITERATIONS

$||J_{n+1} - J_n||2/||J_n||2$	Number of Iterations
0.001	26
0.0001	71
0.00001	150
0.000001	410

Table IV COMPARISON OF THEORETICAL AND MEASURED RESONANT FREQUENCY OF

CONDUCTING SPHERES

Size	Resonant Frequency (GHz)					
(Inch)	Theory	Measured	Theory	Measured	Theory	Measured
1	6.79	6.9	10.37	10.0	13.97	12.8
1 1/8	6.03	6.4	9.22	9.0	12.35	12.0
1 5/16		*	7.90	7.8	10.66	10.8
1 1/2		*	6.90	6.6	9.30	9.0

*Data not available.

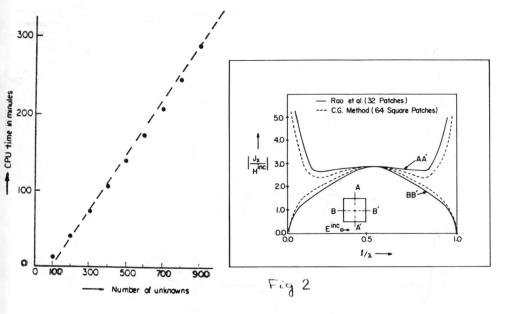

Fig 1

Fig 2

186

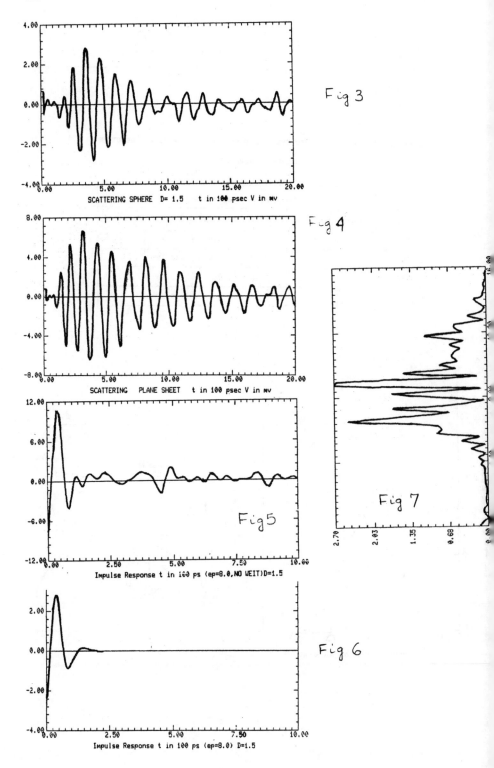

Fig 3

SCATTERING SPHERE D= 1.5   t in 100 psec V in mv

Fig 4

SCATTERING PLANE SHEET   t in 100 psec V in mv

Fig 5

Impulse Response t in 100 ps (ep=8.0,NO WEIT)D=1.5

Fig 7

Fig 6

Impulse Response t in 100 ps (ep=8.0) D=1.5

[3] Harrington, R. F. Matrix methods for field problems. Proc. IEEE, USA
     Feb. 1967, 55, pp. 136–149.

[4] Sarkar, T. K. A study of the various methods for computing electro-
     magnetic fields utilizing thin wire integral equations. Radio Sci.,
     USA Jan. 1983, pp.29–38.

[5] Sarkar, T. K., A note on the variational method, Rayleigh Ritz,
     Galerkin's method and the method of least squares. Radio Sci., USA
     Nov. 1983, pp.1207–1224.

[6] Mikhlin, S. G., The numerical performance of variational methods.
     Walters–Noordhoff, Leyden, Groningen 1971.

[7] Krsnoselskii, M. A., Vainikko, G. M., Zabreiko, P. P., Rutitskii, Ya .
     B., Stetwenko, V. Ya. Approximate solution of operator equations.
     Walters–Noordhoff, Leyden 1972.

[8] Harrington, R. F., Sarkar, T. K., Boundary elements and the method of
     moments, in boundary elements. Proc. of the 5th International Con-
     ference, Hiroshima, Japan, Nov. 1982 [ed, by C. A. Brebbia, T.
     Gutugami and M. Tanaka], pp.21–40.

[9] Sarkar, T. K. A note on the choice of weighting functions in the
     method of moments. IEEE Trans. AP, USA Feb. 1985, 33.

[10] Sarkar, T. K., Siarkiewicz, K. R., Stratton, R. F., Survey of numer-
     ical methods for solution of large systems of linear equations for
     electromagnetic field problems. IEEE Trans. AP, USA Nov. 1981, 29 n
     6, pp.847–856.

[11] Sarkar, T. K., Rao, S. M. An iterative method for solving electro-
     static problems. IEEE Trans. AP, USA July 1982, n 4, pp.611–616.

[12] Sarkar, T. K., Rao, S. M., The Application of the conjugate gradient
     method for the solution of electromagnetic scattering from arbitrary
     oriented antennas. IEEE Trans. AP, USA Apr. 1984, 32, pp.398–403.

[13] Sarkar, T. K. The application of the conjugate gradient method for the
     solution of operator equations arising in electromagnetic scattering
     from wire antennas. Radio Sci., USA Sept. 1984, Vol. 19, Nos.5
     pp.1156–1172.

[14] Sarkar, T. K., On a class of finite step iterative methods (conjugate
     directions) for the solution of an operator equation arising in
     electromagnetics. IEEE Trans. AP, USA, October 1985, vol.33,
     pp.1058–1066.

[15] Sarkar, T. K., Dianat, S. A., Rao, S. M., The application of the con-
     jugate gradient method to the solution of transient scattering from
     wire antennas. Radio. Sci., USA, Sept. 1984, Vol. 19, No. 5, pp.1319–
     1326.

[16] Hestenes, M. R., Applications of the theory of quadratic forms in Hilbert space in the calculus of variations. Pacific Journal of Math. 1951, 1, pp. 525-581.

[17] Hayes, R. M., Iterative methods for solving linear problems in Hilbert space, in O. Taussky (ed.,), Contributions to solution of systems of linear equations and determination of eigenvalues. Nat. Bur. Standards, Appl. Math., Serv. USA 1954, 39, pp. 71-104.

[18] Vandenberg, P. M. Iterative computational techniques in scattering based upon the integrated squared error criterion. IEEE Trans. AP, USA Oct. 1984, Vol. 32, pp. 1063-1071.

[19] Hestenes, M., Steifel, E., Method of conjugate gradients for solving linear systems, J. Res. Nat. Bur. Standards, USA 1952, 49, pp.409-43(

[20] Sarkar, T. K. Application of the conjugate gradient method for numerical deconvolution, in inverse methods in electromagnetic imaging (Edit. by W. M. Boerner), NATO Advanced Study Institute, Bad Windsheim, Germany, set. 1983.

[21] Chen, H., Sarkar, T. K., Dianat, S. A., Adaptive spectral estimation by the conjugate gradient method. IEEE ASSP.

[22] Tseng, F. I. and Sarkar, T. K., Deconvolution of the impulse response of a conducting sphere by the conjugate gradient method, IEEE Trans AP, USA, Jan 1987, vol. 35, pp.105-110.

[23] Rao, S. M. et.al, "Electromagnetic Scattering by Surfaces of arbitrary shape," IEEE Trans. AP, USA, May 1982, Vol.30, pp. 408-418.

[24] Sarkar, T. K., Arvas, E. and Rao, S.M., "Application of the Fast Fourier Transform and the conjugate gradient method for efficient solution of electromagnetic scattering from both electrically large and small conducting bodies," Electromagnetics Journal, Vol. 5, 1985, pp.99-122.

[25] Sarkar, T. K. "The conjugate gradient method as applied to electromagnetic field problems," IEEE Trans AP Newsletter USA Aug. 1986, pp.5-14.

[26] Sarkar, T. K. Arvas, E. and Rao, S.M., "Application of FFT and the conjugate gradient method for the solution of electromagnetic radiation from electrically large and small conducting bodies," IEEE Trans. AP, USA, May 1986, pp.635-640.

[27] Nayanthara, K., Rao, S.M. and Sarkar, T. K., "Analysis of Two Dimensional conducting and Dielectric Bodies Utilizing the Conjugate Gradient Method," IEEE Trans. Ant. & Propagat., vol.35, April 1987, pp.451-453.

[28] Oettli, W. and Prager, W., "Compatibility of Approximate Solution of Linear Equations with given Error Bounds for Coefficients and Right Hand Sides," Num. Math., 6, 1964, pp.405-409.

A SPECTRAL APPROACH TO TRANSIENT PROPAGATION

L.B. FELSEN

Department of Electrical Engineering/Computer Science/Weber Research
Institute, Polytechnic University, Farmingdale, NY 11735 USA

1. INTRODUCTION
    Transient wave propagation and scattering in a rather general environ-
ment, when caused by short pulse illumination and observed over long time
intervals, requires synthesis in terms of harmonic constituents covering
the range from high to low frequencies. Dispersive phenomena (i.e., frequency dependent propagation speeds of plane waves employed in the spectral
synthesis) usually play a minor role for the high frequency contribution
and therefore permit the use of "weakly dispersive" approximations ignoring
these effects. By separating the wide band transient response into weakly
and strongly dispersive portions, one may employ distinct methods better
adapted to analytical and numerical implementation of each rather than
attack the much more complicated overall problem. High frequencies synthesize the initial (early time) response characteristics, whereas the
later time portion behind the wavefront of the initial arrival emphasizes
lower frequencies. When boundaries, scattering centers or medium inhomo-
geneities give rise to multiple wavefront phenomena, each arriving wavefront field has its own causal "turn-on time" at the receiver, with a low
frequency tail trailing behind. These tails cause contributions from early
and later arrivals in the overall received signal to overlap. It is then
preferable, via collective wavefront summation, to restructure the late
time response into resonance fields which sample the global features of the
multiple scattering environment rather than the local features encountered
by each wavefront. The wavefront-resonance alternatives, and their self-
consistent combination into a hybrid form, have been documented in the literature [1] and will not be considered here. Instead, the emphasis is on
the behavior of individual wavefront contributions, assuming that dispersive effects can be ignored. Thus, the problem formulation is exact for
truly nondispersive transient scattering but approximate, and restricted to
the "weakly dispersive" regime, for more general cases.
    Representations involving a continuous superposition of true or local
plane waves characterized by their spatial wavenumbers have been found useful for synthesizing the frequency domain response due to source excitation
in the presence of fairly general, but coordinate separable, propagation or
scattering environments (for descriptions of the method, see [2,3]: local
plane waves result from use of high frequency asymptotic (WKB type) approximations for exact wavefunctions in the spectral integral). In the weakly
dispersive regime, each plane wave in the spatial spectrum is non-dispersive. Therefore, the conventional sequence which recovers source-excited
transient fields from the source-excited time-harmonic fields, can be reversed through performing the frequency inversion before the spatial inversion. This generates a representation of the source-excited transient
field as a spatial superposition of transient true or local plane waves. It
has been customary to perform the integration interchange by keeping both

189

B. de Neumann (ed.), Electromagnetic Modelling and Measurements for Analysis and Synthesis Problems, 189–204.
© 1991 Kluwer Academic Publishers. Printed in the Netherlands.

frequency and wavenumber <u>real</u> [4,5]. This constraint is not detrimental for certain classes of propagation and diffraction environments but can be awkward and indirect for others. This aspect has been pointed out in [6]. Here, a unified theory is presented which, through admission of <u>complex</u> spatial spectra, allows all problems to be treated within the same format. This new Spectral Theory of Transients (STT) has been documented, and applied, in references [6-8]. The discussion which follows is based on these publications. It may also be remarked that for certain special propagation or scattering configurations, the weakly dispersive <u>time-harmonic</u> spectral wavenumber integral can be transformed <u>directly</u> into a Laplace integral, from which the transient field is extracted by inspection. However, this elegant scheme, known as the Cagniard-deHoop method [9,10], cannot attack the broader range of problems accommodated by the spectral approach. Moreover, synthesis by <u>transient</u> plane wave spectra casts a new perspective on the entire mechanism of transient propagation and diffraction.

A final introductory remark refers to recent developments which remove the separability constraint from the formal structure of the local plane wave spectral representation. If the environment deviates <u>weakly</u> from separability in one of the classical separable coordinate systems, one may invoke the high-frequency principle of locality to construct locally adapted spectra. These generalized spectral representations in the frequency domain [5,11-14] can then be inverted into the time domain by the procedure described above. Thus, one may construct a theory of transient fields with great versatility.

II. SPECTRAL INTEGRAL FORMULATION

The <u>non-dispersive</u> local plane wave spectral integral representation of direct, reflected, refracted or diffracted constituents of the source excited time-harmonic field takes the generic form [2,3,6]

$$\hat{u}(\underset{\sim}{r},\underset{\sim}{r}';\omega) = (-i\omega)^{M/2} \frac{i}{2\pi} \int_{C_N} d^N\xi \, A(\underset{\sim}{r},\underset{\sim}{r}';\xi) \, e^{i\omega\tau(\underset{\sim}{r},\underset{\sim}{r}';\xi)} \tag{1}$$

where $\underset{\sim}{r}'$ and $\underset{\sim}{r}$ identify source and observation point locations, respectively, $N = 1$ or 2 for two-dimensional or three-dimensional propagation, and ξ is the corresponding spectral variable (spatial wavenumber) normalized with respect to the frequency ω. The local plane wave field $A(\xi)\exp[i\omega\tau(\xi)]$ has an amplitude A and phase $(\omega\tau)$, which generally result from high frequency (WKB) approximations of exact wavefunctions. The (separable) spectral decomposition occurs along the lateral symmetry coordinate following (for $N = 1$) one of the dashed contours in Fig. 2(a) (for $N = 2$, ξ refers to the lateral plane). Propagation of the local plane wave congruences away from the reference coordinate is specified by the normalized phase function τ, which is real (and therefore defines propagating waves) for ξ in the "visible" spectrum range but is complex (and therefore defines evanescent waves with $\text{Im}\tau > 0$) for ξ in the "nonvisible" spectrum range. By the weakly dispersive assumption, the frequency appears in (1) only as a linear factor in the exponent, or as an algebraic multiplier with possible fractional exponent determined by the integer M. For each ξ, the spectral integrand defines a local plane wave congruence that translates a wavefront passing through the source point into a wavefront passing through the observation point, along ray trajectories with travel time (or time delay) τ (Fig. 2).

The spectral contour C_N follows the real axis in the spectral domain except to avoid, in a manner consistent with the radiation condition, possible singularities located on these axes.

The plane wave congruences with different ξ interfere constructively around central values ξ_j which render the normalized phase $\tau(\underset{\sim}{r},\underset{\sim}{r}':\xi)$ stationary. Each ξ_j selects a wavefront (and therefore a congruence) orientation which ensures that both the source point and the observation point lie on the same ray in the congruence. Asymptotic (saddle point) evaluation of the spectral integral then furnishes the field along that ray as predicted by the geometrical theory of diffraction (GTD ; for a collection of relevant papers, see [15]). The spectral integral in (1) is therefore referred to as the ray integral.

To invert the ray integral into the time domain, we define the Fourier transform

$$u(t) = \underset{\sim}{F}\{\hat{u}\} = \frac{1}{2\pi} \int_{-\infty}^{\infty} d\omega \; e^{-i\omega t} \; \hat{u}(\omega) \tag{2}$$

However, since the plane wave spectra generally depend on the sign of the frequency, it is convenient to restrict the inversion to positive frequencies via the analytic transform

$$u_+(t) = \frac{1}{\pi} \int_0^{\infty} d\omega \; e^{-i\omega t} \; \hat{u}(\omega) \quad , \quad \text{Im } t \leq 0 \tag{3}$$

which render u_+ analytic in the lower half of the complex time plane, with a limit on the real axis given by (see Appendix A)

$$u_+(t) = u(t) + i\bar{u}(t), \quad t \text{ real} \tag{4}$$

The Hilbert transform $\bar{u}$ of u is defined as

$$\bar{u}(t) \equiv \underset{\sim}{H}\{u(t)\} = \underset{\sim}{P} \int_{-\infty}^{\infty} \frac{u(t')}{\pi(t-t')} \, dt' \tag{5}$$

with $\underset{\sim}{P}$ indicating principal value integration. The real transient field follows from u_+ via

$$u(\underset{\sim}{r},\underset{\sim}{r}':t) = \text{Re}\{u_+(\underset{\sim}{r},\underset{\sim}{r}':t)\} \quad , \quad t \text{ real} \tag{6}$$

Applying the analytic Fourier transform (3) to the ray integral (1), with Im $t < 0$, permits interchange of the order of the ξ and ω integrations. Due to the nondispersive integrand, the ω-integration can be done in closed form to yield

$$u(\underset{\sim}{r},\underset{\sim}{r}':t) = D_M(t) * \text{Re}\{U_+(\underset{\sim}{r},\underset{\sim}{r}':t)\} \quad , \quad t \text{ real} \tag{7}$$

where * denotes a convolution, U_+ is the limit on the real axis of the analytic time domain spectral integral with the canonical form

$$U_+(\underset{\sim}{r},\underset{\sim}{r}':t) = \frac{1}{2\pi^2} \int_{C_N} d^N\xi \; \frac{A(\xi)}{t-\tau(\xi)} \quad , \quad \text{Im } t \leq 0 \quad , \tag{8}$$

and

$$D_M(t) = \begin{cases} (\dfrac{d}{dt})^{M/2} \; \delta(t) & , \text{ M even} \quad (9a) \\[3ex] (\dfrac{d}{dt})^{(M+1)/2} \; \dfrac{H(t)}{\sqrt{\pi t}} & , \text{ M odd} \end{cases}$$

Here, $\delta(t)$ represents the Dirac delta function and $H(t)$ the Heavyside function, which equals unity or zero for positive or negative arguments, respectively. In (7) and (9), the differentiation may be commuted to act on U_+. For simplicity, we restrict consideration from here on to geometries in two dimensions (i.e., $N = 1$) so that $\xi \to \xi$ defines a one-dimensional spectral variable, and to $M = 0$ so that from (7) and (9), $u = \text{Re} U_+$.

In (8), the analytic transient source-excited field is synthesized as a spectrum of analytic transient local plane wave congruences defined for each ξ by

$$V_+(\underset{\sim}{r}, t; \xi) = A(\xi) \delta_+ [t - \tau(\xi)], \qquad \text{Im } t \leq 0 \qquad (10)$$

where $\delta_+(t)$ denotes the analytic delta function

$$\delta_+(t) = \begin{cases} \dfrac{1}{i\pi t} & \text{Im } t < 0 \qquad (11a) \\[3ex] \delta(t) + P \dfrac{1}{i\pi t} & \text{Im } t = 0 \qquad (11b) \end{cases}$$

with (11b) representing the distributional limit of (11a) for Im $t = 0$ (see Appendix A). Thus, for ξ in the "visible spectrum", where plane waves propagate along real ray trajectories as in Fig. 2 and $\tau(\xi)$ is real, Re V_+ represents either an impulsive local plane wave with arrival time τ or a local plane wave with a waveform $1/(t - \tau)$, for real or imaginary $A(\xi)$, respectively.

The transient integral U_+ in (8) can be evaluated in terms of the singularities of the integrand. Foremost among these are the <u>time dependent pole</u> singularities $\xi(t)$ defined by

$$\tau[\xi(t)] = t \qquad , \qquad \text{Im } t \leq 0 \qquad (12)$$

which tag those source excited local plane wave constituents that reach the observer at time t. Other, but <u>time-independent</u>, singularities arise from $\tau(\xi)$ and $A(\xi)$. The phase function $\tau(\xi)$ possesses real first order branch points at $\pm \xi_c$ which spectral values separate the visible (propagating) and nonvisible (evanescent) portions where Im $\tau = 0$ and Im $\tau \neq 0$, respectively. The amplitude function $A(\xi)$ may have branch points ξ_b introduced into the visible spectrum by certain interface (for example, critical angle) reflection phenomena in a piecewise continuous medium, and (or) pole singularities ξ_p attributed to shadow boundaries in certain diffracting configurations (for example, a semi-infinite screen). These singularities occur already in the time-harmonic integral in (1). To render the integration in (8) unique in the presence of the branch points, we introduce branch cuts along the real axis (nonvisible spectrum) intervals $|\xi| > \xi_c$ and, if $A(\xi)$ has branch points at $\pm \xi_b$, along the real axis intervals $|\xi| > \xi_b$ (see Fig.3).

The location of the integration contour with respect to the time-independent singularities has already been established in the construction of (1) (see Fig. 3; although C is shown to pass below ξ_p, there are diffrac-

tion problems (see [8]) for which C passes above ξ_p). For the time-dependent pole singularities, the determination is made unique by assuming Im t < 0 so that the real poles $\xi(t)$ in the visible spectrum are displaced from the real ξ axis. Passing to the limit Im t = 0 yields the following rule for the real roots $\xi(t)$ in the visible spectrum:

$$\xi(t) \text{ located } \begin{Bmatrix} \text{above} \\ \text{below} \end{Bmatrix} C \text{ for } \tau'(\xi) \begin{smallmatrix} < \\ > \end{smallmatrix} 0 \tag{13}$$

The limiting process is schematized in Fig. 3 for wave species having two (caustic forming) rays, which are associated with two stationary points ξ_1 and ξ_2 (see discussion in the next section).

Because the transient response u(t) is real, the local plane wave spectral functions in (8) obey certain symmetries in the ξ plane. Since $\tau(\xi)$ is real in the visible spectrum, one may show that

$$\tau(\xi) = \tau^*(\xi^*) \tag{14}$$

in the complex ξ plane, with an asterisk denoting the complex conjugate. The amplitude function A will be either real or imaginary for ξ values belonging to the visible spectrum range $-\xi_\gamma < \xi < \xi_\gamma$, where ξ_γ stands for ξ_b if A has a branch point in the visible spectrum at $\pm \xi_b$, and $\xi_\gamma = \xi_c$ otherwise. Real or imaginary values of A can arise, in this context, from ray phenomena due to turning points and ray caustics in an inhomogeneous medium, and they generate distinct categories of transient wave solutions. The symmetry relations are

$$A(\xi) = A^*(\xi^*) \qquad A \text{ real for } -\xi_\gamma < \xi < \xi_\gamma \tag{15a}$$

$$A(\xi) = -A^*(\xi^*) \qquad A \text{ imag. for } -\xi_\gamma < \xi < \xi_\gamma . \tag{15b}$$

III. The Time-Dependent Spectral Poles

The time-dependent spectral solutions $\xi(t)$ of (12) move in the complex ξ-plane along trajectories that depend on the form of the travel time function $\tau(\xi)$. Initially displaced away from the real ξ-axis when Im t < 0, the trajectories in the limit of real t form real and complex branches, with intersections at the stationary points ξ_j, j = 1,2,..., (if any) of the phase function τ in the visible range (see, for example, in Fig. 3 where the roots tend toward the boundaries between the shaded and unshaded regions as Im t → 0). As noted earlier, the stationary points identify geometrical rays from source to observer, with wavefront arrival times

$$t_j = \tau(\xi_j), \quad \tau'(\xi_j) = 0, \quad j = 1,2,.. \tag{16}$$

where the prime denotes the derivative with respect to the argument. When there is no real stationary point, as on the dark side of a caustic, there are no propagating waves at the observer, and the complex trajectories $\xi(t)$ do not intersect the real axis. Real roots are denoted by $\xi_r(t)$ and complex roots in the upper or lower half planes by $\hat{\xi}(t)$ and $\check{\xi}(t)$, respectively (Fig. 3).

Isolated stationary points are determined by the local maxima or minima of the phase function τ, which can have a form like that sketched in Fig. 4(a). Each ξ_j is at the junction between two real branches and two complex branches, which intersect the real axis at right angles. For $t < t_j$, two roots $\xi(t)$ move towards ξ_j on two opposite branches, and for $t > t_j$,

they move away on the other two (see Fig. 3, near ξ_1 or ξ_2). At $t = t_j$, there are two roots near ξ_j, with behavior inferred by expanding $\tau(\xi)$ to second order near ξ_j,

$$\tau(\xi) = t_j + \frac{1}{2}(\xi - \xi_j)^2 \tau_j'' \quad , \qquad \tau_j'' \equiv \tau''(\xi_j) \qquad (17)$$

and inverting (12) to find

$$\xi(t) = \xi_j \pm [2(t - t_j)/\tau_j'']^{1/2} \quad , \qquad t = t_j \qquad (18)$$

Thus, if $\tau_j'' < 0$ (point ξ_2 in Fig. 3), two roots move toward ξ_j from both sides on the real axis for $t < t_j$, and away from ξ_j on complex trajectories normal to the real axis for $t > t_j$. If $\tau_j'' > 0$ (point ξ_1 in Fig. 3), the movement of the roots with respect to ξ_j is reversed. The complex trajectories generally extend to infinity, with limiting times $t = \pm \infty$ for $\tau_j'' \gtrless 0$, respectively. The explicit expressions in (17) and (18) yield <u>wavefront approximations</u> (see [8]) whose validity is restricted to observation times near the wavefront arrival.

If two (or more) first order stationary points approach one another so that $\tau_j'' \to 0$, it is necessary for a uniform transition to replace the quadratic approximation (17) for the phase $\tau(\xi)$ by a higher order polynomial that accommodates at the point of confluence the vanishing of as many derivations of $\tau(\xi)$ as may occur. The simplest illustration is for two contiguous first order stationary points, representative of observation points near a smooth caustic as in Figs. 1a,b,c. On the caustic, these stationary points coalesce into a single second order stationary point whereas in the shadow they become complex, with one moving above and the other below the real ξ-axis. This behavior is described adequately by a cubic approximation of the phase function in Fig. 4: the wavefront approximation for this case is inferred from the sketch in Fig. 3 when the observer is in the lit region, and from Figs. 5(a) and 5(b) when the observer is on the caustic and in the shadow, respectively. The cubic polynomial has three roots, whose properties can be inferred from these figures. Note that the shadow region (Fig. 5(b)) gives rise to pole trajectories which do not touch the real ξ axis, even for Im $t = 0$.

When the time-dependent spectral poles move into the vicinity of any of the time-independent singularities described earlier, appropriate expansions have to be developed to characterize the transitional behavior under these conditions. For examples, see [8].

IV. Application to a Complex-Source Pulsed Beam

When a line source located at $\rho' = (x', z')$ in a free space with wave propagation speed v is given an impulsive excitation $\delta(t - t')$ at time t', the resulting radiated field at an observation point $\rho = (x, z)$ located at a distance s from the source point consists of a first (causal) response arriving at time $t = t' + s/v$, followed by a decaying tail for $t > t' + s/v$. In fact, the well-known solution for the two-dimensional transient line source Green's function is [6]:

$$G(\underset{\sim}{\rho}, \underset{\sim}{\rho}'; t, t') = \frac{1}{2\pi\Delta} H(t - t' - s/v), \quad \Delta = [(t - t')^2 - (s/v)^2]^{1/2} \qquad (19)$$

where $H(w) = 1$ or 0 for $w > 0$ and $w < 0$, respectively.

It has recently been shown that a new solution of the time-dependent wave equation can be generated by assigning complex $(\bar{\varrho}',\bar{t}')$ values to the source-related coordinates (ϱ',t') [16]. The new solution has the form of a two-dimensional pulsed beam. While time-harmonic beam solutions can be developed from the time-harmonic Green's function solution by direct analytic continuation of ϱ' to $\bar{\varrho}'$ [17-19], this is not possible for the transient case because of the non-analytic behavior of the unit step function in (19). The analytic continuation can be performed uniquely by the spectral theory introduced earlier. The construction below, which closely follows that in [20], illustrates application of the theory, and shows how the spectral integral can be reduced in terms of the transient plane waves contributed by the singularities discussed in Section III.

A. Statement of the problem

We shall look for a solution of the transient-source wave equation

$$\left(\frac{\partial^2}{\partial x^2} + \frac{\partial^2}{\partial z^2} - \frac{1}{v^2}\frac{\partial^2}{\partial t^2}\right) G(\varrho,\varrho':t,t') = - \text{ "}\delta(t - \bar{t}')\,\delta(\varrho-\bar{\varrho}')\text{"} \tag{20}$$

for the case where the source coordinates have the complex values

$$(\bar{x}',\bar{z}') = (0,ib),\ b > 0,\ \text{and}\ \bar{t}' = ib/v \tag{21}$$

The quotation marks on the right-hand side of (20) denote that the delta functions must be appropriately defined. The solution strategy is to employ plane wave spectral analysis and synthesis, as in [20].

B. Frequency domain

Conventional plane wave synthesis of the two-dimensional time-harmonic Green's function $\hat{G}(\varrho,\varrho':\omega)$ for real source points ϱ' yields the form in (1) [21],

$$\hat{G}(\varrho,\varrho':\omega) = \frac{i}{2\pi} \int_C d\xi\, A(\rho,\rho':\xi)\, e^{i\omega\tau(\varrho,\varrho':\xi)} \tag{22}$$

with the (truly non-dispersive) functions τ and A given exactly by

$$\tau(\varrho,\varrho':\xi) = \frac{1}{v}\left[(x - x')\xi \pm (z - z')\sqrt{1 - \xi^2}\right],\ z \gtrless z' \tag{23a}$$

$$A(\varrho,\varrho':\xi) = \frac{1}{2\sqrt{1 - \xi^2}} \tag{23b}$$

The integrand has branch points at the edges $\xi = \pm 1$ of the visible spectrum; branch cuts are chosen along the real axis segments $|\xi| > 1$. In the visible spectrum range $-1 < \xi < 1$, τ is real for real ϱ', whereas it is complex for $\xi < -1$ and $\xi > 1$. The integration contour C runs along the real axis from $-\infty$ to $+\infty$, passing below and above the branch points $\xi = \pm 1$, respectively, so that it proceeds along the side of the cut whereon $\text{Im }\tau > 0$ in accordance with the radiation condition.

Complex source coordinate substitution $\varrho' = (0,ib)$ as in (21) can be performed in (22) by taking the upper or lower signs in (23a) for $z \gtrless 0$, respectively, and requiring that

$$\text{Im }\tau(\xi) \geq - \text{Im }\bar{t}' = - b/v,\ \xi\ \text{on C.} \tag{24}$$

This completes the analytic continuation in the frequency domain.

The resulting time-harmonic field is known to behave like a beam whose strength is maximum on the positive z-axis but decays with transverse displacement (polar angle θ) away from the z-axis. Detailed properties depend on the "beam parameter" b [19]. Large and small (kb) imply strong or weak collimation, respectively. At its narrowest part, the waist, the half-power linear beam width is $(8b/k)^{1/2}$, while in the far zone, its angular width is $(8/kb)^{1/2}$.

C. Time domain

Inversion of $\hat{G}$ into the time domain is performed via the analytic Fourier transform G_+ in (3), from which G is recovered via (6). Then subject to the condition

$$\operatorname{Im}[t - \bar{t}' - \tau(\xi)] < 0 \tag{25}$$

for t in the lower half of the complex t-plane and for all ξ on C, which allows interchange of the order of the ω and ξ integrations, one obtains (cf. (8)),

$$G_+(\rho,\bar{\rho}': t,\bar{t}') = \frac{1}{2\pi^2} \int_C \frac{A(\rho,\bar{\rho}':\xi)}{t - \bar{t}' - \tau(\rho,\bar{\rho}':\xi)}\, d\xi \tag{26}$$

The integral in (26) may be evaluated in terms of the singularities of the integrand in the complex ξ-plane. Denoting by (±) superscripts the results from deforming the integration contour into a semicircle at infinity in the upper and lower half planes, respectively, and noting that these semicircles make no contribution, we get

$$G_+^+(t) = \sum_{\hat{\xi}(t)} -\frac{i}{\pi}\frac{A(\hat{\xi})}{\tau'(\hat{\xi})} + \frac{1}{2\pi^2}\int_1^\infty \left[\frac{A(\xi)}{t-\bar{t}'-\tau(\xi)} - \frac{A(\xi^*)}{t-\bar{t}'-\tau(\xi^*)}\right] d\xi \tag{27a}$$

$$G_+^-(t) = \sum_{\check{\xi}(t)} \frac{i}{\pi}\frac{A(\check{\xi})}{\tau'(\check{\xi})} + \frac{1}{2\pi^2}\int_{-1}^{-\infty} \left[\frac{A(\xi)}{t-\bar{t}'-\tau(\xi)} - \frac{A(\xi^*)}{t-\bar{t}'-\tau(\xi^*)}\right] d\xi \tag{27b}$$

where $\hat{\xi}(t)$ and $\check{\xi}(t)$ are pole solutions satisfying (12) in the upper and lower half planes, respectively, and the integrals result from the branch cuts connected to the branch points $\xi = \pm 1$, with ξ taken along that side a cut whereon the radiation condition is satisfied (i.e., the lower or upper side of the cuts $1 < \xi < \infty$ or $-\infty < \xi < -1$, respectively). These conditions are reversed for the complex conjugate ξ^* of ξ.

With $\tau(\xi)$ given in (23a), there are two roots

$$\xi^\pm(t) = [(x - \bar{x}')(t - \bar{t}') \pm i(z - \bar{z}')\Delta]v/s^2 \tag{28}$$

with

$$\Delta = \sqrt{t^2 - \left(\frac{\rho}{v}\right)^2 - 2i\frac{b}{v}\left(t - \frac{\rho}{v}\cos\theta\right)}, \quad -\pi < \arg\Delta \leq 0 \tag{29}$$

and

$$s = \sqrt{\rho^2 - b^2 - 2ib\,\rho\cos\theta} \quad , \tag{30}$$

and we may choose Im $s \leq 0$. The roots $\xi^{\pm}$ lie on continuous trajectories as a function of t, x or z (see Fig. 6), and one may verify that for $t < t^{\pm}$ defined by

$$t^{\pm} = [\pm x\sqrt{1 + (b/z)^2} - b^2/z]v \quad , \tag{31}$$

the corresponding roots $\xi^{\pm}(t)$ lie on the lower Riemann sheet where $Re\sqrt{1 - \xi^2} < 0$, whereas for $t > t^{\pm}$, the root $\xi^{\pm}(t)$ crosses the branch cut associated with the branch point $\xi = \pm 1$ and moves, respectively, into the upper or lower half of the upper Riemann sheet, never crossing the real ξ-axis again.

In (27), we may therefore identify

$$\hat{\xi}(t) = \xi^+(t), \ \check{\xi}(t) = \xi^-(t) \quad \text{for} \quad t > t^{\pm} \tag{32}$$

and obtain for each pole contribution

$$\frac{i}{\pi} \frac{A[\xi^{\pm}(t)]}{\tau'[\xi^{\pm}(t)]} = \mp \frac{1}{2\pi\Delta} \quad , \ t > t^{\pm} \tag{33}$$

Next, by changing the integration variable to $\xi = \sin\phi$, we may write the branch cut integrals in (27) in the form

$$I^{\pm} = \frac{1}{4\pi^2} \int_{\frac{\pm\pi}{2} + i\infty}^{\frac{\pm\pi}{2} - i\infty} \frac{d\phi}{(t - \bar{t}') - \frac{x - \bar{x}'}{v}\sin\phi - \frac{z - \bar{z}'}{v}\cos\phi} \tag{34}$$

which can be reduced to yield

$$I^{\pm} = \frac{1}{2\pi\Delta} \begin{cases} \frac{1}{\pi i}\log\frac{t - \bar{t}' + \Delta}{s/v} \quad , \quad t > t^{\pm} \\[2ex] 1 + \frac{1}{\pi i}\log\frac{t - \bar{t}' + \Delta}{s/v} \ , \ t < t^{\pm} \end{cases}$$

with

$$\frac{-3\pi}{2} < \text{Im}\log < \frac{\pi}{2} \tag{35a}$$

Then from (6) and either of the representations in (27), and with (33) and (35), one obtains the desired solution of (20), with (21),

$$G(\rho, \bar{\varrho}'; t, \bar{t}') = \text{Re}\left\{\frac{1}{2\pi\Delta}[1 + \frac{1}{\pi i}\log\frac{t - \bar{t}' + \Delta}{s/v}]\right\} \tag{36}$$

which represents the unique analytic continuation of (19) from real to complex source coordinates.

The noncausal transient field in (36) has the characteristics of a pulsed beam, which has maximum strength on the z-axis (the beam axis), decreases away from this axis with increasing transverse distance x, and is

confined in space-time to $(t - \rho/v) \approx 0$. Its shape is influenced by the location of the observer with respect to the Fresnel length $z = b$. Approximating Δ for $t \approx \rho/v$, one gets from the first term in (36) (which yields the dominant contribution),

$$G(t) \approx \frac{1}{\sqrt{8\pi}} \text{ Re } \frac{1}{\sqrt{(t - \rho/v)\rho/v - ib\rho(1 - \cos\theta)/v^2}}, \quad t \approx \rho/v \qquad (37)$$

and in simplified form for the paraxial region $x \ll z$,

$$G(t) \approx \frac{1}{\sqrt{8\pi}} \text{ Re } \frac{1}{\sqrt{(t - z/v)(z - ib)/v - (x^2/2v^2)}} \qquad (38)$$

In (37), ρ and θ are cylindrical coordinates from $\rho = 0$ to the observation point, with the angle $|\theta|$ increasing away from z axis $\theta = 0$. Representative field shapes at various locations are plotted in Fig. 7 [16]. They reveal the strong peaking on the beam axis, and the successively diminished peak amplitude, with weakening precursor, for off-axis locations.

The real pulsed source solution in (19) with $\rho' = t' = 0$ can be recovered by letting $b \to 0$ in (36) (see (21)). From (29),

$$\Delta \to \begin{cases} \sqrt{t^2 - (\rho/v)^2} & t > (\rho/v) & (39a) \\[2ex] -i\sqrt{(\rho/v)^2 - t^2} & -(\rho/v) < t < (\rho/v) & (39b) \\[2ex] -\sqrt{t^2 - (\rho/v)^2} & t < -(\rho/v) & (39c) \end{cases}$$

Then (36) becomes

$$G(t) = \text{Re} \begin{cases} \dfrac{1}{2\pi\sqrt{t^2 - (\rho/v)^2}} \left[1 - \dfrac{i}{\pi} \cosh^{-1}(\dfrac{t}{\rho/v}) \right] & , \ t > \dfrac{\rho}{v} & (40a) \\[3ex] \dfrac{i}{2\pi\sqrt{(\rho/v)^2 - t^2}} \left[1 - \dfrac{1}{\pi} \cos^{-1}(\dfrac{t}{\rho/v}) \right] & , -\dfrac{\rho}{v} < t < \dfrac{\rho}{v} & (40b) \\[3ex] \dfrac{-1}{2\pi\sqrt{(\rho/v)^2 - t^2}} \left[1 - \dfrac{i}{\pi} (-\pi i - \cosh^{-1}(\dfrac{-t}{\rho/v})) \right] & , \ t < -\dfrac{\rho}{v} & (40c) \end{cases}$$

which yields the causal real source field in (2). One may discern from (40a,b,c) how the noncausal spectra cancel in the real source and real time limits.

ACKNOWLEDGEMENT

This work was supported in part by the U.S. Office of Naval Research under Contract NO. N00014-83-K-00214, by the Joint Services Electronics Program under Contract No. F49620-85-C-0078, and by the Strategic Defense Initiative Organization, Innovative Science and Technology Office, under Contract No. DAAL-OE-86-K-0096 managed by Harry Diamond Laboratory.

Appendix A - The Analytic Transforms

Separating the inverse transform in (2) into positive and negative frequency ranges, one may write

$$u(t) = [u_+(t) + u_-(t)]/2 \qquad (A1)$$

where

$$u_\pm(t) = \frac{1}{\pi} \int_{-\infty}^{\infty} H(\pm\omega)\hat{u}(\omega)\exp(-i\omega t)d\omega \qquad (A2)$$

with $H(\omega) = 1$ or 0 for $\omega > 0$ and $\omega < 0$, respectively. For real t, the evaluation of $u_\pm(t)$ can be carried out in terms of the operators

$$\delta_\pm(t) = \frac{1}{\pi} \int_{-\infty}^{\infty} H(\pm\omega)\exp(-i\omega t)dt = \frac{1}{2\pi} \int_{-\infty}^{\infty} [1\pm\text{sgn}(\omega)]\exp(-i\omega t)dt \qquad (A3)$$

$$= \delta(t) \pm P\frac{1}{it\pi} \qquad (A4)$$

with the first and second terms in (A4) contributed by the first and second terms inside the second integrand in (A3). Here, $\delta(t)$ is the conventional delta function, $\text{sgn}(\omega) = +1$ or -1 for $\omega > 0$ or $\omega < 0$, respectively, and the symbol P implies that, in subsequent operations of (A4) on a function u(t), the t integration is understood in the sense of a principal value. To show that $(it)^{-1}$ is the Fourier transform of $\text{sgn}(\omega)$, one observes that $\delta_\pm(t)$ must have the form

$$\delta_\pm(t) = C\delta(t) \pm P(i\pi t)^{-1}, \qquad (A5)$$

where the first term accounts for the singular contribution of the (first) defining integral in (A3) from the "endpoint" $|\omega| \to \infty$, and the second term for the contribution from $\omega = 0$. To determine the weighting coefficient C of the singularity, one recognizes that the Fourier transform S(t) of the real odd function $\text{sgn}(\omega) = [2H(\omega) - 1]$ must satisfy $S(t) = -S^*(t)$, whence Re $S(t) = 0$. But (see (2)),

$$S(t) = F\{2H(\omega) - 1\} = \delta_+(t) - \delta(t) = (C - 1)\delta(t) + (i\pi t)^{-1}, \qquad (A6)$$

therefore requiring $C = 1$. The result in (A4) follows. The functions $u_\pm(t)$ can be represented in terms of the operators $\delta_\pm(t)$ as follows,

$$u_\pm(t) = \frac{1}{2} \int_{-\infty}^{\infty} u_\pm(t')\delta_\pm(t - t')dt' = u_\pm(t) * \delta_\pm(t)/2$$

$$= \frac{1}{2} u_\pm(t) \pm \frac{2}{i} \bar{u}_\pm(t) \qquad (A7)$$

where $\bar{u}(t) \equiv H\{u(t)\}$ is the Hilbert transform in (5), and * denotes convolution.

200

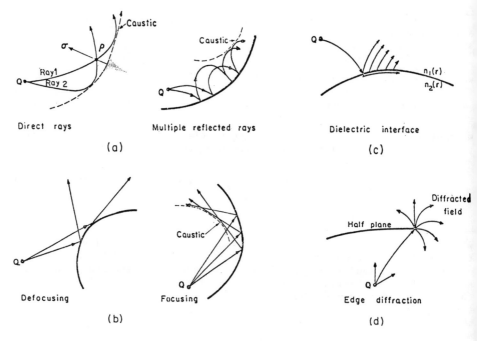

Fig. 1   - Typical configurations treated by STT.   Line source located at Q

    (a) Focusing in inhomogeneous medium.
    (b) Reflection from smooth curved boundary.
    (c) Reflection (not shown) and lateral wave diffraction, when
         relevant, due to interface between two media. $n_1(\underset{\sim}{r})$ and
         $n_2(\underset{\sim}{r})$ are spatially dependent refractive indeces.
    (d) Edge diffraction in inhomogeneous medium.

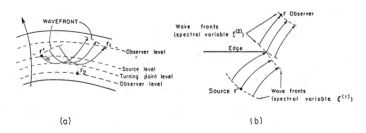

(a)   (b)

Fig. 2 – Spatial spectral local plane wave congruences. The spectral
variable is $\xi = \cos\alpha$, where α is the departure angle at the source
level. The phase $\tau(\xi)$ in (1) is the time delay (measured along
rays in a congruence) from the wavefront which passes through the
source point $\underline{r}'$ to the wavefront which passes through the observation point $\underline{r}$. For ξ_j that render $\tau(\xi)$ stationary, a ray in the
congruence passes through <u>both</u> $\underline{r}'$ and $\underline{r}$.

(a) Ray congruence with turning point. Observer at $\underline{r}_1$ is
     reached by real rays in the visible spectrum while observer
     at $\underline{r}_2$ is reached by complex rays belonging to the nonvisible
     (evanescent) spectrum.

(b) Double spectral decomposition, due to the actual source
     ($\xi^{(1)}$) and the virtual source at the edge ($\xi^{(2)}$).

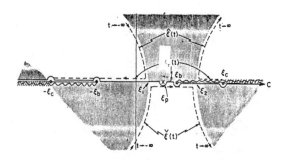

Fig. 3 – Singularities in complex ξ-plane for phase function $\tau(\xi)$ as in
Fig. 4(a), having two stationary points ξ_1 and ξ_2. Of the
corresponding two caustic forming rays (see Fig. 1(a)), ray 1
arrives first at the observer at time $t_1 = \tau(\xi_1)$ before touching
the caustic, whereas ray 2 arrives second at time $t_2 = \tau(\xi_2) > t_1$
after touching the caustic (note also that $\tau''(\xi_{1,2}) \gtrless 0$).
Im $\tau(\xi) > 0$ and Im $\tau(\xi) < 0$ in the shaded and unshaded regions,
respectively. With Im $t = 0$, the roots $\xi(t)$ in the u.h.p., l.h.p.
and on the real axis are denoted by $\hat{\xi}$, $\check{\xi}$ and ξ_r, respectively.
In the limit of real t, their trajectories (shown dashed), tend
to the boundary between the shaded and the unshaded regions, and
intersect at ξ_j as explained in (18). $\pm \xi_c$ and the corresponding
shaded segments of the real axis represent the branch points and
branch cuts of $\tau(\xi)$ associated with the nonvisible spectrum. $\pm \xi_b$
and the corresponding wiggly lines represent possible branch
points and branch cuts of $A(\xi)$. ξ_p represents a possible pole of
$A(\xi)$, with C passing above or below, depending on the diffraction
problem at hand.

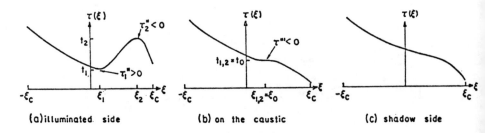

Fig. 4 – Phase function $\tau(\xi)$ in the visible spectrum $(-\xi_c < \xi < \xi_c)$ for a caustic forming wave species (see also Figs. 3 and 5).

    (a)   Observer away from the caustic on the lit side.
    (b)   Observer on the caustic: $\xi_{1,2} \to \xi_0$, $\tau''_{1,2} \to 0$.
    (c)   Observer in the shadow.

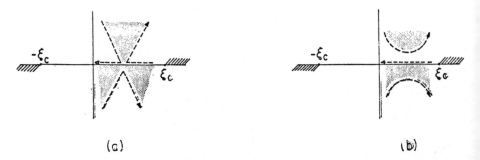

Fig. 5 – Local third order behavior of phase function $\tau(\xi)$ of Fig. 3 for an observer near a caustic. As in Fig. 3, shaded area represents $\mathrm{Im}\tau > 0$, and the dashed trajectories are the roots $\xi(t)$, with t having a small negative imaginary part.

    (a)   Observer on the caustic.
    (b)   Observer in the shadow.

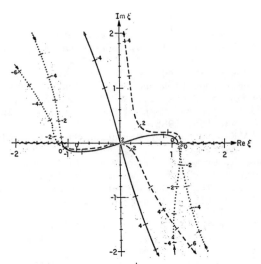

Fig. 6 – Trajectories of the roots $\xi^{\pm}(t)$ in (28) in the complex ξ-plane (after [20]). The full and dashed trajectories are for $\rho/b = 2$, $\theta = 0$ and $\rho/b = 2$, $\theta = 20°$, respectively. For $t < t^{\pm}$, the roots lie on the lower Riemann sheet $Re(1 - \xi^2)^{1/2} < 0$, and are described by the dotted trajectories. Branch cuts are indicated by wiggly lines. The normalized time $(v/b)t$ is indicated along the trajectories.

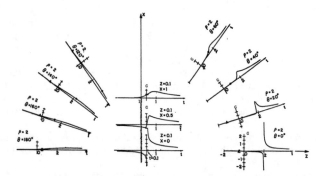

Fig. 7 – Waveforms of the pulsed beam corresponding to parameters in Fig. 6. Observation points are near the initial plane (Fresnel zone): $z/b = 0.1$, and $x/b = 0, 0.5, 1$; in the "far" zone: $\rho/b = 2$, and in the forward and backward directions ($\theta = 0°$, 20°, 40°, 60° and $\theta = 180°$, 160°, 140°, 120°, respectively). In the figure, the field as well as the space and time coordinates are normalized to $(b/v)G$, $(x/b, z/b)$ and $(v/b)t$, respectively.

REFERENCES

1. E. Heyman and L.B. Felsen, "A Wavefront Interpretation of the Singularity Expansion Method", IEEE Trans. on Antennas Propagat., AP-33, pp. 706-718, 1985.
2. L.B. Felsen, "Progressing and Oscillatory Waves for Hybrid Synthesis of Source Excited Propagation and Diffraction", Invited Paper, IEEE Trans. on Antennas and Propagat., AP-32, pp. 775-796, 1984.
3. L.B. Felsen, "Real Spectra, Complex Spectra, Compact Spectra", J. Opt. Soc. Am. A3, pp. 486-496, 1986.
4. C.H. Chapman, "A New Method for Computing Synthetic Seisograms", Geophys. J. Roy. Astr. Soc., 54, pp. 481-518, 1978.
5. C.H. Chapman and R. Drummond, "Body-Waves in Inhomogeneous Media Using Maslov Asymptotic Theory", Bull. Seism. Soc. Am., 72 pp. 277-317, 1982.
6. E. Heyman and L.B. Felsen, "Non-Dispersive Closed Form Approximation for Transient Propagation and Scattering of Ray Fields", Wave Motion, 7, pp. 335-358, 1985.
7. E. Heyman and L.B. Felsen, "Weakly Dispersive Spectral Theory of Transients-I. Formulation and Interpretation", IEEE Trans. Antennas Propagat., AP-35, pp. 80-86, 1987.
    "--=II. Evaluation of the Spectral Integral", IEEE Trans. Antennas Propagat., to be published.
8. E. Heyman, "Weakly Dispersive Spectral Theory of Transients-III. Applications, IEEE Trans. Antennas Propagat., to be published.
9. L. Cagniard, "Reflection and Refraction of Progressive Seismic Waves", (translation from French by Flinn and Dix), McGraw Hill, New York, 1962.
10. A.T. DeHoop, "A Modification of the Cagniard Method for Solving Seismic Pulse Problems", Appl Sci. Res. B8, pp. 349-356, 1960.
11. G. Mazak, I.B. Bernstein, and T.M. Smith, "Integral Representation for Geometric Optics Solutions", Phys. Fluids 26, pp. 684-688, 1983.
12. I.T. Lu and L.B. Felsen, "Adiabatic Transforms for Spectral Analysis and Synthesis of Weakly Range Dependent Shallow Ocean Green's Functions", J. Acoust. Soc. Am. 81(4), pp. 897-911, 1987.
13. J.M. Arnold, "Geometrical Theories of Wave Propagation: A Contemporary Review", IEE Proc. 133, pp. 165-187, 1986.
14. J.M. Arnold and L.B. Felsen, "Spectral Reconstruction of Uniformized Wavefields from Nonuniform Ray or Adiabatic Mode Forms for Acoustic Propagation and Diffraction", submitted to J. Acoust. Soc. Am.
15. R.C. Hansen, ed. "Geometric Theory of Diffraction", IEEE Press, New York, 1981.
16. E. Heyman and L.B. Felsen, "Propagating Pulsed Beam Solutions by Complex Source Parameter Substitution", IEEE Trans. Antennas Propagat. 8, pp. 1062-1065, 1986.
17. G.A. Deschamps, "Gaussian Beams as a Bundle of Complex Rays", Electronic Lett., 7, pp. 684-685, 1971.
18. L.B. Felsen, "Complex-Source-Point Solutions of the Field Equations and their Relation to the Proagation and Scattering of Gaussian Beams", Symp. Matemat. Istituto Nazionale di Alta Matematica, XVIII, pp. 40-56, Acad. Press. London, 1976.
19. L.B. Felsen, "Geometrical Theory of Diffraction, Evanescent Waves, Complex Rays and Gaussian Beams", Geophys. J. Roy. Astron. Soc., 79, pp. 77-78, 1984.
20. E. Heyman and B.Z. Steinberg, "Spectral Analysis of Complex-Source Pulsed Beams", J. Opt. Soc. Am. A4, pp. 474-480, 1987.

MODEL-BASED PARAMETER-ESTIMATION APPLICATIONS IN ELECTROMAGNETICS

E. K. Miller
General Research Corporation
Santa Barbara, CA 93111

SUMMARY
Numerical modeling in electromagnetics, whether based on integral, differential, or other formulations, inevitably involves a discretized, sampled-field description of the physical reality which is being modeled. These descriptions exhibit a commonality with signal processing and model-based parameter estimation in terms both of sampling requirements, information content and system rank, and the role of difference-equation linear predictors. This discussion surveys some aspects of computational electromagnetics from a parameter-estimation viewpoint, emphasizing the variety of ways in which an exponential-series model can arise.

The discussion begins by examining the various field observables whose behavior has exponential-series form, devoting special attention to the physical significance of the parameters involved. Next, the sampling of such observables in order to extract physically useful information from processing of the resulting data is considered. The occurrence of similar functional descriptions originating in other mathematical and numerical forms such as recursion and finite-difference approximations is also explored. Finally, an attempt is made to unify the various concepts that have been raised in terms of their implications for both the direct (analysis) and the inverse (synthesis) problems.

B. de Neumann (ed.), Electromagnetic Modelling and Measurements for Analysis and Synthesis Problems, 205–256.
© 1991 Kluwer Academic Publishers. Printed in the Netherlands.

1 INTRODUCTION
1.1 Radar Scattering as an Encoding Process

When an object (target) is illuminated by an electromagnetic field, the energy it scatters to a given observation point depends upon the target's geometry and material characteristics, the conditions of illumination [angle of incidence, polarization, frequency(ies), etc.] and the medium through which the field propagates. It is intuitively obvious, and demonstrated by analysis and measurement, that the scattered field (which will also contain, in general, interference and/or noise energy) has impressed or encoded on it information about the target which might somehow be extracted for purposes of: 1) detection (i.e., is there a target?); 2) classification (does the target belong to a class of interest?); 3) identification (if in the class of interest, which one of that class it is?); and 4) imaging or inversion (can the target's geometry and/or material properties be reconstructed?). An interesting view of this encoding process is described by Fiddy and Ross (1979) and Fiddy et. al (1982).

The particular problem of deducing the geometry and material characteristics of a radar target from its scattered fields is one of continuing interest. Generally speaking, the amount of data which is needed, and the amount of processing that data will require, can be expected to grow commensurately with the specificity and confidence desired in the answer being sought. Thus, the general inverse problem is by far the most difficult of those applications listed as (1) to (4) above. Fortunately, many problems of practical importance do not require the most general answer. For example, airborne radar targets naturally fall into one or more of a few sets, i.e., friend or foe, missile or aircraft, etc. In such circumstances, target classification and identification (C/I) can be based less on a rigorous inverse approach and more on whether their radar signatures match prestored information and expectations about the targets of potential interest.

But the quality and amount of target-scattered energy that are available place limits on how much information an observer can extract solely from measurement. Rarely, however, is the observer dependent wholly on the measurement alone to accomplish the goal of target identification. He can, in addition, make use of a priori information (e.g., atmospheric noise data, knowledge that airborne targets are planes, missiles, or helicopters, etc.) and/or modify the measurement itself (change frequencies, polarization, pulse shape, etc.). These factors suggest that the observer should process the available data in such a way as to utilize all available information completely, and to the extent possible, adapt the controllable measurement conditions to the characteristics of the target. In both, knowledge of at least some of the characteristics of expected targets is an essential requirement for improving the signal processing.

1.2 Target Characterization

Target characteristics may be recognized to belong to (at least) two domains. One, the target domain, is that in which the target is a basic entity having either directly measurable properties (such as

mass, volume, shape) independent of its electromagnetic response, or properties which are derivable from those measurements (moment of inertia and higher-order moments, harmonic expansion of the surface profile, etc.), either of which can be used to uniquely describe the target within some degree of accuracy. The other, which we term the data domain, is that in which the target exists only indirectly, in terms of its effect upon a measured observable (a scattered EM field, for example), and from which various properties may be derived. When these derived, or data-domain, properties can be uniquely related to the basic, or target-domain, properties, then it can be said that an inverse problem has been solved. For many applications, however, classification or identification alone may suffice or even be all that is possible to achieve, in which case the data-domain properties may be most directly relevant. In the former application, the problem is more absolute in nature, i.e., (determine the target) whereas in the latter, it is more comparative or relative (i.e., differentiate between targets). In either case, the way in which target-domain and data-domain properties interrelate is crucial to a successful outcome.

Another key to characterizing a target is the number of independent properties that are needed to accomplish either of these objectives. This number, which we call the target rank, can depend on the characterization being used and the availability of any prior information. For example, if all targets are known to be penetrable, homogeneous spheres, then the only basic properties of concern are radius and composition. We might also expect that a desirable property of the characterization is that it be parsimonious, i.e., that with minimum rank it include all independent properties over some set of measurements. This statement implies that the characterization in either the target domain or data domain is probably not unique, and that alternative property sets are available.

1.3 Information Content and Feature Sets

Besides the issues of which target or data properties may be best, and what their associated ranks may be, there is the closely related question of the information content of the measurement. This might be loosely defined as the total number of bits required to represent the maximum number of independent properties (basic or derived) obtainable from the data. Implicit in this definition is a tradeoff between the number of properties and their complexity, as we speculate that the total information represented by the data is fixed. A priori knowledge, however, can greatly influence how the data are best used, changing the problem from one of system identification (determining a model for the data) to one of parameter estimation (given a model, estimating the model parameters). In other words, by reducing the number of unknowns via the prior knowledge (a model), the data can yield either more complete or more accurate information about the model parameters (the properties).

In radar terminology, the EM properties of a target are sometimes called its features, the ad hoc nature of which is demonstrated by the

features that have at one time or other been used, some examples being: the maximum value of target cross section at a given frequency, the aspect-averaged cross section over a specified frequency band, and the ratio of cross sections for two orthogonal polarizations of the incident field. The success of such an approach will depend in part on the degree to which the features span the space of target-radar and target-geometry characteristics, individually and as a set. When the features chosen are target-radar characteristics or electromagnetic-observables data, we observe that it is the data domain in which C/I is being attempted. On the other hand, when the features are derived from the radar signature to yield target-geometry characteristics, then C/I is being pursued in the target domain. Clearly, target-geometry features would be more desirable, everything else being equal, because these are features closer to describing the target in ways which are recognizable to human observers. However, it is more straightforward to work with target-radar features which can unfortunately vary due to the aspect, frequency and polarization dependence of the target's scattering properties. It would be beneficial to have a feature set that is a fundamental property of the target and is independent of how the target is excited.

1.4 Complex Resonances and other Singularities

An example of such a feature set is the complex source-free resonances, or poles, popularized in recent years in connection with the Singularity Expansion Method (SEM) [Baum (1972, 1976)]. These frequency-domain poles are dependent only on the target geometry (and impedance, if it is imperfectly conducting) with oscillatory or imaginary parts established by the lengths(s) of propagation path(s) on the object, and the damping or real parts associated with energy loss (radiative and dissipative) over that path. While significant analytical questions continue to be studied in connection with SEM (the implications of higher-order layer poles, branch cuts, singularities at infinity, etc.), from a pragmatic viewpoint, the poles seem to be unique properties of a target and, therefore, provide an eminently attractive mechanism for target identification. In some sense, too, they seem to represent a kind of minimum-parameter target description which has interesting implications for the inverse problem from an information-theoretic viewpoint. Finally, they may provide useful insight concerning other feature sets that have been or might be used for target identification.

Considering the use of frequency-domain poles for target identification presumes the availability of wideband or impulsive scattering data for a target. An alternate approach is to make the measurement as a function of angle in a bistatic or monostatic scattering mode at a fixed frequency. In this case, a different kind of pole (or more properly a singularity) arises due to the equivalent or actual point sources which give the scattered field. Such sources can correspond to scattering centers and points of radiation from the target, and thereby might also serve as a feature set. To the extent that they are related to the target geometry, these space singularities

could provide a more direct link between the target and data domains than do the complex-frequency poles. For example, the maximum physical extent of the target might be obtained directly from extrema of the space-singularity distribution, but could only be inferred from the wavelength of the smallest-value pole.

Space singularities can also be associated with measurements made as a function of position (e.g., along a line or circular arc) or as a function of frequency. In the former situation, the singularities correspond to the direction angles of the plane-waves which make up the total field or to other physical observables of the problem [Bleszynski (1987). The result is that a spatial array can image a distribution of sources located on the far-field sphere. An analogous situation arises in measuring the field of a set of point sources as a function of frequency, but at a fixed point in space, where the singularities also correspond to the source positions. By making measurements from three orthogonal (or linearly independent) directions, the possibility arises of imaging a three-dimensional source distribution. Another example is provided by measuring the far-field of a linear array of point sources wherein the space singularities correspond to the source locations.

Alternate approaches are described by Fiddy and Ross (1979) using complex zeros as information sources of the field and by Inagaki and Garbacz (1982) in terms of the eigenfunctions of operators with respect to discrete and radiating systems. In the latter case, the fact has been demonstrated that the width of the eigenfunction spectrum of the field decreases upon moving away from the source until a steady state is reached in the far field, due to the non-radiating, evanescent near fields.

An interesting property of these representations is their close connection with widely used signal-processing techniques. For example, a linear-predictor filter implicitly incorporates a series-exponential model, as can be demonstrated by applying Prony's method to a set of uniformly spaced (in x) samples of f(x). Prony's method and related techniques provide a way to estimate values of the amplitudes (residues) and exponents (poles) and residues from the sampled data. Thus, there is the possibility of not only obtaining parameters which may be useful EM feature sets, but also that of processing EM data for target identification.

In the next section, we compare direct and inverse models in the context of an integral-equation formulation to emphasize their similarities and differences.

2 DIRECT AND INVERSE MODELING COMPARED

It is generally recognized that inverse problems are more difficult to solve than are direct problems because the direct problem can usually be reduced to the solution of a linear system whereas inverse problems are intrinsically transcendental and nonlinear. These

issues are discussed briefly below.

2.1 Numerical Approximations to Operator Relationships

Linear problems in electromagnetic, acoustic and similar applications can be expressed in operator form as [Harrington (1968), Poggio and Miller (1973)]

$$L(\vec{r},\vec{r}\,')f(\vec{r}\,') = g(\vec{r})$$

(2.1)

where:

$g(\vec{r})$ is a field,

$f(\vec{r})$ is a source, and

$L(\vec{r},\vec{r}\,')$ is an integral operator determined by the problem parameters and governing equations.

Eq. (1) is a general statement relating sources and fields in the frequency domain. An analogous treatment can also be developed for the time domain [Poggio and Miller (1973)]. When the source position vector $\underline{r}'$ and field vector $\underline{r}$ lie on a locus of points B over which some known relations hold due to boundary conditions (e.g., vanishing of tangential electric field on a perfect electric conductor, or vanishing of the normal particle velocity on a rigid body), then Eq. (2.1) becomes an integral equation for the source, i.e.,

$$L_B(\vec{r},\vec{r}\,')f(\vec{r}\,') = g(\vec{r}); \; \vec{r},\vec{r}' \, \varepsilon \, B$$

(2.2)

with L_B denoting the operator L with the position vectors constrained to B.

A numerical representation of Eq. (2.2) can be derived using the moment method [Harrington (1968)], leading to a linear system of the form

$$B_{ij}x_j = y_i$$

(2.3)

where B_{ij} is a matrix of N^2 interaction coefficients relating the vectors of N sampled-field values y_i and the N sampled-source values x_j over the boundary B, and repeated indices are summed over N terms. We use the term "sampled" in a generalized sense because x_j and y_i represent actual spatial samples only for certain numerical treatments. The relationships in Eq. (2.3) can be inverted to obtain

$$x_i = B_{ij}^{-1}y_j$$

(2.4)

where B_{ij}^{-1} is a matrix of inverse interaction coefficients (assumed nonsingular) and

$$B_{ij}^{-1}B_{ij} = I_{ij}$$

with I_{ij} the identity matrix.

The matrices B_{ij} and B_{ij}^{-1} have straightforward physical interpretations, representing a constrained source-field, sampling relationship for a given problem. In electromagnetics, they are referred to most often as the impedance and admittance matrices,

respectively.

A numerical representation of Eq. (2.1) can also be expressed in terms of sampled values as

$$z_i = A_{ij}x_j = A_{ij}B_{jk}^{-1}y_k \qquad (2.5)$$

where z_i is a sampled field on some locus of points in space other than B_{ij}. For far-field samples of z_j, the matrix A_{ij} is especially simple and can be expressed in terms of plane-wave projections onto B, but in general, A_{ij} involves samples of the operator L.

2.2 Direct and Inverse Solutions Using Sampled Operators

The preceding discussion was presented to set the stage for clarifying the difference between direct and inverse problems. A common approach to both of these problems is to use an input-output model which leads to a generalized transfer function in which two basic problem types become apparent. For the analysis or direct problem, the input is known and the transfer function is derivable from the problem specification, with the output or response to be determined. For the case of the synthesis or inverse problem, two problem classes may be identified. The "easier" synthesis problem involves finding the input given the output and the transfer function, an example of which is that of determining the source voltages which produce an observed pattern for a known antenna array. The "harder" synthesis problem itself separates into two problems. One entails finding a source distribution that produces a given far field. The other and still more difficult is that of finding the object geometry which produces an observed scattered field from a known exciting field. The latter problem is the most difficult of the three synthesis problems to solve because it is intrinsically transcendental and nonlinear. Furthermore, such problems are subject to uniqueness constraints which can impose difficulties and uncertainties in developing their solutions.

In essence, both the direct and inverse problems can involve an Nth order, sampled, approximated operator, and differ mainly in terms of how the operator (or its equivalent) is obtained. In the direct problem, the N^2 coefficients of this operator are derived from the known governing equations and problem parameters. In the inverse problem, an operator or its equivalent must also be obtained for a rigorous solution, where the absence of known problem parameters must be compensated by known input and output quantities.

Two approaches might be considered for the latter problem. One would be to obtain the N^2 coefficients directly from the known input and output quantities, which requires in principle solution of up to an N^2th order linear system. The other would be to exploit the known governing equations as well to reduce the number of coefficients requiring solution directly from the data.

If an inverse solution is being sought only to predict the target's response to a new incident field, then an operator matrix is all that is

needed. Furthermore, this matrix need not be unique since the tangential field distribution on any closed surface which contains the target could be used to define it. On the other hand, the target geometry and electrical properties could be derived only from that matrix which represents the true boundary-value problem. This presumes that the object geometry can be related unambiguously to the coefficients of B_{ij}. The unknowns actually being sought would be the geometrical parameters needed to specify the Nth order decomposition of the object's geometry, from which the coefficients in B_{ij} could be computed, possibly using an iterative approach. An example using a time-domain integral equation is provided by Bennett (1981) with an earlier example of approximating transient responses given by Kennaugh and Moffett (1965). While the direct problem is a linear one when the governing equations are linear, the inverse problem is not because the object's boundary and/or its constitutive parameters are unknown, which is why inverse problems (using our definition) are so much more difficult.

It should be noted that this is the most general form of the inverse problem, and that it entails reconstructing the object's geometry and constitutive parameters. A simpler inverse problem is to estimate model or physical parameters from observational data to decide whether the object (process) belongs to a class of interest (the classification problem) and, if so, which one of that class it is most likely to be (the identification problem). These latter problems are amenable to resonance or pole-zero parameterization.

2.3 Implications of Sampling
Inherent in all analysis of field problems is the idea of sampling. Sampling is involved in representing sources and fields whether the problem geometry permits solution via classical separation of variables, or requires numerical techniques. For numerical solutions, sampling functions can exist over all of B, or be defined on subdomains of B. In either case, some lower limit on the number of samples is required to achieve a given degree of convergence in the solution.

The number of samples needed to represent a solution for a given problem represents a measure of the problem dimensionality, which is also referred to as rank or the number of degrees of freedom. Problem rank is a fundamental property for both direct and inverse problems, in terms of solution difficulty and information content. It appears that the geometrical (target-domain) rank and electromagnetic or acoustic (data-domain) rank of a problem are not necessarily the same, a situation that requires consideration in treating inverse problems.

2.4 A Unified Approach
We have included the above discussion to emphasize that direct and inverse problems involve aspects of not only field theory, but also information theory and signal processing. The literature in these separate areas is extensive. By comparison, relatively little

evidence is available to show that a unified approach is likely to be considered when treating inverse problems. Unless these closely related aspects of direct and inverse problems are addressed in a unified way, progress towards achieving practical solutions is likely to be slower than would otherwise be possible.

An area where information-theory concepts are being pursued in inverse problems is in optical imaging. Representative examples in this area can be found in the work of Winthrop (1971), Ross et al (1978), Gori and Ronchi (1981), and Bortero and Pike (1982). The central theme of these and related works is that the information content of a field is a finite quantity that is relatable to the field's rank or number of degrees of freedom, based for example on an eigenvalue measure. This concept is one of paramount importance in any inverse treatment, since it evidently establishes the number of independent parameters that can be extracted from the data, whether these are eigen vectors, complex-frequency poles, or other representations.

Object poles represent another way by which data rank can be quantified and object identification or even reconstruction might be attempted. These possibilities have not been very actively pursued, partly because estimating poles from noisy data is not easily or reliably accomplished, and the accompanying question of how poles can be used for identification has no clear answer. The results of an initial study of some of these questions are presented by Miller and Lager (1983).

One disadvantage of pole estimation is that only the source-free part of an object's response can be used in a straightforward way for this purpose. Otherwise, the model used and subsequent pole estimation can become much more complex because of the need to model the driven response as well. Unfortunately this means that information contained in the early-time, forced response might not be exploitable using only a pole-based scheme. The early-time ramp response, however, can be used to estimate an object's cross-sectional shape and area [Kennaugh and Moffatt (1965)]. By combining both ideas into a unified treatment, more effective utilization of the entire object response might be achieved as discussed by Miller et al (1984).

In the next section we discuss one approach to finding the poles of objects from their EM response in either the time domain or frequency domain using Prony's Method or straightforward variations thereof.

3 PROCESSING POLE-BASED REPRESENTATIONS

In a wide variety of physical problems, especially those which involve wave phenomena such as in electromagnetics, acoustics, and seismology, a behavior results that can be described by systems of linear, partial differential equations. Solutions to such problems often can be expressed simply in exponential form (series or integrals), the occurrence and use of which we explore more fully below.

These exponential forms are of interest not only because they provide solutions but also because the parameters they contain, the exponents in particular, possess physical information basic to a given problem. In transient phenomena for example, the exponents (or poles) contain resonant frequencies and damping constants for the problem. Many other kinds of exponential forms which contain other kinds of physically relevant information also exist. Moreover, the poles (and associated zeroes or residues) can be extracted from a wide variety of data, as we shall demonstrate below.

Exponential-form representations not only include parameters relevant to a problem's physics, but are also valuable because they:

1)   provide a way to condense the data needed to describe a problem's physical behavior;

2)   offer insight into the physical phenomena involved;

3)   provide a measure of the information content of the data;

and

4)   suggest a conceptual framework for solving the inverse problem.

Some example applications are given below in Section 5. Before discussing these specific issues however, we present some general background material in Section 3 and summarize some of the wide variety of problems which exhibit exponential-form solutions in Section 4.

3.1 OBTAINING POLES

Many kinds of mathematical transformations are used in solving physical problems. One of the more useful of these, and one which is ideally suited to the exponential forms of interest here, is the Laplace transform. For the generic waveform-domain exponential series

$$f(x) = \sum_{\alpha=1}^{W} R_\alpha \exp(s_\alpha x) \qquad (3.1)$$

a Laplace transform leads to the generic spectral-domain pole series

$$F(X) = \sum_{\alpha=1}^{W} R_\alpha / (s_\alpha - X) \qquad (3.2)$$

where we use the terms waveform and spectral domains as the most commonly encountered physical examples given by the forms in Eqs. (3.1) and (3.2) respectively. We note that knowledge of the W s_α's (the poles) and the W R_α's (the residues) provides complete information both for $f(x)$ and $F(X)$. This means that rather than needing many sampled values to reconstruct a waveform or spectrum, we need only store the two sets of numbers (generally complex) s_α and R_α as depicted in Fig. 1. The result can be a greatly reduced number of bits required for storing such data while also

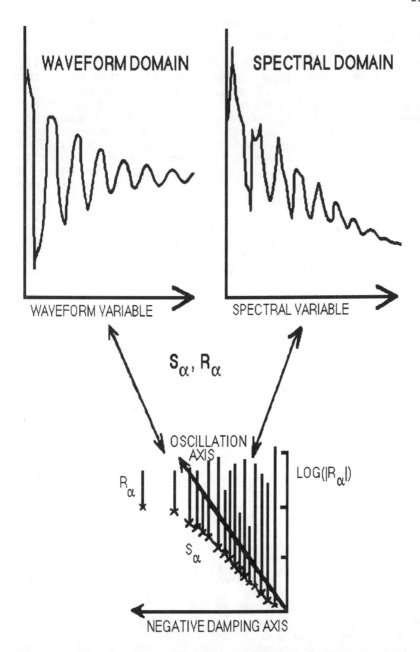

Figure 1. Waveform-domain and spectral-domain representations for an impulsively-excited straight wire together with a perspective plot of the complex frequency plane showing the pole locations s_α and residue magnitudes R_α on a three-decade log scale. These generic results illustrate the connection between the responses in the two domains and their relationship via the poles and residues.

providing an extrapolation function for f(x) or interpolation function for F(X), continuous in the independent variable.

Practical implementation of Eqs. (3.1) or (3.2) requires that s_α and R_α be derivable from whatever analytical, computed, or measured data or information is available. A procedure for doing this from waveform data was developed in 1795 by Prony [Hildebrand (1965)]. Prony's method in various modified and extended forms may be recognized in many signal-processing methods in present use [see Kay and Marple (1981) for example]. As outlined below, it can be used as originally developed to process waveform data, and in a slightly different, but closely analogous version, to process spectral data. We derive first the waveform-domain version of Prony's method, followed by its spectral-domain counterpart.

3.2 Waveform-Domain Prony

Determination of the poles from Eq. (3.1) using the Prony technique involves sampling f(x) at D data points $x_i = i\delta$, $i = 0, 1, \ldots,$ D-1 over a window of width $T = \delta(D-1)$. We thus rewrite Eq. (1) as

$$f_i = f(x_i) = \sum_{\alpha=1}^{P} \hat{R}_\alpha \hat{X}_\alpha{}^i; \quad i = 0, 1, \ldots, D - 1 \tag{3.3}$$

where

$$\hat{X}_\alpha = e^{\hat{s}_\alpha \delta},$$

P is the number of poles being sought and the caret denotes a parameter-value estimate.

Eq. (3.3) can be written

$$\hat{R}_1 + \hat{R}_2 + \cdots + \hat{R}_P = f_0,$$

$$\hat{R}_1 \hat{X}_1 + \hat{R}_2 \hat{X}_2 + \cdots + \hat{R}_P \hat{X}_P = f_1,$$

$$\hat{R}_1 \hat{X}_1{}^2 + \hat{R}_2 \hat{X}_2{}^2 + \cdots + \hat{R}_P \hat{X}_P{}^2 = f_2,$$

$$\vdots$$

$$\hat{R}_1 \hat{X}_1{}^{D-1} + \hat{R}_2 \hat{X}_2{}^{D-1} + \cdots + \hat{R}_P \hat{X}_P{}^{D-1} = f_{D-1}, \tag{3.4}$$

where each data sample can be seen to generate an equation involving higher-order polynomials in the exponential terms as a consequence of using uniformly spaced data samples. The requirement that a polynomial form be generated from uniform sampling of the waveform-domain signal model arises because if the $\exp(s_\alpha \delta)$ represent the roots of a P-th order polynomial, i.e.,

$$\sum_{\alpha=0}^{P} \hat{a}_{\alpha} X^{\alpha} = (X - \hat{X}_1)(X - \hat{X}_2) \cdots (X - \hat{X}_P)$$

$$= \prod_{\alpha=1}^{P}(X - \hat{X}_{\alpha}); \; a_P = 1 \tag{3.5}$$

it is then possible to reduce Eq. (3.4) to a linear-predictor form.

This can be done by first multiplying the equation series (3.4) respectively in turn by

$$\hat{a}_0, \hat{a}_1, \ldots, \hat{a}_P$$

and adding the first $P + 1$ of these equations together. The result is that a new equation can be found, which upon using Eq. (3.5) can be reduced to

$$\hat{a}_0 f_0 + \hat{a}_1 f_1 + \cdots + f_P = 0$$

If this operation is repeated another $D - P - 1$ times by successively beginning with the second equation of Eq. series (3.4), . . . , up to the $D - P - 1$ equation, the following $D - P$ linear equations will result:

$$\overrightarrow{\underset{t}{M}} \hat{a} = -\overrightarrow{f}, \tag{3.6a}$$

where

$$\overrightarrow{\underset{t}{M}} = \begin{bmatrix} f_0 & f_1 & \cdots & f_{P-1} \\ f_1 & f_2 & \cdots & f_P \\ \vdots & \vdots & \vdots & \vdots \\ f_{D-P-1} & f_{D-P} & \cdots & f_{D-2} \end{bmatrix}, \tag{3.6b}$$

$$\overrightarrow{f} = \begin{bmatrix} f_P \\ f_{P+1} \\ \vdots \\ f_{D-1} \end{bmatrix} \tag{3.6c}$$

and the vector of unknowns represents the estimated coefficients of the polynomial in Eq. (3.5) (sometimes called the characteristic equation). Observe that Eq. (3.6a) has the form of a linear predictor or difference equation, which occurs commonly in various signal processing procedures, and that the data matrix of Eq. (3.6b) is of Toeplitz form. The linear system in Eq. (3.6) provides the basis for obtaining the P coefficients in the polynomial in Eq. (3.5), from whose roots the P pole values are found as

$$\hat{s}_{\alpha} = (1/\delta) \ln(\hat{X}_{\alpha}) \tag{3.7}$$

If D = 2P, the system is square of order P and can be solved directly for the P real a's. For D>2P, the method of least squares can be used. In either case, the residues can be found by various means, such as fitting the first P equations of Eq. series (3.4) or by applying the least-squares technique to the entire set.

It is important to recognize that for the method to work the data must be available in uniform steps of the exponent, so that Eq. (3.1) can be written as in Eq. (3.3). This limitation will be encountered in various situations discussed below in Section 4. Note also that the technique just outlined can be used for data whose exponential terms are products of two or more exponents depending on different variables, e.g., where

$$f(x,y) = \sum_{\alpha=1}^{W_x} \sum_{\beta=1}^{W_y} R_{\alpha\beta} \exp(s_\alpha^{(x)}) \exp(s_\beta^{(y)}) \tag{3.1}'$$

as discussed below in section 4.2.2 in connection with rectangular-mesh sampling.

We observe that a constraint or auxiliary equation is needed to obtain an inhomogeneous system of equations. It is conventional to impose condition (3.5) which leads to the linear-predictor form seen in Eq. (3.6a). Other forms could also be used, e.g. [Moffatt and Mains (1975)],

$$\sum_{\alpha=0}^{P} \hat{a}_\alpha{}^2 = 1.$$

3.3 Spectral-Domain Prony

The treatment of Eq. (3.2) to derive the pole values proceeds in an analogous fashion. First, rewrite Eq. (3.2) as

$$F(X) = \sum_{\alpha=1}^{P} R_\alpha \prod_{\beta=1}^{P} {}^\alpha (X - \hat{s}_\beta) / \prod_{\gamma=1}^{P} (X - \hat{s}_\gamma) \tag{3.8}$$

where the superscript α on the first product signifies omission of the $\beta = \alpha$ term and again the carat denotes an estimated parameter. Upon recognizing that

$$\prod_{\gamma=1}^{P} (X - \hat{s}_\gamma) = \sum_{\gamma=0}^{P} \hat{a}_\gamma X^\gamma; \; a_P = 1 \tag{3.9a}$$

$$\prod_{\beta=1}^{P} {}^\alpha (X - \hat{s}_\beta) = \sum_{\beta=0}^{P-1} \hat{b}_\beta {}^{(\alpha)} X^\beta; \; b_{P-1}^{(\alpha)} = 1 \tag{3.9b}$$

Eq. (3.8) can be expressed as

$$\sum_{\alpha=0}^{P-1} [\hat{a}_\alpha X^\alpha F(X) - \tilde{b}_\alpha X_\alpha] = -F(X) X^P \tag{3.10a}$$

where

$$\tilde{b}_\beta = \sum_{\alpha=1}^{P} \hat{R}_\alpha \hat{b}_\beta \, (\alpha),$$

and it should be noted that characteristic-equation coefficients are real for spectra where $F(X) = F^*(X^*)$. This fact can be used to reduce the number of real unknowns in Eq. (3.10a) from 4P to 2P, and is assumed to be the case here.

If D>2P samples are available for $F_i = F(X=X_i)$, where $i = 0,1,...,D-1$, Eq. (3.10a) provides the basis for obtaining the 2P real coefficients from

$$\overset{\Rightarrow}{M_s} \overset{\wedge}{\vec{c}} = -\vec{F} \tag{3.10b}$$

where

$$\overset{\Rightarrow}{M_s} = \begin{bmatrix} F_0 & X_0 F_0 & \cdots & X_0^{P-1}F_0 & -1 & -X_0 & \cdots & -X_0^{P-1} \\ F_1 & X_1 F_1 & \cdots & X_1^{P-1}F_1 & -1 & -X_1 & \cdots & -X_1^{P-1} \\ \vdots & \vdots & & \vdots & \vdots & \vdots & & \vdots \\ F_{D-1} & X_{D-1}F_{D-1} & \cdots & X_{D-1}^{P-1}F_{D-1} & -1 & -X_{D-1} & \cdots & -X_{D-1}^{P-1} \end{bmatrix}, \tag{3.10c}$$

$$\overset{\wedge}{\vec{c}} = \begin{bmatrix} \hat{a}_0 \\ \hat{a}_1 \\ \vdots \\ \hat{a}_{P-1} \\ \tilde{b}_0 \\ \tilde{b}_1 \\ \vdots \\ \tilde{b}_{P-1} \end{bmatrix}, (3.10d); \quad \vec{F} = \begin{bmatrix} X_0^P F_0 \\ X_1^P F_1 \\ \vdots \\ \vdots \\ \vdots \\ X_{D-2}^P F_{D-2} \\ X_{D-1}^P F_{D-1} \end{bmatrix}, (3.10e)$$

The P pole frequencies are then obtained as the roots of a Pth-order homogeneous polynomial from the characteristic equation (3.9a). Similarly, the residues can be found using Eq. (3.9b) or, more directly, by returning to Eq. (3.2) and matching P points or using all D points and using a pseudo inverse to obtain a least-squares solution.

3.4 Discussion

Note that Prony's Method linearizes the problem (except for finding the roots of a Pth-order polynomial) of estimating the s_α and R_α in Eq. (3.1) or (3.2) by separating their solutions. The above steps (in Time-Domain Prony) comprise Prony's original method of 1795 as described by Hildebrand (1956). It is worth noting that Eq. (3.6) is a

difference equation whose form in the Z-domain is that of a pole-zero model.

In its simplest form, Prony's Method requires that the user specify only two parameters for the time-domain case, say δ and P, from which D (=2P) and T [=δ(D-1)] follow automatically. The result is a square system of linear equations with P unknowns, the coefficients of which are the sampled data where the general coefficient A_{ij} is given by f_{i+j-2}, where $i,j=1,...,P$. A unique solution is possible only when $P \geq W$, and when $\delta < 1/2f_{max}$ where f_{max} is the highest frequency component of f(t). It should also be noted that only W of the P poles can be valid parameters of the process which produced the sampled data. In general, because W may not be known a priori, experimentation using different values for P may be required to establish an acceptable choice for it. Similarly, in order to establish the best value of δ, its variation to the extent permitted by the data may be necessary. While δ must be small enough to avoid aliasing (i.e., δ must satisfy the Nyquist criterion), it cannot be made too small without risking an illconditioned data matrix as a result of decreasing the linear independency of neighboring rows and columns. Clearly, application of even the basic version of Prony's Method requires care if acceptable results are to be obtained.

If more data points (a total of D') were available than the minimum $2W$ values required to compute the W poles of f(t), this would clearly result in an overdetermined system because the number of equations would exceed the number of unknowns. Two approaches to handling this situation might be considered. In the first, we could continue to use D' = 2P', i.e., compute as many poles as permitted by the number of data samples available. This number of poles would, of course, be greater than the number actually contributing to f(t) and the data matrix in Eq. (3.6) should then be singular unless, as is invariably the case, the data are noisy or of otherwise limited accuracy. This approach has been followed in much of the early work in electromagnetics that employed Prony's Method.

Alternatively, we could use the original value of P and reduce the number of equations from P' to P by multiplying Eq. (3.6) by the transpose of the data matrix to obtain a least-squares solution. This results in

$$\overset{\Rightarrow}{M_t}{}' \overset{\wedge}{\vec{c}} = - \vec{f}{}' ,$$ (3.11a)

where

$$\overset{\Rightarrow}{M_t}{}' = \begin{bmatrix} C_{00} & C_{01} & \cdots & C_{0,P-1} \\ C_{10} & C_{11} & \cdots & C_{1,P-1} \\ \vdots & \vdots & \vdots & \vdots \\ C_{P-1,0} & C_{P-1,1} & \cdots & C_{P-1,P-1} \end{bmatrix} ,$$ (3.11b)

$$\vec{f}\,' = \begin{bmatrix} C_0 \\ C_1 \\ \vdots \\ C_{P-1} \end{bmatrix}, \tag{3.11c}$$

$$C_{ij} = \sum_{k=0}^{P'-1} f_{i+k} f_{j+k}, \tag{3.11d}$$

and

$$C_i = \sum_{k=0}^{P'-1} f_{i+k} f_{P+k}. \tag{3.11e}$$

The two approaches could also be combined by choosing any value of P' between P and D'/2.

The matrix of Eq. (3.11b) has the form of an autocovariance matrix of data samples which is a common starting point for digital signal processing of transient waveforms. This form results directly if the minimization of the square of the error between the model and the data is realized by varying the model parameters. Prony's Method does not consider this error explicitly, however, and as originally formulated it is essentially a point-matching procedure. The process of developing the new data matrix in Eq. (3.11b) through multiplying the original matrix by its transpose also minimizes the squared error; this is sometimes called regularization or is also referred to as generating the pseudo inverse.

There are additional ways to handle an overdetermined system. One of the more useful in our experience is the so-called moving-window technique which involves processing the data using several sequential time windows that "move" along the waveform by $\Delta = n\delta$ (n an integer ≥ 1). In this way, the data are not used all at one time and several sets of estimates are obtained for the poles, a particularly helpful procedure when the data is noisy, as Prony's method can be very susceptible to noise [Dudley (1977)]. If the noise is random these individual windows will yield independent estimates for the pole sets, and authentic poles should cluster around an average value while spurious poles due to noise and curve fitting (caused by P>W) will be scattered rather randomly in the s-plane. Superimposing all pole-set estimates on one graphical plot can then help to identify the valid poles. An interesting alternate procedure [Van Blaricum et al (1984)] is to process the data in both forward and reverse directions with respect to time. The result is that poles actually present in the waveform are reflected about the $j\omega$ axis while noise and curve-fitting poles remain predominantly in the negative half-plane. It is worth mentioning that Kay and Marple (1981) found that, when compared with a number of other more recently developed spectral-estimation procedures for handling noisy, time-series data, a variant of Prony's method yielded the best estimate for a noise-contaminated waveform

having discrete and broadband components.

It should also be emphasized that Prony's Method is but one of a variety of techniques that might be considered for use in estimating poles from EM transient or spectral data. Our concern is not Prony's Method per se, but in the potential of EM poles for various applications. Since the need for polynomial forms to be generated by the observed data seems to be a central one however the pole estimation is accomplished, we continue to focus on EM observables that are amenable to such representations.

4   PHYSICAL PROBLEMS HAVING EXPONENTIAL FORM
Our interest in Prony's Method and other related signal processing techniques, as emphasized above, stems from the wide range of problems for which an exponential or a pole series provides a rigorous description. A generic listing of some exponential representations is given in Table 1. Examples of each type are briefly discussed further below to highlight the physical information represented by the pole set and residues for a given problem. Note that we include here only those phenomena whose representation makes them suitable for parameter estimation using Prony's Method, i.e. whereby a polynomial data model applies. We omit the carat superscript from estimated quantities in the following discussion to simplify the notation. Discussed in turn are time waveforms and frequency spectra, space waveforms and angle spectra, angle waveforms and space spectra, and frequency waveforms and space spectra.

4.1   Time Waveforms and Frequency Spectra
Transient electromagnetics provided the original motivation for field representations using Prony's Method. For linear transient problems (electromagnetics, acoustics, structural mechanics, etc.) we have in general

$$f(t) = \sum M_\alpha \exp(s_\alpha t) \tag{4.1}$$

with:

observation variable t=time,

poles $s_\alpha = \pm i\omega_\alpha \angle \sigma_\alpha$ where ω_α is the oscillation rate (radian frequency) and σ_α is the decay rate,

and residues $R_\alpha = M_\alpha$, the complex amplitude of the αth mode or pole.

Note that for realtime functions, s_α and M_α occur either in conjugate pairs or are pure real. The corresponding spectral response on the $i\omega$ axis is given by

$$F(i\omega) = \sum M_\alpha / (s_\alpha - i\omega) \tag{4.2}$$

where we observe that Eq. (4.2) can be written more generally in terms of the complex frequency s.

The value of representation (4.1) for transient phenomena is that the poles are independent of the excitation (for linear problems) and are determined instead solely by the object geometry. This principle is

Table 1
VARIOUS WAVEFORM- AND SPECTRAL-DOMAIN FIELD EXPRESSIONS

DOMAIN	VARIABLES		MODEL	PARAMETERS
WAVEFORM (Exponential)	x	g	$f(x)$	Residues
SPECTRAL (Pole)	X		$F(X)$	Poles
Time	t	1	$f(t)=\sum M_\alpha \exp(s_\alpha t)$	M_α=amplitude of αth mode.
Complex Frequency	s		$F(s)=\sum M_\alpha/(s-s_\alpha)$	$s_\alpha=i\omega_\alpha \pm \sigma_\alpha$=complex-resonance frequency of αth mode.
Frequency source.	ω	i/c	$f(\omega)=\sum S_\alpha \exp(g\omega R_\alpha)$	S_α=amplitude of αth
Space	R		$F(R)=\sum S_\alpha/[g(R-R_\alpha)]$	R_α=position of αth source along line of view.
Space	x	ik	$f(x)=\sum P_\alpha \exp\{gx[\cos(\varphi_\alpha)]\}$	P_α=amplitude of αth incident plane wave.
Angle	φ		$F(\varphi)=\sum P_\alpha/\{g[\cos(\varphi)-\cos(\varphi_\alpha)]\}$	φ_α=incidence angle of αth plane wave with respect to line of observation (x).
Angle	φ	ik	$f(\varphi)=\sum S_\alpha \exp[gx_\alpha \cos(\varphi)]$	S_α=amplitude of αth source.
Space	x		$F(x)=\sum S_\alpha/[g(x-x_\alpha)]$	x_α=position of αth source along linear array.

illustrated by the behavior of a tuning fork. For such problems, the s_α are the complex frequencies at which a source-free solution exists. The pole set may extend over essentially an infinite range in frequency, but for practical problems the data will be band-limited and only a finite number of poles will contribute measurably to the waveform. We briefly consider below the cases of pure real, pure imaginary and complex poles in transient waveforms.

4.1.1. Pure-Real Poles
In this case, the waveform is given by

$$f(t) = \sum M_\alpha \exp(-\sigma_\alpha t) \qquad (4.1)'$$

where only the damped or convergent case is considered. The corresponding spectral response is

$$F(i\omega) = \sum M_\alpha/(\sigma_\alpha - i\omega) \tag{4.2}'$$

As one example, this is the time behavior of radioactive decay problems. It was for this kind of application that Prony's Method was "re-discovered" in the 1950s [Householder (1950)]. Little success was evidently achieved for this application at that time. More recent work on this problem [Wiscombe and Evans (1977)] to characterize radiative transmission through an absorbing medium has been more successful. The observability of the poles is in this case an especially important problem, as the decaying waveform limits the useful observation time, and furthermore, all of the poles in the complex-frequency plane are aligned on the negative damping axis. The question of pole observability is discussed further in Section 5.2.1.

Another situation exhibiting the time behavior of Eq. (4.1)' is found in the scattering of short optical pulses from a fluid medium containing a particulate suspension. In this case, the autocorrelation function of the transient scattered field is given by a Laplace transformation of the particle-size number density, which may be approximated as in Eq. (4.1)' [Novotny (1982)].

4.1.2. Pure-Imaginary Poles

Pure-imaginary poles are associated with purely sinusoidal behavior for which the waveform becomes

$$f(t) = \sum M_\alpha \exp(-i\omega_\alpha t) \tag{4.1}''$$

This undamped oscillatory behavior occurs in only continuous-wave or time-harmonic fields, and could be used to represent cavity or waveguide problems, for example. The spectral response is then given by

$$F(i\omega) = \sum M_\alpha/(i\omega_\alpha - i\omega) \tag{4.2}''$$

In both of these special cases, it would be helpful if the characteristic equation could be so constrained as to produce pure-real or pure-imaginary roots, respectively. Otherwise, the numerical result can generally be expected to be a complex root whose unwanted component may at best provide an indication of solution accuracy, while at worst it can invalidate the solution. This question has been discussed by Miller and Goodman (1983) for the pure-imaginary case and is considered further in Section 4.2.1a.

4.1.3. Complex Poles

This is the most general case and the situation commonly encountered in acoustics, structural mechanics, electromagnetics, and other areas. The imaginary component of the pole accounts for the oscillation rate of the phenomenon being modeled, while the real component accounts for the rate of decay. In most applications, these individual components provide different, complementary information concerning the underlying physics which

produced the observed waveform.

For this reason, the poles can be of broader interest than merely permitting an efficient way of representing the data. In electromagnetics for example, the oscillatory pole component conveys information concerning the object's size while the damping component similarly reveals information about its shape. This is so because standing-wave or resonance effects are determined mostly by the length(s) of the path(s) on the object which contribute to the resonance condition. Aside from dissipative loss, the damping that occurs due to radiation of energy is primarily due to geometrical shape which causes charge acceleration. Because of these relationships, and the fact that the accuracy with which the damping component can be estimated is generally less than that of the oscillatory component [Dudley (1977)], it can be concluded that object size is more easily determined than object shape using waveform-domain data.

4.1.4 A Special Case Exponential In Both the Time and Frequency Domains

The field reflected from an infinite planar slab due to an incident impulsive plane (or equivalently the current-voltage wave on a transmission line) can be expressed in exponential form in both the time domain and the frequency domain [Lytle and Lager (1976)]. The time-dependent reflection coefficient for an impulsive, normally incident plane wave can be written as

$$r(t) = R_{12}\delta(t) + T_{12}T_{21}R_{23}(R_{21}R_{23})^{-2} \sum_{\alpha=-\infty}^{\infty} \exp(i\omega_\alpha t) \qquad (4.3a)$$

where the ω_α are the roots of

$$1 - R_{21}R_{23}\exp\left[-2i\omega_\alpha\sqrt{\mu_2\varepsilon_2}D_2\right] \qquad (4.3b)$$

with the slab thickness D and electrical properties denoted by the subscript "2", while "1" and "3" respectively indicate the upper and lower half spaces. The reflection and transmission coefficients are R and T, with the first subscript indicating the incident medium and the second the medium reflected from, or transmitted into, respectively. It is also assumed that all the media are non-dispersive.

The corresponding frequency-dependent reflection coefficient is similarly given by

$$R(\omega) = R_{12} + R_{23}T_{12}T_{21} \sum_{\alpha=1}^{\infty} (R_{21}R_{23})^{\alpha-1}\exp\left[-2i\omega_\alpha\sqrt{\mu_2\varepsilon_2}D_2\right] \qquad (4.4)$$

The time- and frequency-domain poles are thus given respectively by
$$s_\alpha(t) = \omega_\alpha = [-i\ln|R_{21}R_{23}| + Ang(R_{21}R_{23} \pm \alpha 2\pi]/[2D_2\sqrt{(\mu_2\varepsilon_2)}]$$
and $s_\alpha(\omega) = -i2\alpha D_2\sqrt{(\mu_2\varepsilon_2)}$
Similarly, the corresponding residues are
$$R_\alpha(t) = T_{12}T_{21}R_{23}(R_{21}R_{23})^{-2}$$

and $\quad R_\alpha(\omega) = T_{12}T_{21}R_{12}(R_{21}R_{23})^{\alpha-1}$

4.2 Space/Angle Waveforms and Angle Spectra

There are a variety of problems involving the space sampling of fields whose data structure makes them suitable for analysis using Prony's Method. Consider a set of W plane waves of strength P_α having direction angles θ_α and ϕ_α in a spherical coordinate system, being sampled at the point in space whose position vector is given by

$$\vec{r} = x\hat{x} + y\hat{y} + z\hat{z},$$

with $\hat{x}$, $\hat{y}$, and $\hat{z}$ unit vectors.

The total field can then be written

$$\vec{f(r)} = \sum_{\alpha=1}^{W} P_\alpha \exp\{ik[(\sin\theta_\alpha\cos\varphi_\alpha)x + (\sin\theta_\alpha\sin\varphi_\alpha)y + (\cos\theta_\alpha)z]\} \quad (4.5)$$

with $k=2\pi/\lambda$. Eq. (4.5) forms the basis for several interesting applications as outlined below. The first three cases, 4.2.1a, 4.2.1b, and 4.2.2a are pole-based only in that the direction angle(s) are obtained entirely from the exponents or poles. The latter applications illustrate sampling arrays which require the residues in addition to poles to find the desired information.

4.2.1 Sampling on Lines

The problem of sampling a field made up of incident plane waves is directly analogous to time-series analysis. In spatial field sampling however, we deal with complex data samples, or phase and amplitude, where the poles provide the incidence angles of the incoming waves. These angles are defined, for example, with respect to the line along which the sampling occurs (straight-line sampling) or with respect to angular arc along a circular sampling path. Some examples of line sampling are given here.

a. A straight line--The total field along a line due to a set of incident plane waves can be written from Eq. (4.5) by equating x=y=0 to obtain

$$f(z) = \sum_{\alpha=1}^{W} P_\alpha \exp[ikz\cos(\theta_\alpha)] \quad (4.6)$$

with:

observation variable z=linear coordinate along a straight line, with $P \geq 2W$ in total,
poles $s_\alpha = ik\cos(\theta_\alpha)$,
and   residues $R_\alpha = P_\alpha$, the amplitude of the αth plane wave.

For this problem, $F(\theta)$, given by

$$F(\theta) = \sum_{\alpha=1}^{W} \frac{P_\alpha}{[ik\cos(\theta_\alpha) - ik\cos(\theta)]} \quad (4.7)$$

provides the plane-wave spectrum (as a function of incidence angle) that produces the sampled data f(z). The incidence angles of the waves being sampled by the array are thus obtained from the poles to provide a procedure for direction finding, an application discussed earlier by Benning (1969) and Kelso (1972). More recently, this basic application was explored by Haykin and Reilly (1980) and Aurand (1987), among others, as a general signal-processing problem.

One peculiar attribute of using Prony's Method for this and similar kinds of data is that the computed pole values can be complex in general, whereas the actual poles in many cases are either pure real or pure imaginary. For uncontaminated or low-noise data, the size of the incorrect component will usually be small relative to that which properly models the data, a fact that can be used to separate valid poles from curve-fitting and noise poles as in the moving-window technique discussed above. But in some cases, pattern synthesis for example, the specified pattern may not be possible to produce with pure imaginary poles, in which case the real pole component can be interpreted to imply a source directivety. An alternative to this would be to constrain the calculation procedure to produce only imaginary poles as discussed by Goodman and Miller (1983).

b. A circular arc--In this case a Fourier transformation is needed to render the original data suitable for processing since it is not otherwise possible to sample in uniform steps of the exponent. We first write the total field, which is obtained from Eq. (3.13) by letting z=0 and $\rho=\sqrt{x^2+y^2}$, the radius of the circular arc along which the sampling is being performed to get

$$f(\varphi) = \sum_{\alpha=1}^{W} P_\alpha \exp[ik\rho\sin(\theta_\alpha)\cos(\varphi-\varphi_\alpha)] \tag{4.8}$$

where φ_α is the angle of incidence in azimuth of the αth plane wave. As written, Eq. (4.8) is not amenable to Prony processing since a polynomial form cannot be obtained directly. An intervening Fourier transformation can be used however to produce the desired form.

If samples of $f(\varphi)$ are taken at D equispaced points around the circle, they are given by

$$f_n = f(\varphi_n) = \sum_{\alpha=1}^{W} P_\alpha \exp[ik\rho\sin(\theta_\alpha)\cos(\varphi_n-\varphi_\alpha)], \quad n=1,...,D, \text{ and}$$

$$\varphi_n = 2\pi n/D$$

where we note that use of a circular arc less than 2π radians in angle is also possible if the f_n are set to zero outside the sampling region. For convenience, let $D = 2^N \geq 2W$ with N integer, and take the FFT of the sequence f_n to get [Moody (1980)]

$$\tilde{f}_q = \frac{1}{D}\sum_{n=1}^{D} f_n \exp(-i2\pi nq/D)$$

$$= \frac{1}{D}\sum_{\alpha=1}^{W}\sum_{n=1}^{D} P_\alpha \exp\{ik[\rho\sin(\theta_\alpha)\cos(\varphi_n-\varphi_\alpha)]\}\exp(-i2\pi nq/D)$$

$$= \sum_{\alpha=1}^{W} P_\alpha J_q[k\rho\sin(\theta_\alpha)]\exp(-iq/2)\exp(iq\varphi_\alpha) \qquad (4.8a)$$

The data sequence in Eq. (4.8a) is still not amenable for processing using Prony's method because of the presence of the $J_q[k\rho\sin(\theta_\alpha)]$ term which creates an additional and unknown α-dependent multiplier of the sought-for polynomial form represented by the $\exp(iq\varphi_\alpha)$ term. However, if either: i) all $\theta_\alpha = \pi/2$; or ii) all θ_α are equal and known so that in either case $k\rho\sin(\theta_\alpha) = kP$ is a known constant dependent only on q, then the $\exp(-iq/2)/J_q(kP)$ term can be taken outside the α summation of Eq. (4.8a) to obtain

$$\tilde{\tilde{f}}_q = \frac{\exp(iq/2)}{J_q(kP)}\tilde{f}_q = \sum_{\alpha=1}^{W} P_\alpha\exp(iq\varphi_\alpha) \qquad (4.8b)$$

It can be seen that Eq. (4.8b) now has the form of Eq. (3.1), with q the independent (observation) variable, and whose parameters may therefore be estimated using Prony's Method. Thus, proceeding from samples of an incident plane-wave spectrum, uniformly-spaced in angle for convenience, a subsequent Fourier transformation does lead to the required polynomial form as given by Eq. (4.8b).

We thus have:
        observation variables q, φ_n = index of Fourier-transformed data
            and angular coordinate on circle of nth sample of
            $2N\geq2W$ original samples,
        poles $s_\alpha=i\varphi_\alpha$,
and   residues $R_\alpha = P_\alpha$.

As in the case of sampling on a line, sampling on a circle also provides a procedure for direction finding, but with the incidence angles now occurring as poles of the Fourier-transformed data rather than of the data originally sampled. The Fourier transform of the sampled data can be regarded as a preconditioning step needed to obtain the kind of exponential, waveform series to which Prony's method is applicable. The spectra-domain form of the transformed data then becomes

$$F(\varphi) = \sum_{\alpha=1}^{D} \frac{P_\alpha}{[i\varphi_\alpha - i\varphi]} \qquad (4.9)$$

where we observe that angle φ is the variable in both domains, a result of the pre-conditioning Fourier transform.

Because the $\sin(\theta_\alpha)$ in Eq. (4.81) may not be known in general, the procedure just described is not always applicable and its utility for azimuthal direction finding is limited. However, by combining azimuth and axial sampling as described further below in 4.2.2c, it may be possible to circumvent this obstacle.

4.2.2 Sampling over Surfaces

By sampling over a surface, we encounter the possibility of estimating both the elevation and azimuth direction angles of incoming plane waves. This can be done using either two polynomials whose simultaneous solutions are needed, or two separate polynomials whose solutions must be determined in a self-consistent fashion. In the discussion that follows we explore some applications involving surface or area sampling.

a. Rectangular-mesh sampling--The total field of a set of plane waves incident on the x-y plane is given from Eq. 4.5) by setting z=0 to get

$$f(x,y) = \sum_{\alpha=1}^{W} P_\alpha \exp\{ik[\sin(\theta_\alpha)\cos(\varphi_\alpha)x + \sin(\theta_\alpha)\sin(\varphi_\alpha)y]\} \qquad (4.10)$$

whose spectral-domain form can then be written as

$$F(\theta,\varphi) = \sum_{\alpha=1}^{W} \left[\frac{P_\alpha}{\{ik[\sin(\theta_\alpha)\cos(\varphi_\alpha) - \sin(\theta)\cos(\varphi)]\}} \right] \cdot$$

$$\cdot \left[\frac{1}{\{ik[\sin(\theta_\alpha)\sin(\varphi_\alpha) - \sin(\theta)\sin(\varphi)]\}} \right] \qquad (4.11)$$

with:

> observation variables x,y = position coordinates in the x-y plane for P≥4W samples in total,
> poles $s_\alpha{}^{(x)} = ik\sin(\theta_\alpha)\cos(\varphi_\alpha)$ and $s_\alpha{}^{(y)} = ik\sin(\theta_\alpha)\sin(\varphi_\alpha)$,

and residues $R_\alpha = P_\alpha$.

This case is an extension of sampling the fields along a line considered above. Now, because of the area sampling however, the incident-wave directions are obtained both in azimuth φ_α and elevation θ_α. A planar array can thus be used for two-dimensional direction finding, the simplest situation being two orthogonal linear arrays each having at least 2W samples.

We note that estimation involving Eq. (4.10) can be approached using various pairs (or more) of polynomials in two variables, depending upon the array configuration [Benning (1969)]. Two polynomials are needed because two poles are associated with each plane wave.

Upon denoting the sampled values of f(x,y) as

$$f_{ij} = f(x_i, y_j)$$

where $x_i = i\delta x$, $y_j = j\delta y$, we observe that the simplest configuration that might be used could employ orthogonal sampling along the x and y axes respectively. This could be done by using f_{i1} and f_{1j}, where i,j =

0, . . . , D-1 and amounts to developing two independent polynomials in the x and y directions. If $D \geq 2W$ samples are obtained in each direction, then separate estimates of $s_\alpha^{(x)} = ik\sin(\theta_\alpha)\cos(\varphi_\alpha)$ and $s_\alpha^{(y)} = ik\sin(\theta_\alpha)\sin(\varphi_\alpha)$ can be obtained from the corresponding roots X_α and Y_α. However, unless the estimated residue amplitudes are different enough to permit organizing the separate x and y poles into ordered pairs, an ambiguity can remain in determining the individual θ_α and φ_α of each incoming plane wave.

A more robust way of resolving this ambiguity is developing a polynomial whose roots simultaneously involve both the x- and y-axes poles. One way of doing this is by performing one additional row of sampling that bisects the x and y axes, to obtain f_{ii}, where $i = 0$, . . . , D-1. The poles in this case will now be $s_\alpha^{(xy)} = ik\sin(\theta_\alpha)\cos(\varphi_\alpha) + ik\sin(\theta_\alpha)\sin(\varphi_\alpha)$ from the product root $(XY)_\alpha = X_\alpha Y_\alpha$. The separate poles $s_\beta^{(x)}$ and $s_\gamma^{(y)}$ whose sum equals $s_\alpha^{(xy)}$ are then the ordered pair which belongs to the same incident wave. The total number of samples needed for these schemes is 4W and 6W respectively.

Benning (1969) discusses some other possibilities for using planar arrays for two-dimensional direction finding. In one case, a square array of W by W samples is employed, with an additional two samples being added to the ends of the first two x-rows of the matrix. Two polynomials are then developed by using all W samples in the x direction and the next-column y sample in row number 1. A second polynomial is developed in the same way except by using the next-column y sample in row number 2. The polynomials that arise can then be shown to have the forms

$$Y + a_{W-1}X^{W-1} + a_{W-2}X^{W-2} + \cdots + a_2X^2 + a_1X + a_0, \quad (4.12a)$$
and
$$XY + b_{W-1}X^{W-1} + b_{W-2}X^{W-2} + \cdots + b_2X^2 + b_1X + b_0. \quad (4.12b)$$

These equations can be combined to obtain

$$(a_0 - b_1)X^1 + (a_1 - b_2)X^2 + \cdots + (a_{W-2} - b_{W-1})X^{W-1} = b_0 - a_{W-1}X^W \quad (4.13)$$

whose roots provides the poles $s_\alpha^{(x)}$. The corresponding y poles $s_\alpha^{(y)}$ can then be found from either of the original polynomials. This approach requires $W^2 + 2$ samples.

Benning points out still another approach which requires 2W samples along the x axis at y-coordinate 1 and another W samples along x at y-coordinate 2. The resulting polynomials can then be written as

$$a_{W-1}X^{W-1} + a_{W-2}X^{W-2} + \cdots + a_2X^2 + a_1X + a_0, \quad (4.14a)$$
and
$$Y + b_{W-1}X^{W-1} + b_{W-2}X^{W-2} + \cdots + b_2X^2 + b_1X + b_0 \quad (4.14b)$$

whose roots are X_α and $(XY)_\alpha$. In this particular approach, 3W samples are required. As mentioned by Benning, numerous other possibilities exist for other implementations in terms of the surface-sampling configuration employed. The specific nature of the polynomial is determined by the sampling sequence employed in developing the data matrix used in Prony's method.

b.   Circular-mesh sampling--Area sampling

can also be done using non-rectangular coordinate systems, one example being polar coordinates where the sampling variables are ρ and φ. Again letting $z = 0$ in Eq. (4.5) we obtain

$$f(\rho,\varphi_n) = \sum_{\alpha=1}^{W} P_\alpha \exp[ik\rho\sin(\theta_\alpha)\cos(\varphi - \varphi_n) \qquad (4.15)$$

where we have:

observation variable ρ = linear coordinate along radius of sampling grid at angular position $\varphi_{n=1}$ of $P \geq 2W$ radial samples in total,

poles $s^{(1)}_\alpha = ik\sin(\theta_\alpha)\cos(\varphi_1-\varphi_\alpha)$,

and   residues $R_\alpha = P_\alpha$,

with the corresponding spectral form given by

$$F(\theta,\varphi_n) = \sum_{\alpha=1}^{W} \frac{P_\alpha}{\{ik[\sin(\theta_\alpha)\cos(\varphi_n - \varphi_\alpha) - \sin(\theta)\cos(\varphi_n - \varphi)]\}} \qquad (4.16)$$

Radial sampling thus produces information about the combined influence of θ_α and φ_α about which further data is needed in order to estimate their separate values. Upon obtaining further radial-line samples at $D \geq 1$ additional angles, there is available in principal the additional needed information. Two or more values for $s^{(n)}_\alpha$ obtained at respective angles φ_n in the form above provide the equations required. For example, with

$s^{(1)}_\alpha = ik\sin(\theta_\alpha)\cos(\varphi_1-\varphi_\alpha)$,

$s^{(2)}_\alpha = ik\sin(\theta_\alpha)\cos(\varphi_2-\varphi_\alpha)$

and assuming the independent estimates $s^{(1)}$ and $s^{(2)}$ belong to the same term in α, it is possible to solve for θ_α and φ_α. This approach could be used when the residues differ enough so that they can be used to identify the same-α poles that arise from two (or more) estimates. If this cannot be assured, then some ambiguity is likely to result. In the special case where $\varphi_1 = \varphi_2 + \pi/2$, rectangular-mesh sampling along two lines is recovered.

       c. Cylindrical-surface sampling--If we let $\rho = \sqrt{x^2 + y^2}$ in Eq. (3.13) with ρ the radius of the cylindrical surface over which the field is sampled and with $x = \rho\cos(\varphi)$, $y = \rho\sin(\varphi)$, we obtain

$$f(\varphi,z) = \sum_{\alpha=1}^{W} P_\alpha \exp\{ik[\rho\sin(\theta_\alpha)\cos(\varphi - \varphi_\alpha) + \cos(\theta_\alpha)z]\} \quad (4.17)$$

By sampling in z at azimuth angle φ_1, we can obtain a set of estimates for the elevation incidence angles $\theta^{(1)}_\alpha$ with the result:

observation variables z, φ_1 = z coordinate at azimuth angle φ_1 and radius ρ for a total of $P \geq 2W$ samples,

poles $s^{(1)}_\alpha = ik\cos(\theta_\alpha)$,

and   residues $R^{(1)}_\alpha = P_\alpha \exp[ik\rho\sin(\theta_\alpha)\cos(\varphi_1-\varphi_\alpha)]$,

where we see that θ_α can be found directly. This is not unexpected considering that sampling along the z axis has already been shown in Section 4.2.1 to provide the incidence angle in elevation.

A set of $D \geq 1$ additional z samples at other azimuths provides the needed additional equations from which estimation of the φ_α can be attempted using the new residue values in a fashion analogous to that discussed for the case of the circular mesh. Thus, if at azimuth angles φ_n we obtain poles $s^{(n)}{}_\alpha = ik\cos(\theta_\alpha)$, and residues $R^{(n)}{}_\alpha = \{P_\alpha \exp[ik\rho \sin(\theta_\alpha)\cos(\varphi_n - \varphi_\alpha)]\}^{(n)}$, we could seek solutions of the same-α residue terms satisfied simultaneously by $\theta_\alpha = \cos^{-1}[s^{(n)}{}_\alpha/ik]$, φ_α, and P_α.

This is certainly not as simple as finding the incidence angles directly from the poles as was possible in previous cases. The difficulty arises from the fact that upon Fourier transforming the original data, given by

$$f_{nm} = f(\varphi_n, z_m) = \sum_{\alpha=1}^{W} P_\alpha \exp\{ik[\rho\sin(\theta_\alpha)\cos(\varphi_n - \varphi_\alpha) + \cos(\theta_\alpha)z_m]\} \quad (4.17a)$$

$$= \sum_{\alpha=1}^{W} P_{\alpha m}\exp\{ik[\rho\sin(\theta_\alpha)\cos(\varphi_n - \varphi_\alpha)]\},$$

where

$$P_{\alpha m} = P_\alpha \exp[ik\cos(\theta_\alpha)z_m] \quad (4.17b)$$

we find

$$\tilde{f}_{mq} = \frac{1}{D}\sum_{n=1}^{D} f_{nm}\exp(-i2\pi nq/D) \quad (4.18)$$

$$= \sum_{\alpha=1}^{W} P_{\alpha m}\exp(-iq/2)J_q[k\rho\sin(\theta_\alpha)]\exp(iq\varphi_\alpha)$$

where an α- and q-dependent $J_q[k\rho\sin(\theta_\alpha)]$ term again appears under the summation, as in Eq. (4.8a). Furthermore, even though all the quantities which appear in Eq. (4.15) except for P_α and φ_α are known or computable, this knowledge does not permit rewriting the equation in the form needed to employ Prony's method. This appears to be possible only for the uninteresting (in terms of 2D direction finding) special case where all θ_α are equal as already discussed in Section.4.2.1b.

d. Spherical-surface sampling--
The requirement that the successive samples produce a uniform change in the exponents of the model was satisfied for a circular arc by Fourier-transforming the azimuthally sampled data prior to using the Prony model. This basic idea does not appear to be extendible to the case of a spherical surface since expressions of the form

$$f(\theta, \varphi) = \sum_{\alpha=1}^{W} P_\alpha \exp\{ik\rho[\cos(\theta)\cos(\theta_\alpha) + \sin(\theta)\sin(\theta_\alpha)(\cos(\varphi)\cos(\varphi_\alpha)$$

$$+ \sin(\varphi)\sin(\varphi_\alpha))]\} \quad (4.19)$$

where equal changes cannot be produced in all the terms of the

exponent sampling in either θ or φ. Thus, a linearizing polynomial cannot be developed and Prony's method is not directly applicable to spherical-surface sampling of incident plane waves.

4.2.3 Sampling in Three Dimensions
The basic approaches outlined above can be implemented in three dimensions in various ways as well. Here we briefly consider some possibilities.

a. Rectangular sampling--From Eq. (4.5) we observe that by sampling separately along the x,y, and z axes in a rectangular coordinate system, three sets of poles can be obtained. This is also a straightforward extension of rectangular-mesh sampling on a plane as previously discussed. With the following form for the sampled field

$$f(x,y,z) = \sum_{\alpha=1}^{W} P_\alpha \exp\{ik[\sin(\theta_\alpha)\cos(\varphi_\alpha)x + \sin(\theta_\alpha)\sin(\varphi_\alpha)y + \cos(\theta_\alpha)z]\}$$

$$(4.20)$$

we thus have:

observation variables x,y,z = position coordinates in the x-y-z sampling volume,

poles $s_\alpha^{(x)} = ik\sin(\theta_\alpha)\cos(\varphi_\alpha)$, $s_\alpha^{(y)} = ik\sin(\theta_\alpha)\sin(\varphi_\alpha)$, and $s_\alpha^{(z)} = ik\cos(\theta_\alpha)$

and residues $R_\alpha = P_\alpha$.

We note that although residue matching might provide one way of identifying the 3-tuple of poles that correspond to a single incident plane wave, this is not particularly dependable and an alternative is desirable. After factoring out the ik multiplier, the sum of the squares of the poles should satisfy the following equation

$$[s_\alpha^{(x)}/ik]^2 + [s_\alpha^{(y)}/ik]^2 + [s_\alpha^{(z)}/ik]^2 \approx 1 \qquad (4.21)$$

for the 3-tuple of valid poles all belonging to the same plane wave. As another alternative, we could employ a polynomial in x, y, and z as an extension of the procedure outlined above for planar sampling.

b. Other sampling strategies--
Other three-dimensional sampling might also be employed to acquire the needed information in a more interpretable format. Among the possibilities worth considering might be sampling over an area combined with sampling on a line normal to that surface, or other combinations of the cases already discussed. No further discussion is devoted here to such special cases.

4.3 Angle Waveforms and Space/Angle Spectra
Another set of field-sampling problems, but involving sampling over the far-field sphere rather than the space sampling just discussed, is also suitable for Prony-type analysis. These problems have as their starting point

$$f(\theta,\varphi) = \sum_{\alpha=1}^{W} S_\alpha \exp\{ik[\sin(\theta)\cos(\varphi)x_\alpha + \sin(\theta)\sin(\varphi)y_\alpha + \cos(\theta)z_\alpha]\} \quad (4.22)$$

with

$$\vec{r}_\alpha = x_\alpha \hat{x} + y_\alpha \hat{y} + z_\alpha \hat{z}$$

the space position of source α of strength S_α, and θ and φ the spherical-coordinate observation angles. Not surprisingly, this class of problems is closely related to that just discussed. Some examples of this problem class follow.

4.3.1 Sampling the Far Field of a Linear Source Distribution

For an array of isotropic, point sources along the z axis (i.e., $x_\alpha = y_\alpha = 0$) we obtain from Eq. (4.22)

$$f(\theta) = \sum_{\alpha=1}^{W} S_\alpha \exp[ikz_\alpha \cos(\theta)] \qquad (4.23)$$

with:

observation variable θ = elevation angle with respect to the array axis on the far-field sphere

poles $s_\alpha = ikz_\alpha$,

and residues $R_\alpha = S_\alpha$, the strength of the αth source,

It can be seen that Eq. (4.23) is the reciprocal application of Eq. (4.6) in that space sampling to determine angles of incidence in the former has been replaced by angle sampling to determine space positions in the latter.

It can be seen that spectral-domain representation of the pattern which is given by

$$F(z) = \sum_{\alpha=1}^{W} S_\alpha/(ikz - ikz_\alpha) \qquad (4.24)$$

provides the aperture distribution which produces $f(\theta)$. But the sources are poles rather than the delta functions which is implicitly assumed in writing $f(\theta)$. Again note that it is necessary to sample the waveform domain data, in this case $f(\theta)$, in uniform steps of the exponent of Eq. (3.23a) in order to produce the polynomial form needed in Prony's Method. This means that the observation angle must change incrementally such as to produce uniform steps of $\cos(\theta)$ rather than θ itself.

This kind of far-field pattern sampling can be useful for EM, acoustic, and seismic problems. One application is pattern synthesis, since upon specifying $f(\theta)$ it then becomes possible to find the set of point sources required to produce it [Goodman and Miller (1983)]. In general, this procedure leads to non-uniformly spaced sources, with no a priori specification other than the pattern itself and the number of sources chosen to approximately produce it. This case can be seen to be the inverse of the line-sampling problem discussed above, in that sampling over the far-field sphere of sources along a line is performed rather than sampling the fields along a line of sources located in the far field. Other approaches to the problem of synthesizing non-uniformly spaced arrays are described by Maffett

and Curtz (1967), Sahalos (1974), and Steinberg (1982).

Another application for $f(\theta)$ is imaging, i.e., finding the sources that produced a given far-field [Miller (1983)]. This particular problem is equivalent to performing one-dimensional inversion. This basic Prony technique can be extended to two-dimensional source distributions as well (see below), but apparently cannot be used for general three-dimensional source distributions because it is then not possible to sample the data in terms of uniform steps in the exponent of each exponential. It is possible in the latter case however, to solve the inverse-source problem using nonlinear optimization of an exponential model, a procedure that can be very time consuming computationally. Furthermore, a global minimum cannot be guaranteed to be found and the procedure can be quite sensitive to noise and the initial values used for the unknowns.

4.3.2 Sampling the Far Field of a Circular Source Distribution

The far field of a set of isotropic point sources located on a circle of radius ρ in the x-y plane as obtained from Eq. (4.22) in the x-y plane is given by

$$f(\varphi) = \sum_{\alpha=1}^{W} S_\alpha \exp[ik\rho_\alpha \cos(\varphi - \varphi_\alpha)]; \quad \rho_\alpha = \rho, \text{ for all } \alpha \quad (4.25)$$

with φ_α the angular position of the αth source. This is precisely the expression for the field samples on a circle due to a set of incident plane waves. Thus, the Fourier-transform technique applies here as well to get

$$f_n = \sum_{\alpha=1}^{W} S_\alpha \exp[ik\rho \cos(\varphi_n - \varphi_\alpha)], \text{ and}$$

$$\tilde{f}_q = \sum_{\alpha=1}^{W} S_\alpha \exp(iq\varphi_\alpha) = \exp(iq/2)/[J_q(k\rho)D] \sum_{n=1}^{D} f_n \exp(-i2\pi nq/D) \quad (4.25a)$$

with:

observation variables q, φ_n = index number of Fourier-transformed data and angular coordinate (in azimuth) of nth sampling point on far-field sphere,

poles $s_\alpha = i\varphi_\alpha$,

and residues $R_\alpha = S_\alpha$

Observe that the radius of the circle on which the sources are located needs to be known if the actual source distribution is to be accessible. Otherwise, only an equivalent distribution will be obtained. As in the case of linear array of sources, this approach also provides a procedure for synthesizing the circular-source distribution that produces a specified radiation pattern, a problem also discussed by Graham, Johnson, and Elliott (1978).

The spectral-domain result is given by

$$F(\varphi) = \sum_{\alpha=1}^{W} \frac{S_\alpha}{[i\varphi_\alpha - i\varphi]} \tag{4.26}$$

where we observe that as in previous cases involving the Fourier transform, the waveform-domain and spectral-domain variables are both azimuth angle.

4.3.3 Sampling the Far Fields of a Planar Source Distribution

If the sources are distributed over the x-y plane (i.e., z_α = 0), then from Eq. (4.22) we have

$$f(\theta,\varphi) = \sum_{\alpha=1}^{W} S_\alpha \exp\{iks\sin(\theta)[\cos(\varphi)x_\alpha + \sin(\varphi)y_\alpha]\} \tag{4.27}$$

It is clear that this field can be sampled in θ at fixed φ-planes to produce uniform exponential changes, and is therefore suitable for analysis using Prony's method. We obtain then

$$f(\theta,\varphi_p) = \sum_{\alpha=1}^{W} S_\alpha \exp\{iks\sin(\theta)[\cos(\varphi_p)x_\alpha + \sin(\varphi_p)y_\alpha]\} \tag{4.27a}$$

with:

observation variable θ = angular elevation coordinate on far-field sphere for sampling in the φ_p plane,

poles $s_\alpha^{(p)} = ik[\cos(\varphi_p)x_\alpha + \sin(\varphi_p)y_\alpha]$,

and residues $R_\alpha = S_\alpha$, the strength of the αth source.

Note that each azimuth sampling plane yields a set of poles. If the source strengths are all resolvably different, then the pole estimates having common residues from different sampling planes can be employed to estimate source locations using

$$\cos(\varphi_1)x_j + \sin(\varphi_1)y_j = s_j^{(1)}$$
$$\cos(\varphi_2)x_j + \sin(\varphi_2)y_j = s_j^{(2)}$$

where $S_j^{(1)} \approx S_j^{(2)}$ and $\varphi = \varphi_1, \varphi_2$ respectively. For the special case where $\varphi_1 = 0$ and $\varphi_2 = \pi/2$, then the x coordinates and y coordinates are obtained separately in turn, since each sampling plane provides source positions projected onto its base in the x-y plane. It may happen, however, that two or more of the source strengths are equal, within the numerical precision available from the data and the processing. In that case, an ambiguity exists, so that with no additional measurements or information, nothing further could be concluded about the actual source locations.

We note that the spectral representation can be written as

$$F(x,y) = \sum_{\alpha=1}^{W} \frac{S_\alpha}{\{ik[\cos(\varphi_p)x_\alpha + \sin(\varphi_p)y_\alpha] - ik[\cos(\varphi_p)x + \sin(\varphi_p)y]\}} \tag{4.28}$$

4.3.4 Sampling the Far Field of a Cylindrical Source Distribution of Known Radius

From Eq. (4.22) we obtain, in a fashion analogous to Eq. (4.25),

$$f(\theta,\varphi) = \sum_{\alpha=1}^{W} S_\alpha \exp\{ik[\rho\sin(\theta)\cos(\varphi-\varphi_\alpha) + \cos(\theta)z_\alpha]\} \quad (4.29)$$

with φ_α and z_α the position coordinates of sources on a cylindrical surface of radius ρ. The Fourier transform can again be used to put the azimuth variation into the required polynomial form. There is then obtained

$$f_n(\theta_p) = \sum_{\alpha=1}^{W} S_\alpha \exp\{ik[\rho\sin(\theta_p)\cos(\varphi_n-\varphi_\alpha) + \cos(\theta_p)z_\alpha]\} \quad (4.29a)$$

$$\tilde{f}_q(\theta_p) = \sum_{\alpha=1}^{W} S_\alpha \exp[iq\varphi_\alpha + ikz_\alpha\cos(\theta_p)]$$

$$= \exp(iq/2)/\{J_q[k\rho\sin(\theta_p)]D\} \sum_{n=1}^{D} f_n \exp(-i2\pi qn/D) \quad (4.29b)$$

with:

observation variables q, φ_n, θ_p = index number of Fourier-transformed data and observation angle (in azimuth) on far-field sphere of pth elevation cut,

poles $s_{\alpha p} = i\varphi_\alpha$,

and residues $R_{\alpha p} = S_\alpha \exp[ikz_\alpha\cos(\theta_p)]$.

Because the Bessel-function $J_q[k\rho\sin(\theta_p)]$ which appears in Eq. (4.29b) involves a known quantity, it can be factored from the transformed-data summation. Consequently, tit becomes feasible to synthesize a cylindrical source distribution needed to produce a specified θ,φ pattern using Prony's method.

4.3.5 Sampling the Far Field of a Spherical Source Distribution of Known Radius

Using Eq. (4.22), we obtain

$$f(\theta,\varphi) = \sum_{\alpha=1}^{W} S_\alpha \exp\{ikR[\sin(\theta_\alpha)\sin(\theta)\cos(\varphi-\varphi_\alpha) + \cos(\theta)\cos(\theta_\alpha)]\} \quad (4.30)$$

where the sources are assumed to be located on a sphere of radius R at angle coordinates θ_α and φ_α.

Upon sampling the far field in the θ_p plane at equal angular increments φ_n, there is obtained

$$f_n(\theta_p) = \sum_{\alpha=1}^{W} S_\alpha \exp\{ikR[\sin(\theta_\alpha)\sin(\theta_p)\cos(\varphi_n-\varphi_\alpha) + \cos(\theta_p)\cos(\theta_\alpha)]\} \quad (4.31a)$$

$$\tilde{f}_q(\theta_p) = \frac{1}{D}\sum_{n=1}^{D} f_n\exp(-i2\pi nq/D)$$

$$= \sum_{\alpha=1}^{W} S_\alpha J_q[kR\sin(\theta_\alpha)\sin(\theta_p)]\exp(-iq/2)\exp[iq\varphi_\alpha + kR\cos(\theta_\alpha)\cos(\theta_p)]$$

$$(4.31b)$$

with:

> observation variables q, φ_n, θ_p = index number of Fourier-transformed data and observation angle (in azimuth) on far-field sphere of pth elevation-angle cut.

It can be seen that the problem of locating the positions of sources on a spherical surface is not amenable to Prony's method because of the $J_q[kR\sin(\theta_\alpha)\sin(\theta_p)]$ in the transformed-data series. As an alternative, the far field instead might be represented as a spherical-wave expansion whose projection onto the sphere could approximate the source distribution.

4.3.6 Sampling the Far Fields of a Three-Dimensional Source Distribution

The general inverse problem requires finding the 3-dimensional source distribution which gives rise to a measured near or far field. As demonstrated above, Prony's method is applicable to some 2-dimensional subsets of this general problem. However, the method unfortunately cannot be used to derive a 3D source distribution from its field, at least not from sampling in angle only. This is due to the requirement of Prony's method that polynomial data forms be obtained, an outcome that cannot be achieved from angle sampling alone. We might speculate that this outcome reflects the more basic need of obtaining 3D-equivalent information to derive 3D sources, whereas angle sampling alone is intrinsically capable of providing only 2D information. Another approach that does work for the 3D problem is discussed in the next section.

4.4 Frequency Waveforms and Space Spectra

Consider again Eq. (4.22) for the far field of sources with position coordinates x_α, y_α, and z_α. In contrast to the previous section where various kinds of angle sampling were examined, here we consider sampling in frequency f (or wave number k) at fixed angles $\theta = \theta_p$, $\varphi = \varphi_p$. We then have

$$f(k,\theta_p,\varphi_p) = \sum_{\alpha=1}^{W} S_\alpha(k)\exp\{ik[\sin(\theta_p)\cos(\varphi_p)x_\alpha$$

$$+ \sin(\theta_p)\sin(\varphi_p)y_\alpha + \cos(\theta_p)z_\alpha]\} \tag{4.32}$$

where the possible frequency dependence of the source strengths is acknowledged in $S_\alpha(k)$. If either: 1) $S_\alpha(k) = S_\alpha f(k)$ for all α, where $f(k)$ is known; or 2) $S_\alpha(k) = S_\alpha$, a constant independent of k, then it is clear that Eq. (4.32) provides a way to find the source coordinates. One possibility is given below.

4.4.1 Sampling the Far Field of a 3D Source in Frequency Along Three Lines

If Eq. (4.32) is sampled over frequency in three different directions we get

$$f(k,\pi/2,0) \quad = \sum_{\alpha=1}^{W} S_\alpha(k)\exp(ikx_\alpha)$$

$$f(k,\pi/2,\pi/2) \quad = \sum_{\alpha=1}^{W} S_\alpha(k)\exp(iky_\alpha)$$

$$f(k,0,0) \quad = \sum_{\alpha=1}^{W} S_\alpha(k)\exp(ikz_\alpha) \tag{4.33}$$

with:

observation variable k = wavenumber at fixed point(s) in space,
poles $s_\alpha = ix_\alpha, iy_\alpha,$ and $iz_\alpha,$
and   residues $R_\alpha = S_\alpha.$

Eq. (4.33) provides the possibility of separately estimating the x, y, and z coordinates of the sources. We thus see that frequency sampling at a fixed point on a line provides the projections of the source coordinates onto that line. However, unless the source amplitudes are all different, determination of the actual source coordinates cannot be easily accomplished. The corresponding spectral forms are given by

$$\vec{F}(r,\theta_p,\varphi_p) = \sum_{\alpha=1}^{W} \frac{S_\alpha}{i[\sin(\theta_p)\cos(\varphi_p)x_\alpha + \sin(\theta_p)\sin(\varphi_p)y_\alpha + \cos(\theta_p)z_\alpha] - [\sin(\theta_p)\cos(\varphi_p)x +}$$

$$\cdots \frac{1}{\sin(\theta_p)\sin(\varphi_p)y + \cos(\theta_p)z]} \tag{4.34}$$

4.5 Discussion

Only brief attention was given above to the process by which the signal-processing problem is linearized to one of finding the set of coefficients defining one or more polynomials, and from whose roots

the poles are obtained. For one-dimensional data, the choice of an harmonic polynomial is fairly obvious (but not unique), and leads to the rows of the linear system being formed from sequential sets of data samples. When two-dimensional data is involved, as in sampling on a two-dimensional plane for example, the polynomial choice is no longer so straightforward. This general problem has been discussed in some detail by Benning (1969) in connection with the design of direction-finding arrays.

The most important aspect of Prony's method in terms of its utility for field-sampling applications of the kinds illustrated here is the need to develop a linearizing polynomial which transforms the original exponential equation for the data samples to a linear system whose solution gives the coefficients of a linear predictor or characteristic equation. For some problems, as typified by time-series data, this requirement leads to the simple strategy of sampling at uniform time intervals. A similar situation holds for field sampling along a straight line. But the relatively simple reciprocal problem of sampling the far fields of a linear array requires that the samples be spaced in uniform increments of $\cos(\theta)$ rather than the observation angle θ.

Extension to sampling on a circular arc was found to be even more complicated, requiring the preprocessing step of Fourier transforming the data to make possible development of a polynomial form. However, as was illustrated in Sections 4.2.1b and others where the Fourier-transform step is required, special conditions are necessary if the resulting exponential model is then suited for Prony modeling. If, as can happen, the coefficients in the transformed data are dependent on both the index number of the individual wave components as well as the transform variable [see Eq. (4.8a) for example], then Prony's method may still be inapplicable.

The possibility that the coefficients in the individual terms of the series can include a dependence on the observation variable in addition to the exponential itself is thus one that deserves additional consideration. The modified model takes the form

$$f(x) = \sum R_\alpha(x) \exp(s_\alpha x)$$

where the question to be examined is what kinds of $R_\alpha(x)$ are needed for Prony's method be applicable. For the cases of circular-arc sampling (Section 4.2.1b) and frequency sampling in three dimensions (Section 4.4.1), it was pointed out that if $R_\alpha(x) = R(x)$ independent of α, then the extra x dependence can be factored and a new observation variable $g(x) = f(x)/R(x)$ can be defined. Aside from this simple, special case however, no other straightforward examples have been found which permit realizing the polynomial form needed for the f(x) samples to be Prony processed.

5 SOME REPRESENTATIVE APPLICATIONS

We have demonstrated above that exponential forms, as discrete series and continuous integrals, arise in many areas of physics, not least of all acoustics and electromagnetics. In this concluding section we review briefly some of the computational issues that may arise and summarize some representative applications of pole-based parameter estimation.

5.1 SEM (Singularity Expansion Method) Motivation

The idea of using poles for characterizing radar targets was pursued in electromagnetics shortly after introduction of the Singularity Expansion Method (SEM) [Baum (1972,1976)]. Examples of studies in this area include Miller, et al (1975), Moffatt and Mains (1975), Lytle and Lager (1976), Moffatt et al (1981) , Miller and Lager (1982), and Tijhuis and Blok (1984a, 1984b). For the most part, except for the simpler one-dimensional problems, these studies generally do not address the possible physical interpretability of the poles. Instead, they attempt to use poles in various ways, for example to estimate whether an unknown waveform belongs to a particular class or target in that class, or to formulate transient problems. A notable exception to this kind of approach is due to Hoenders (1981) who showed that the pole sets for a radially inhomogeneous sphere uniquely can determine its index of refraction as a function of r. Other work along this line is due to Gaunaurd and Uberall (1981), Miller (1981), Barber et al (1982), and Conwell et al (1984).

It can be demonstrated for data similar to that which occurs in sampling EM fields that Prony's method can preserve information originally present in the waveform; that is it gives pole values whose accuracy on the average equals the accuracy of the input data [Miller (1981)]. It is generally true however, that the imaginary (oscillatory) components are almost always found with greater accuracy than the real (damping) components, where the accuracy of the latter seems more closely correlated to that of the predictor coefficients. This observation relates to the issue of pole observability which we discuss further below, as well as to the influence of object geometry and size on the pole values, which is also discussed.

5.2 Computing Poles

Interest in poles in the electromagnetics and acoustics communities originated from development and application of time-domain modeling and the Singularity Expansion Method [Baum (1972), (1976)]. The poles were originally studied for their possible use in obtaining transient results using the framework of the SEM. One way of getting the poles needed for SEM was provided by processing computed or measured transient responses, using a technique such as Prony's method. We found that, for wirelike objects at least, as many as 20 to 25 pole pairs could be found accurately from computed currents or scattered fields when an object was excited impulsively. A variety of examples are presented by Miller (1981).

Present literature that deals with poles in electromagnetics and acoustics generally focusses on either of two problems. One is that of obtaining poles from measured or computed data. The other is that of exploring how poles can be most effectively exploited for various applications, or what conceptual implications or utility poles might have for the inverse-type problem.

Both problems are relevant. Our tentative conclusion is that pole-based schemes for target classification and identification in acoustics and electromagnetics may be limited intrinsically by the difficulty of obtaining poles under even the ideal circumstances of using noise-free, computed data. The basic problem for this situation seems to be one of the decreased observability of low-Q resonances in time-series data. It is worth noting, however, that poles can be obtained from other than response data. For example, the zeros of the B_{ij} matrix define the pole locations when a frequency-domain formulation is used. In a time-domain formulation, the poles can be found as the eigen-values of a state-transition matrix, the coefficients of which are given by the time-domain equivalent of B_{ij} [Cordaro and Davis (1980)].

Two particular issues associated with pole-based representations are discussed here. First we consider the question of pole observability, i.e., how can poles be found indirectly from either time-domain or frequency <u>observations</u> as opposed to analytical formulations. The other issue is that of whether the coefficients of a pole-zero model themselves might not be modeled in a hierarchical or nested fashion.

5.2.1 Pole Observability
In an attempt to quantify in a generic way the degree of difficulty in computing poles, we have conducted a series of computer experiments. Parameters varied in these experiments were the number of poles P and the locus of points described by the poles in the complex-frequency "s" plane. Since the pole locations for various objects thus far investigated generally tend to lie along smooth paths or trajectories in the s-plane, we parameterized test pole sets using a straight line parallel to the jω axis. A still more realistic experiment might be conducted by varying the angle of this pole trajectory relative to the jω axis. The poles were spaced evenly along the imaginary axis, with the constant loss component systematically increased. Time-domain waveforms were constructed from the test-pole set thus established and the poles were estimated from Prony's method for varying levels of added white noise using a Monte-Carlo approach. Expected values for the accuracy of the extracted poles were obtained as function of the number of poles, the constant damping-component value, and the noise level.

The goal of this experiment is to establish regions in the s-plane within which the poles can be extracted to a given degree of accuracy for selected noise levels. Upon superimposing on such a plot known pole sets for various objects of different shape (obtained in various

ways but as accurately as possible), it should be feasible to establish the number and accuracy of poles that might be estimated from actual transient data of varying signal-to-noise (S/N) ratios. These parametric results can then be used to sort ultrasonic and electromagnetic targets into three sets, viz. those for which the poles can generally be estimated from transient data, those for which the poles can generally not be estimated, and those which are borderline. Subsequent work on pole-based classification and identification schemes could then concentrate on those target classes where there is a reasonable probability of success.

This work continues and results and implications thereof are still evolving, but some preliminary findings are presented in Fig. 2. There we plot the average pole accuracies (in bits) obtained from the simulated data, with all calculations performed to computer accuracy (8-10 digits). The rectangles have lengths along the $j\omega$ and σ axes proportional to the average bit accuracies of the respective pole components, and are centered at s-plane coordinates given by the parameters of the pole set they represent. Thus for example, a rectangle at $s_\alpha = \sigma_\alpha + j2\pi f_\alpha$, the pole-set parameters are a fixed real component σ_α, and imaginary components $\pm j2\pi f_\alpha$, $\pm j2\pi(f_\alpha - 2)$, . . ., $\pm j2\pi(2)$, with the number of poles P equal to the numerical value used for $f_\alpha = 2, 4, . . . , 22$. Unit-amplitude residues were used for all terms in the transient waveforms.

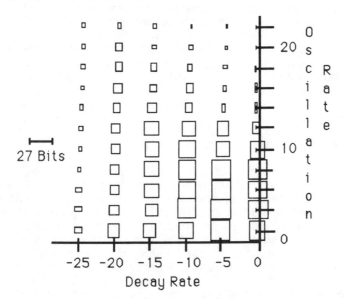

Figure 2. Average pole-estimation accuracy as a function of number of poles and their real and imaginary components.

That the possibility of accurate pole estimation depends on the relative sizes of the real and imaginary pole components is easily

explained. For example, if a two- (conjugate) pole waveform is only slightly damped, it can be expected that its decay rate is more difficult to estimate than its frequency over one period. Conversely, if it is highly damped, the frequency would be more difficult to obtain. Maximum accuracy should be expected for some intermediate ratio. This has been demonstrated in a previous computer simulation by Miller (1978) who also used Prony's method. While Prony's method may not be the most accurate way to estimate the poles, the result of using other more robust techniques should only change the overall attainable accuracy, but not the basic implications.

5.2.2 Higher-Order, Pole-Based Models

As mentioned, our experience thus far has been mixed in terms of estimating the poles of acoustic targets from their transient, scattered fields. Even when using computer-simulated data without additive noise, consistent pole sets were not obtained as the number of model poles was increased, although the mismatch error monotonically decreased. Because the s-plane pole trajectories of smooth objects are also found to be "smooth", a constrained estimation procedure was tried as an alternative as demonstrated in Fig. 3 [Miller, et al (1984)].

The approach used involved forcing the pole trajectory to satisfy certain prescribed conditions or constraints. The simpler approach required that the damping components increase monotonically with increasing frequency. A more sophisticated procedure placed the poles on a trajectory described by a low-order polynomial. Both techniques greatly improved the pole estimates judged by our expectations based on the target shape and compared with the results obtained without using a constraint. Constrained estimation can be viewed as a second-order model as it models the parameters of the data model (the exponential series) itself. Success of such an approach demonstrates that the target rank is of lower order than the data rank for such targets, and represents a way to use a priori information which is available about the target.

5.3 Pole Dependence on Target Size and Shape

Given the results of various electromagnetic studies, we conclude that the oscillatory components of an object's poles contain information primarily related to its size and the damping components contain information related to its shape and electrical impedance. This conclusion is based on observations of pole locations in wires as a function of bends, junctions, and lumped impedance loading as demonstrated for representative cases as presented by Miller (1981). Our conjecture is that $d\omega_\alpha/dL$ will be much larger than $d\sigma_\alpha/dL$ for L/a constant (i.e., a wire of fixed shape) where L is the wire's length and a is its radius. Conversely, $d\omega_\alpha/da$ will be much smaller than $d\sigma_\alpha/da$ for L constant with $s = \sigma_\alpha + \omega_\alpha$ representing the αth pole. We suspect that these relationships hold for more general objects as well.

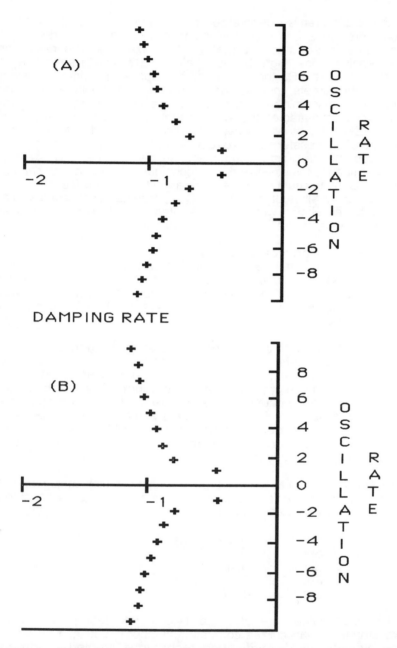

Figure 3. Examples of pole results obtained using higher-order pole models based on constrained estimation procedures applied to transient responses for a conducting sphere. In part (A) the constraint is that the real components increase in value as a function of frequency. The constraint in (B) is that of fitting a fourth-order polynomial to the real components of the poles [Miller et al (1984)].

If this basic premise generalizes to nonwire and acoustic targets, it could provide a basis for developing classification and identification schemes in specific situations. For example, the resonance frequencies alone might be adequate to discriminate among targets of different size. On the one hand, if shape is the primary difference among the targets, then the damping constants would be needed to provide a discriminant. Given that damping constants are less accurately estimated on the average than are the resonances, it might be concluded that pole-based classification and identification would be more successful for size discrimination. On the other hand, because the early-time ramp response gives target cross-sectional area or shape, it is possible that a technique that unifies early and late time responses could yield both size and shape. This issue has been discussed by Miller et al (1984). Our presentation concludes with a computer simulation which demonstrates use of several pole-based procedures for target classification/identification (C/I).

5.4 Using Poles for C/I

A fundamental question concerning the possible use of poles in C/I involves just what might be the most effective way to exploit pole-related information. Various schemes have been explored in this connection. We summarize here some results of a simple computer experiment.

The problem to be explored concerns testing to which pole set in a library of pole sets an unknown transient waveform most likely belongs. The unknown waveform, which has been analytically generated from one of the pole sets in the library, is sampled at a total of D evenly spaced data points, and there are P pole in each of the S sets in the library. Monte Carlo experiments are performed to compare the various techniques outlined below, by adding zero-mean, uniformly distributed noise have a peak signal-to-peak-noise ratio S_P.

The study is limited to ten poles per set (five complex-conjugate pairs) and a total of five pole sets all having constant real parts equal to -1 with the imaginary parts given by:

$$\text{Set } 1: \pm j2, \pm j4, \pm j6, \pm j8, \pm j10$$
$$\text{Set } 2: \pm j3, \pm j5, \pm j7, \pm j9, \pm 11$$
$$\text{Set } 3: \pm j4, \pm j6, \pm j8, \pm j10, \pm 12$$
$$\text{Set } 4: \pm j2, \pm j3, \pm j6, \pm j9, \pm j10$$
$$\text{Set } 5: \pm j2, \pm j5, \pm j6, \pm j7, \pm j10.$$

All residue values are equal to 1+j0 in generating the transient waveforms. A normalized correlation number is obtained for each of the classification/identification techniques studied. Individual correlation numbers are recorded to estimate false alarm rates, and average values are computed over several (usually 10) noise-contaminated waveforms to obtain performance estimates. The four classification/identification techniques for which results are presented here are outlined below.

5.4.1 Linear Predictor

A source-free linear predictor takes the form

$$f_N(t) = \sum_{\alpha=1}^{N} A_{N,\alpha} f(t - \alpha \Delta t) \tag{5.1a}$$

where $A_{N,\alpha}$ is the Nth order set of prediction coefficients. Here we see that the present value of $f(t)$ is given as a linear combination of the previous N samples uniformly spaced Δt apart. Of course, for Eq. (5.1a) to be valid, the coefficients $A_{N,\alpha}$ must belong to the function on which the prediction is being done.

For our problem, we use the notation

$$f_P^{(i,j)}(t_k) = \sum_{\alpha=1}^{} A_{P,\alpha}^{(i)} f_M^{(j)}(t_k - \alpha \Delta t) \tag{5.1b}$$

where $k = P, \ldots, D-1$ with $t_k = k\Delta t$. Here the superscript i denotes the pole set to which the prediction coefficients belong, and j the pole set from which the "measure" waveform $f_M^{(j)}(t)$ is derived. The procedure used is to compute $f_P^{(i,j)}(t_k)$ at $K = P, P+1, \ldots, D-1^*$ to obtain $D-P$ predicted values of the waveform from the total of D measured values. A mean-squared error between the two sets of waveform samples is then given by

$$r_{ij}^2 = 1/(D-P) \sum_{\beta=P}^{D-1} [f_P^{(i,j)}(t_\beta) - f_M^{(j)}(t_\beta)]^2 \tag{5.2a}$$

If D=2P, as is used here unless otherwise indicated, the P predicted values are used in the correlation. Note that in applying Prony's method to compute the poles of a given waveform having P poles, D must be at least as large as 2P. A normalized correlation is defined from Eq. (5.2a) as

$$R_{ij} = \frac{1}{1 + r_{ij}} \tag{5.2b}$$

and an average value is obtained from

$$<R_{ij}> = \frac{1}{R} \sum_{\alpha=1}^{R} R_{ij}(\alpha) \tag{5.2c}$$

where R is the number of separate Monte Carlo runs (nominally 10) and α denotes the run number. Note that using Eq. (5.1b) once requires on the order of $P(D-P)$ multiplies (multiplication steps), whose value is P^2 for the minimum D=2P samples required if Prony's method were to be used to directly compute the poles. And with W waveforms to be compared against S pole sets (we have used W = W = 5 in our calculations) in the library, the total number of multiplies is $\sim WS(P^2+P)$, of which the first term comes from the calculation in Eq. (5.2a).

5.4.2 Residue Calculation

The residue calculation is based explicitly on the

exponential character of the waveforms with which we are dealing [in Eq. (5.1) this dependence is implicit]. It proceeds from

$$f_M^{(j)}(t_k) = \sum_{\alpha=1}^{P} \hat{R}_\alpha{}^{(i,j)} \exp(s_\alpha^{(i)} t_k); \ k=0,1,\ldots,P-1,\ldots, \tag{5.3}$$

where the carats indicate estimates of the P residue values which match waveform j to pole set i. If P waveform samples are used in Eq. (5.3), then a square system of equations results. Otherwise, for K=P, . . . , D-1, an overdetermined system is produced, for which a pseudo inverse can be obtained by regularization.

A correlation value follows from the residue estimates as

$$r_{ij}^2 = \frac{1}{D}\sum_{\beta=1}^{D}\left\{ \sum_{\alpha=1}^{P} \hat{R}_\alpha{}^{(i,j)} \exp(s_\alpha^{(i)} t_\beta) - f_M^{(j)}(t_\beta)\right\} \tag{5.4}$$

from which a normalized value is found by using Eq. (5.2b).

The minimum number of multiplies involved in using Eq. (5.3) once is on the order of $(P^3/3) + P^2$ due to the linear system solution. This assumes the terms $\exp[s_\alpha^{(i)} t_k]$ are computed once and stored for use as needed. The total number of computations involved in using W waveforms and S pole sets is then on the order of $S\{[(P^3/3)+P^2W]+2P^2W\}$, the last term being due to Eq. 5.4.

5.4.3   Pole Calculation

Proceeding in a fashion similar to that used for the residue calculation, we write

$$f_M^{(j)}(t_k) = \sum_{\alpha=1}^{P} \hat{R}_\alpha{}^{(j)} \exp(\hat{s}_\alpha{}^{(j)} t_k); \ k=0,1,\ldots,D-1 \tag{5.5}$$

where the carats indicate residue and pole estimates for waveform j. Prony's method is used for this calculation and requires a minimum of 2P data points for its implementation.

A correlation value is then obtained from

$$r_{ij}^2 = \frac{1}{P}\sum_{\alpha=1}^{P}[|\hat{s}_\alpha{}^{(j)} - s_\alpha^{(i)}|_{MIN}]^2 \tag{5.6}$$

Eq. (5.6) provides the minimum distance between the pole set computed from waveform j and poles of set i. Again, a normalized correlation is obtained from Eq. (5.2b).

The minimum number of multiplies required to use Prony's method once is on the order of $(P^3/3)+P^2+ePI$, where the last accounts for the polynomial solution which Prony's method necessitates, and I is the average number of iterations required per root. The total associated with W waveforms and S pole sets is then on the order of $W\{[(P^3/3)+P^2+3PI]+Ps\}$, where the last term is due to Eq. (5.6).

5.4.4 Waveform Correlation

Here, we compute R_{ij} directly from

$$R_{ij} = \frac{\displaystyle\sum_{\beta=1}^{D} f_M^{(j)}(t_\beta) f_L^{(i)}(t_\beta)}{\displaystyle\sum_{\beta=1}^{D} [f_M^{(j)}(t_\beta)]^2} \tag{5.7}$$

where $f_L^{(i)}(t_\beta)$ denotes waveform i from the library of stored waveforms. While the poles are source independent, the residues are not; hence, if the library is to be complete, many more waveforms than the number of targets of interest need to be stored. For simplicity in our calculations, we stored only the original (noise-free) waveforms, i.e., those for which all residues are equal to 1+j0. In using this approach, the minimum number of multiplies is on the order of 2WSP+2WP.

For comparison, the numbers of multiplies for the various techniques are summarized in Table 2. Also included there are the totals for the parameter values used in this study: W=S=5, P=10, and I~50. Note that these numbers do not account for all of the computer time needed to exercise a given technique, as this can also be strongly influenced by other factors and is algorithm-dependent as well.

TABLE 2
COMPARISON OF OPERATION COUNT FOR VARIOUS POLE-BASED TECHNIQUES USED FOR WAVEFORM IDENTIFICATION

TECHNIQUE	Number of Multiplies/Divides	
	EXPRESSION	Value relative to waveform correlation for I=50, W=S=5, P=10
Waveform correlation	2WSP + 2WP	1
Linear predictor	WSP(P + 1)	4.6
Residue calculation	$SP^2[P/3 + 3W]$	15
Pole calculation	$WP[P^2/3 + P + 3I + S]$	17

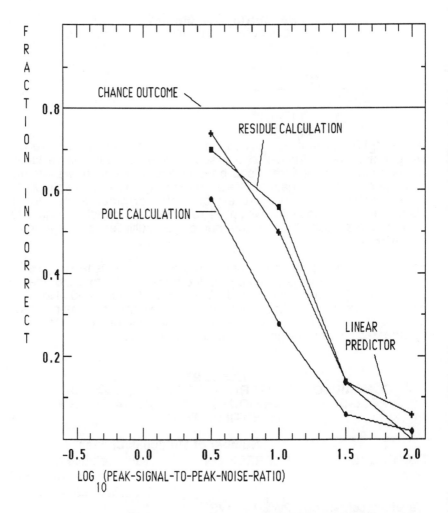

Figure 4.   Results for transient-waveform identification using three pole-based techniques for 10-pole transients. The basic idea is to test whether the pole set to which a noise-contaminated waveform belongs can be identified using stored sets of poles in a library. In this graph are displayed the average false-alarm rates for 10 runs as a function of peak-signal-to-peak-noise ratio S_P.

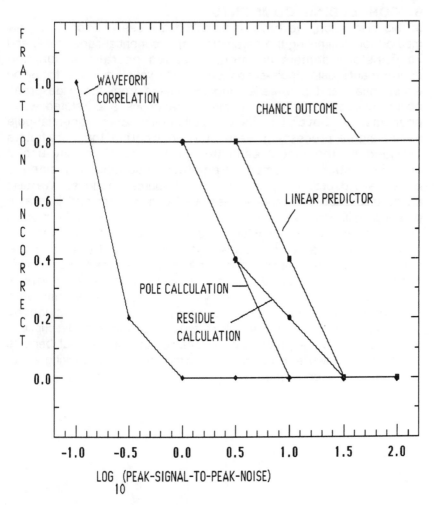

Figure 5.   Results for transient-waveform identification using three pole-based techniques and waveform correlation for 10-pole transients. The basic idea is to test whether the pole set to which a noise-contaminated waveform belongs can be identified using stored sets of poles in a library. In this graph are displayed the false-alarm rates based on averages of the correlation coefficients obtained from 10 runs as a function of peak-signal-to-peak-noise ratio S_P.

6 CONCLUDING COMMENTS

In this discussion, we have summarized the use of Prony's original method for estimating the parameters of exponential-series data, and its transform-domain counterpart based on rational function approximants using pole-series data. Motivation for this focus on exponential- and pole-series models arises from the widespread occurrence of such forms in electromagnetics and related wave phenomena. Several kinds of applications for exponential-pole based signal processing were outlined wherein the parameters provided by the model estimates pertain to physically useful quantities such as complex-frequency resonances, incidence plane-wave direction angles, and spatial source positions. We also briefly discussed the important issue of observability and illustrated potential limitations with some computations. The possibility of using higher-order models in which the coefficients themselves are modeled was also discussed as one way by which the noise-sensitivity and observability problems associated with Prony's method might be reduced. Finally, a simple application of various pole-based schemes for identifying transient waveforms was illustrated with some sample computations. We conclude by observing that although the utility of using pole-based descriptions may not always be practically exploitable, the conceptual benefits alone are worthwhile in providing an alternate way of thinking about electromagnetic phenomena.

7 REFERENCES

Anton, J. J., and A. J. Rockmore (1976), "A unified approach to array-factor synthesis for line arrays with nonuniformly positioned elements", *IEEE J. of Oceanic Engineering*, OE-1(1), pp. 14-21.

Aurand, J. F. (1987), "An antenna array processing system for multiple source bearing estimation", PhD Dissertation, Iowa State University, Department of Electrical and Computer Engineering, Ames, Iowa.

Barber, P. W., J. F. Owen, and R. K. Chang (1982), "Resonant scattering for characterization of asymmetric dielectric objects", *IEEE Trans. on Antennas and Propagation*, 29(2), pp. 165-172.

Baum, C. E. (1972), "Electromagnetic transient interaction with objects with emphasis on finite sized objects and some aspects of transient pulse production", in *Digest of Spring URSI Meeting*, Washington, DC.

Baum, C. E. (1976), "The singularity expansion method", Chapter 3 in *Transient Electromagnetic Fields*, Springer-Verlag, New York.

Bennett, C. L. (1981), "Time Domain Inverse Scattering", *IEEE Trans. on Antennas and Propagation*, 29(3), pp. 213-219.

Benning, C. J. (1969), "Adaptive array beamforming and steering using Prony's method", Proceedings 19th Annual Symposium on USAF Antenna Research and Development Program, U. of Illinois, Allerton.

Bertero, M., and E. R. Pike (1982), "Resolution in diffraction-limited imaging, a singular value analysis. I. The case of coherent illumination", *Optica Acta,,* 29(6), pp. 727-746.

Bleszynski, Marek (1987), Private Communication

Brittingham, J. N., E. K. Miller, and J. L. Willows (1980), "Pole extraction from real-frequency information", *Proc. IEEE*, 68(2), pp. 263-273.

Conwell, P. R., P. W. Barber, and C. K. Rushforth (1984), "Resonant scattering for characterization of asymmetric dielectric objects", *IEEE Trans. Antennas and Propagation*, AP-30, pp. 165-172.

Cordaro, J. T. and W. A. Davis (1980), "Time domain techniques in the singularity expansion method", *IEEE Transactions on Antennas and Propagation*, 23, pp. 358-367.

Dudley, D. G. (1977), "Fitting noisy data with a complex exponential series", Lawrence Livermore Laboratory Rept. UCRL-52242.

Fiddy, M. A., and G. Ross (1979), "Analytic Fourier optics: The encoding of information by complex zeros", *Optica Acta*, 26(9), pp.

1139-1146.

Fiddy, M. A., G. Ross, M. Nieto-Vesperinas, and A. M. J. Huiser (1982), "Encoding of information in inverse optical problems", *Optica Acta*, 29(1), pp. 23-40.

Gaunaurd, G. C., and H. Uberall (1981), "Solution of the inverse electromagnetic scattering problem in the resonance case", *IEEE Trans. on Antennas and Propagation*, AP-29, pp. 293-297.

Gori, F., and L. Ronshi (1981), "Degrees of freedom for scatterers with circular cross section", *Journal of the Optical Society of America*, 71(3).

Graham, O., R. M. Johnson, and R. S. Elliot (1978), "Design of circular apertures for sum patterns with ring sidelobes of individually arbitrary heights", *Alta Frequenza*, 47(1), pp. 21-25.

Harrington, R. F. (1968), *Field Computation by Moment Methods*, Macmillan Co., New York, NY.

Haykin, S., and J. Reilly (1980), "Mixed autoregressive-moving average modeling of the response of a linear array antenna to incident plane waves", *IEEE Proceedings*, 68(5), pp. 622-623.

Hildebrand, F. (1956), *Introduction to Numerical Analysis*, McGraw-Hill, New York, NY.

Hoenders, B. J. (1983) "The unique determination of an object with a radially dependent index of refraction by its natural frequencies", *Optica Acta*, 29(1), pp. 55-62.

Householder, A. S. (1950), "On Prony's method of fitting exponential decay curves and multiple-his survival curves", Oak Ridge National Laboratory, Oak Ridge, TN, Rept. ORNL-455.

Inagaki, N., and R. J. Garbacz (1982), "Eigenfunctions of composite Hermitian operators with application to discrete and continuous radiating systems", *IEEE Trans. Antennas and Propagation*, AP-30, pp. 571-575.

Kay, S. M., and S. L. Marple (1981), "Spectrum analysis--a modern perspective", *IEEE Proceedings*, 69(11).

Kelso, J. M. (1972), "Measuring the vertical angles of arrival of HF skywave signals with multiple modes", *Radio Science*, 11(3), pp. 199-209.

Kennaugh, E. M., and D. L. Moffatt (1965), "Transient and impulse response approximations", *Proc. IEEE*, pp. 893-901.

Lytle, R. J. and D. L. Lager (1976), "Using the natural-frequency

concept in remote probing of the earth", *Radio Science*, **11**, pp. 199-209.

Maffett, A. L., and T. B. Curtz (1967), "Moment synthesis of array factors with nonuniform spacing and amplitude parameters", *Radio Science*, 2(7), pp. 721-728.

Miller, E. K., F. J. Deadrick, H. G. Hudson, A. J. Poggio, and J. A. Landt (1975), "Radar target classification using temporal mode analysis", Lawrence Livermore National Laboratory Report UCRL-51936.

Miller, E. K. (1978), "Prony's method revisited", Lawrence Livermore National Laboratory Report UCRL-52590.

Miller, E. K. (1981), "Natural mode methods in frequency and time domain analysis", in *Theoretical Methods for Determining the Interaction of Electromagnetics Waves with Structures*, Sijthoff and Noordhoff, pp. 173-212.

Miller, E. K., and D. L. Lager (1982), "Inversion of one-dimensional scattering data using Prony's method", *Radio Science*, 17(1), pp. 211-217.

Miller, E. K. (1983), "Imaging of linear source distributions", *Electromagnetics*, 3(1), pp. 21-40.

Miller, E. K., and D. M. Goodman (1983), "A pole-zero modeling approach to linear array synthesis 1. The unconstrained solution", *Radio Science*, 18(1), pp. 57-69.

Miller, E. K., G. A. Clark, G. D. Poe, B. D. Cook, and J. A. Jackson (1984), "Combining pole and ramp-based techniques for target identification", NATO Advisory Group for Aerospace Research & Development Meeting on Target Signatures, London, UK.

Moffatt, D. L., and R. K. Mains (1975), "Detection and discrimination of radar targets", *IEEE Trans. Antennas and Propagation*, AP-23, pp. 358-367.

Moffatt, D. L., J. D. Young, A. A. Ksienski, H-C. Lin, and C. M. Rhoads (1981), "Transient response characteristics in identification and imaging", *Trans. Antennas and Propagation*, AP-29, 192-205.

Moody, M. P. (1980), "Resolution of coherent sources incident on a circular antenna array", *IEEE Proceedings*, 68(2), pp. 276-277.

Novotny, V. (1982), "Particle size distribution by optical transients", Program of the First Annual Conference of the American Association for Aerosol Research, Santa Monica, CA.

Poggio, A. J. and E. K. Miller (1973), "Integral equation solutions of three-dimensional scattering problems", in *Computer Techniques for*

Electromagnetics, Pergamon Press, pp. 159-264.

Poggio, A. J., M. L. Van Blaricum, E. K. Miller, and R. Mittra (1978), "Evaluation of a processing technique for transient data", *IEEE Trans. on Antennas and Propagation*, 26(1), pp. 165-173.

Ross, G., M. A. Fiddy, M. Nieto-Vesperinas, and I. Manolitsakis (1978), "The propagation and encoding of information in the scattered field by complex zeros", *Optica Acta*, 26(2), pp. 229-238.

Sahalos, J. (1974), "A solution of the general nonuniformly spaced antenna array", *Proc. IEEE*, 62, pp. 1292-1294.

Steinberg, B. D. (1982), "Adaptive microwave holography", *Optica Acta*, 29(4), pp. 363-369.

Tijhuis, A. J., and H. Blok (1984a), "SEM approach to the transient scattering by an inhomogeneous, lossy dielectric slab; Part 1: the homogeneous case", *Wave Motion*, 6, pp. 61-78.

Tijhuis, A. J., and H. Blok (1984b), "SEM approach to the transient scattering by an inhomogeneous, lossy dielectric slab; Part 2: the inhomogeneous case", *Wave Motion*, 6.

Van Blaricum, M. L., J. R. Auton, and T. L. Larry (1984), "Target identification via complex natural resonance extraction from radar signatures", NATO AGARD Electromagnetic Wave Propagation Panel Symposium on Target Signatures, London, UK.

Winterfeld, Chr. v. (1987), "Directional analysis using vector spherical waves", IEEE AP-S 1987 International Symposium Digest, Vol. II, pp.1074-1076.

Winthrop, J. T. (1971), "Propagation of structural information in optical wave fields", *Journal of the Optical Society of America*, 61(1), January.

Wiscombe, W. J., and J. W. Evans (1977), "Exponential sum fitting of radiative transmission functions", *J. Comp. Physics*, 24.

SELECTED APPLICATIONS OF METHOD OF MOMENTS MODELLING

Professor Stanley J. Kubina
and
Professor Christopher W. Trueman
Concordia University
Montreal, Canada

1. INTRODUCTION

The renaissance of interest in the electromagnetic scattering from complex bodies has brought about a re-examination of the potential and the limitations of numerical methods in electromagnetics including integral equation methods for the analysis of complex geometries. These techniques had been surveyed in the 1979 NATO ASI[1]. An indication of this renewed interest are the several sessions on Numerical Techniques and a special review session on Matrix Methods in Electromagnetics at the 1987 AP-S International Symposium and URSI Radio Science Meeting in Blacksburg Virginia[2]. A directed survey and a compendium of case studies of practical structures can be found in the book edited by Moore and Pizer[3].

In the latter work some comparisons are made between several computer codes including the Numerical Electromagnetic Code or NEC[4]. The experiences with these codes are translated into some guidelines for the creation of wire-grid models that should assure a reasonable degree of agreement with measured results. The degree of agreement which is desirable varies with the application. In the present paper, experiences with the use of NEC are described which have a more stringent requirement on the extent of the agreement with measured results. At the same time, aspects of the corresponding scale model measurements are discussed which have a bearing on the closeness of the agreement that can be expected. The applications to be considered are for radiation rather than scattering problems.

The morphology of the numerical process that is being used can be considered to consist of three distinct but interlinked portions. The central, mathematical portion, consists of the appropriate integral equation formulation, its conversion into a matrix form whose solution can be used to produce the desired results - far fields, RCS, near fields or impedance. Essentially linked with this core is the selection of basis and weighting functions and the description of the source excitation. The rate of convergence of the method is effected by the selection of the basis and weighting functions and the source modelling effects the reliability of the impedance computations. More efficient solution methods than matrix decomposition are described in this Institute by Sarkar[5].

The third important element in the process is the methodology for the creation of the wire-grid model itself. In this process the user asks, where shall I place the wires? How dense shall I make the grid? What radius shall I use? How small should the wire segments be? These questions are directed at seeking the best equivalent structure in the electromagnetic sense. This practical problem is

257

B. de Neumann (ed.), Electromagnetic Modelling and Measurements for Analysis and Synthesis Problems, 257–310.
© 1991 Kluwer Academic Publishers. Printed in the Netherlands.

not unrelated to the sampling considerations discussed by Miller[6], except that the sampling of the yet to be established current distribution must be determined heuristically with some awareness of the consequences imposed by the mathematical formulation.

The time scale for practical analysis and design of complex real-world vehicles seldom allows the development of new computer codes. Rather means are sought for the most productive use of existing codes which are versatile and which have already been "de-bugged" and where extensive usage has made guidelines for their successful application available to the user community. The exchange of such application information, as well as information on new codes, is one of the main aims of the newly-formed Applied Computational Electromagnetics Society[7]. In this paper, the use of NEC is described from this purview and the following section provides a summary of its features.

The error-free generation of a three-dimensional wire grid to be a close replica of the complex physical structure is a time-consuming and tedious task which is made less formidable if done by computer-aided methods. One such process via the DIDEC[8] computer code is described in the third section. The remaining sections describe the results with computer models of HF antennas on aircraft, helicopters and ships and some of the corresponding scale model measurement results.

2. THE NUMERICAL ELECTROMAGNETICS CODE

The "Numerical Electromagnetics Code"[4,9] or "NEC" is a computer program for analysing a general class of radiating structures constructed from either continuous surfaces or from straight wires of cylindrical cross-section. This section describes the methods used by NEC to determine the current flow on a wire structure.

To use the NEC program, the antenna and the the ship or aircraft on which it is mounted is represented or "modelled" as a mesh of interconnected wires usually coincident with the surfaces. Techniques for deriving such models are discussed below. NEC accepts a list of the rectangular coordinates of the endpoints of the wires making up the model, the radius of each wire, the location of the feed point, and the frequency of operation. In a NEC solution, each wire must be sub-divided into "segments" shorter than one-tenth or preferably one-twentieth of a wavelength. The antenna is excited by specifying the voltage impressed across the segment coinciding with the feed point. The NEC program finds the magnitude and phase of the current flowing at the center of each segment, using a "moment method" solution[10] of Pocklington's Equation, as discussed below. If there are N_t segments in the model, then an N_t by N_t square matrix must be formulated and solved.

Once the currents flowing on the wires of the model have been found, they are readily integrated to find the radiation patterns of the antenna on the ship or aircraft. In the following some of the details of the NEC formulation are briefly described.

2.1 Pocklington's Integral Equation

The NEC is based on Pocklington's Integral Equation, which states that for perfectly conducting wires, the total axial electric field must be equal to zero at any point on any of the wires of the

antenna. The "primary" field is the field exciting the antenna, which could be an incoming plane wave, but in radiation problems is usually associated with a voltage applied a the feed point of the antenna. If the antenna has N_w wires in total, then the "secondary" field at any point is the sum of the contribution of each of the wires,

$$\overline{E}_{sec} = \sum_{i=1}^{N_w} \overline{E}_i \qquad \cdots 1$$

where the subscript "sec" stands for "secondary". If $\hat{s}_k$ represents a unit vector which is parallel to the axis of wire #k, then the component of the field parallel to the axis of wire #k can be expressed as

$$\overline{E}_{sec} \cdot \hat{s}_k = \sum_{i=1}^{N_w} \overline{E}_i \cdot \hat{s}_k \qquad \cdots 2$$

A fundamental assumption must be made, namely that the wires are "thin" so the the current flow on each of the wires of the antenna is purely axial, with no circumferential component. Also, the current must be uniformly-distributed over the wire surface. A "thin" wire is one whose diameter is much smaller than the wavelength. Also, wires must be farther apart than their diameters. If s_i measures distance along wire #i, then the field axial to wire #k due to the current $I_i(s_i)$ on wire #i can be expressed as

$$\overline{E}_i \cdot \hat{s}_k = \frac{-j\eta}{4\pi\beta} \int_L I_i(s_i) \left\{ \beta^2 \hat{s}_k \cdot \hat{s}_i - \frac{\partial^2}{\partial s_k \partial s_i} \right\} g(s_k, s_i) ds_i \qquad 3$$

where β is the free-space wave number, η is the free-space intrinsic impedance, and the i-subscripts denote the "source wire" which is causing the field. The Green's Function or "reduced kernel" is given by

$$g(s_k, s_i) = \frac{e^{-j\beta|s_k - s_i|}}{|s_k - s_i|} \qquad \cdots 4$$

where $\backslash s_k - s_i \backslash$ is the distance from point s_k on wire #k to point s_i on wire #i. This "reduced kernel" is derived by assuming that the current effectively flows on the axis of the source wire, and that the observer is located on the periphery of wire #k. A better approximation, called the "extended kernel", is sometimes used for parallel, connected segments, in which the current is assumed to be distributed uniformly over the surface of the source wire, and integration is used to account for its distribution.

To state Pocklington's Equation, the total axial field on wire number k must be made equal to zero. The primary field is that which is applied to the wires to excite the wire antenna model. radiating antenna problems the primary field is applied to a segment coincident with the feed point, which is excited with a specified voltage. The primary field is then that voltage divided by the length of the feed segment. The total field is the sum of the primary field plus the secondary field given by Equation 2,

$$\overline{E}_{primary} \cdot \hat{s}_k + \overline{E}_{sec} \cdot \hat{s}_k = 0 \qquad \cdots 5$$

which by substitution of Equation 2 obtains

$$\sum_{i=1}^{N_w} \overline{E}_i \cdot \hat{S}_k = -\overline{E}_{primary} \cdot \hat{S}_k \qquad \cdots 6$$

Together with Equation 3, this is Pocklington's Integral Equation. Ideally, the unknown current $I_i(s_i)$ should be determined such that it satisfies Pocklington's Integral Equation at every point s_k on each wire of the wire antenna model. In practice, an approximate solution must be used, as described in the following.

2.2 Junction Constraints

Pocklington's Equation states a boundary condition at any point s_k on wire #k, in terms of the currents $I_i(s_i)$ flowing on all of the N_w wires of the model. The NEC finds the currents $I_i(s_i)$ by expanding these unknown functions in terms of simple "basis functions" with unknown, complex-valued coefficients. This section deals with the current expansion and the junction constraints, and the next discusses the "moment method".

In the NEC solution, the unknown current function $I_i(s_i)$ is found by subdividing wire #i into N_i "segments", and then expanding the current on segment #j on wire #i as a constant plus a sine plus a cosine. If the length of wire #i is L_i, then the "segment length" for wire #i is

$$\Delta_i = \frac{L_i}{N_i} \qquad \cdots 7$$

The current on wire number i, segment number j is expressed as

$$I_{ij}(s_i) = A_{ij} + B_{ij}\sin\beta(s_i - s_{ij}) + C_{ij}\cos\beta(s_i - s_{ij}) \qquad \cdots 8$$

where s_{ij} is the coordinate of the center of segment #j. This introduces three unknowns per segment. If the total number of segments on the model is

$$N_t = \sum_{i=1}^{N_w} N_i \qquad \cdots 9$$

then there is a total of 3 N_t unknowns. To obtain the values of these unknowns, linear equations are derived from continuity conditions for the current and its derivative, and from Pocklington's Equation using the "moment method".

Along the length of any wire, the current is required to be continuous from one segment to the the next. Hence

$$I_i\left(s_{ij} + \frac{\Delta_i}{2}\right) = I_i\left(s_{i,j+1} - \frac{\Delta_i}{2}\right) \qquad \cdots 10$$

which must hold for j = 1, ... (N−1). Also, the charge density on the wire,

$$q(s_i) = \frac{-1}{j\omega} \frac{\partial I_i}{\partial s_i} \qquad \cdots 11$$

is required to be continuous across segment boundaries. With a total
of 3 N_i unknowns for wire #i, there are (N_i-1) segment boundaries and
so continuity of current and charge density provide $2(N_i-1)$ equations
for each wire.

At any wire junction, Kirchoff's Current Law is required to
hold, providing one equation per junction. Also, at a wire junction,
the King-Wu constraint is imposed on the charge density[11]

$$\frac{\partial I_\iota}{\partial s_\iota}\bigg|_{at\,junction} = \frac{Q}{\ln\left(\frac{2}{\beta a_\iota}\right) - \gamma} \qquad \cdots 12$$

where a_i is the radius of wire #i, γ = 0.5772 is Euler's Constant,
and Q is the total charge in the vicinity of the junction. In
effect, at the junction of n wires, a set of $(n-1)$ relations is
obtained, of the form

$$\left(\ln\left(\frac{2}{\beta a_1}\right) - \gamma\right)\frac{\partial I_1}{\partial s_1} = \left(\ln\left(\frac{2}{\beta a_2}\right) - \gamma\right)\frac{\partial I_2}{\partial s_2} = \cdots 13$$

These $(n-1)$ charge density constraints plus Kirchoff's Current Law
give a set of exactly n equations per junction. Where a wire has a
"free end" not connected to any other wire, the wire "endcap" is
permitted to store charge, and the current flowing onto the endcap is
thus related to the charge stored there. Free-end conditions yield
one equation per free end.

It may be shown that this set of continuity conditions on the
current and charge density amount to 2 N_t equations, and a further N_t
equations are needed to solve for the A, B and C current amplitudes
on each segment of the model. These are obtained from Pocklington's
Equation via the "moment method".

2.3 Moment Method Solution

In a wire antenna model with N_t segments, there are a total of 3
unknown current amplitudes per segment for a total of 3 N_t unknowns
in all. The junction conditions discussed above provide a set of
2 N_t relations, and a further N_t are required to determine the values
of the current coefficients.

Pocklington's Equation given by Eqn. 6 above states the boundary
condition that at any point on any wire, the axial component of the
total field must be zero. The "moment method" enforces this boundary
condition at the center of each segment, and thus "point matches" the
secondary field to the excitation field at segment centers, that is,
at the set of "match points" $s_k = s_{kj}$ for $j=1,\ldots,N_k$ on each wire.
Thus for each wire of the model, N_k "point matching" equations are
obtained, for a total of N_t equations.

The internal details of the solution are so arranged in NEC that
the full set of 3 N_t equations are never explicitly generated or
stored. Rather, the junction conditions are imposed as the "point
matching" equations are assembled, and so the matrix size in NEC is
never larger than N_t by N_t.

The NEC solution accounts for the interaction of every segment
of the model with every other segment, both by direct connection such
as at wire junctions, and by coupling via the near fields of the
segments. Once the current on each segment has been found by this

"moment method" solution, the fields of the antenna can be computed by integration of the currents on the wires. NEC provides for the evaluation of "far field" radiation patterns, and of the "near fields" in the vicinity of the model.

2.4 Ground Options

The NEC program is often used to model antennas which operate above a ground plane, which may represent real ground or the surface of the sea. A useful approximation has been to consider the ground plane to be highly-conducting or "perfect", in which case the NEC program uses the method of images to obtain an efficient solution.

NEC provides two alternatives to "perfect" ground conductivity. If the antenna is located well above the ground plane, such as a helicopter or aircraft in flight, then Fresnel reflection coefficients can be used to account for the reflection of the signal from the surface of lossy ground. However, if parts of the structure are directly connected to the ground, as is the case for a model of a ship, then the Sommerfeld-Norton ground model can be used. In this model, Sommerfeld Integrals are used to express the exact interaction of each segment with lossy ground of a specified conductivity and relative permittivity. NEC uses numerical integration and interpolation to construct an efficient method for approximating the interaction terms for any distance between the source segment and the observer[12,13]. The Sommerfeld-Norton ground option is costly in terms of increased running time for the NEC program.

2.5 Geometrical Restrictions

The user of the NEC code must observe certain guidelines in constructing a wire antenna model for analysis by the program. The guidelines are necessary so that the basic assumptions upon which NEC is based not be violated.

The "thin wire" assumption requires that the wire diameter must be much less than the wavelength. In subdividing each wire into "segments", the length of each "segment" must be shorter than one-tenth wavelength, and segments shorter than one-twentieth are often used. Segments must not be shorter than one-thousandth of the wavelength. The segment should be much longer than the diameter of the wire. Segments longer than 8 times the wire radius are safe, and segments as short as the wire diameter are sometimes used. With the "extended kernel", segments should be longer than the wire diameter, although segments as short as half the wire radius are sometimes used. Parallel wires must be much farther apart than their diameter.

Restrictions apply at wire junctions[9]. When wires of different radii are connected, the ratio of the larger to the smaller radius must be less than 5. The segment length must not be too dissimilar when two wires are connected. Differences larger than a factor of five are to be avoided. A more subtle restriction requires that the match point on a segment at a junction be located outside the volume of all the wires connected to the junction. Non-physical results can be seen if this restriction is violated.

These restrictions are local, relating segment length and wire radius within one wire, and on adjacent wires making up a junction. The NEC program is used in the following to analyse complex geometries, in which a solid body of complex shape is replaced by a "grid" or mesh of wires describing its shape. The wires of the grid must satisfy the guidelines set out above. The remainder of this paper discusses the derivation of complex wire-grid models, and

illustrates the results that can be obtained using them.

3. THE MODEL CREATION ART

The basic code requirements are that the start and end co-ordinates and the diameter of each wire be specified. For an aircraft model with 350 wires, some 600 vertices must be scaled and tabulated from the respective views of the working drawings. Errors of overlap and discontinuity can frequently occur and these are not necessarily detected by the computer code itself. The computer-aided model creation system called DIDEC was designed to simplify and accelerate this process, to make it less error-prone and to add an interactive dynamic dimension to the exercise which makes the validity criteria and model features more visible to the engineer. The acronym DIDEC[8] means Digitize, Display, Edit and Convert.

The necessary first step in both the manual as well as the computer-aided process is the marking of the vertices on acccurate dimensioned three-view drawings of the structure. The model designer then proceeds to "digitize" the vertices on a digitizing tablet. The drawing is first calibrated by digitizing two known reference points and then the co-ordinates of each vertex are determined sequentially: X,Y co-ordinates from a top-view and Y,Z co-ordinates from a side-view. Points can be linked automatically to form wires, or their linkage can be specified by listing their vertex numbers. At each stage of the digitization of any section of the structure, the vertices and wires can be displayed in multi-port displays with selected views corresponding to the source drawings and/or isometric 3-D displays.

Wires and vertices can be deleted or added as desired. For large symmetrical structures commands such as "REFLECT" and "MERGE" exploit symmetry and allow progressive build-up of a complex model from individual, manageably small sections. Colour-coded 3-D displays can show each wire in a colour that corresponds to a coded value of diameter or length. This provides meaningful visualization of these parameters as they relate to modelling criteria for variable sections of the model. If the model is found to be suitable, a final command will automatically format the internal files into an input data set that conforms to the format requirements of the computer code that is being used.

The DIDEC system was intended to have features and to use software standards that would allow its evolutionary development and to minimize the danger of its early obsolescence. Changes in graphics standards have now eroded the value of this characteristic.

3.1 Implementation and Use

The graphics software used in DIDEC is a portion of package called DIGRAF[14]. It conforms to the Core Graphics Standard[15] and is readily available to the university environment. The program itself is written in RT-11 Fortran IV in a highly structured way so that it could be effectively overlayed to fit within the limited memory space of PDP-11 systems.

The major portions of the model development process are illustrated in the first two figures. Fig.3.1 shows the individual portions of the aircraft that were digitized separately. They are displayed as separate files in one of the many viewport configurations that DIDEC allows. The fuselage section in the upper left window was obtained by digitizing one half of the fuselage in

top and side-view drawings and by using the REFLECT command to generate the whole section. The horizontal stabilizers, shown in the lower right window were formed also by using the REFLECT command.

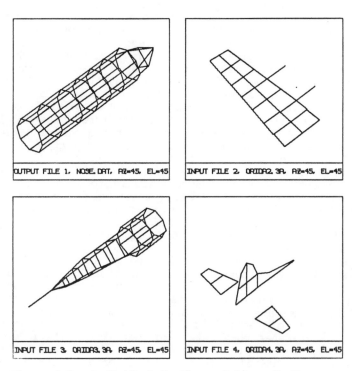

Figure 3.1.   Individual Sections of Aircraft Structure

By a subsequent set of operations, the wing was reflected and merged with the fuselage and then the remaining sections were merged to form the complete aircraft model shown in Fig.3.2.   A scaling command was used when necessary to keep the display of the merged sections within the limits of the display of the port.   Fig.3.2 shows the separate views of the aircraft simultaneously with the isometric display.   Such views are useful for comparisons with the original aircraft drawings.   If desired, they can be displayed individually in single ports over the entire screen.   Readers should also visualize the colour coding of these displays for diameter and length.   Finally the NEC command translates the model description data into the proper format as an input data file for NEC execution.

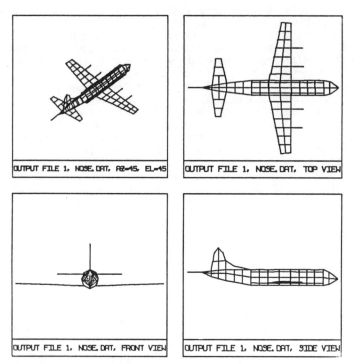

OUTPUT FILE 1, NOSE.DAT, AZ=45, EL=45

OUTPUT FILE 1, NOSE.DAT, TOP VIEW

OUTPUT FILE 1, NOSE.DAT, FRONT VIEW

OUTPUT FILE 1, NOSE.DAT, SIDE VIEW

Figure 3.2. Merged Sections and Separate Views

3.2 Consequences

The DIDEC interactive model building system accelerates the process of developing wire-grid models by factors exceeding 20:1 over any manual process. It does so without mechanical measurement errors other than the tablet tolerances. These errors can be removed at will by numerical input. In the process of model creation it provides the medium for the creator to have better insight into the significant parameters that are important to the success of the model simulation. DIDEC is also being augmented for surface patch modelling purposes.

Firms that now have access to CAD systems have also been able to adapt them to the generation of wire-grid models for electromagnetic and other simulations. The package called IGUANA[16] is also now available for model creation using a personal computer.

This section has dealt with the computer-aided model creation process only. The questions of model parameters and detail are commented upon in the application sections that follow.

4. CASE STUDY NO.1 - HF ANTENNAS ON LONG RANGE AIRCRAFT

Results of the modelling of the long wire HF antennas on the Canadian CP-140/AURORA aircraft are presented next. This aircraft is a modern version of the P-3C airframe used by many nations.

What is desired from a computer model is that it produce radiation pattern data of sufficient accuracy to be useful for system

performance calculations. As a second objective, it is also useful that the impedance of the antennas be calculated in order to anticipate tuning system requirements, although it is generally recognized that these calculations are less reliable than pattern calculations. With multi-antenna systems, it is also desired that the coupling between antennas be estimated.

For reasons of economy, it is desirable to use a single computer model to cover the 2-30MHz HF frequency band. The first complex model(#3D) that was developed using DIDEC is shown in Fig.3.2 and in Fig. 4.1 with the antennas in place. It consists of 327 segments. The main portion of the fuselage has an octagonal section(i.e. 8 longitudinal wires) with the two upper longitudinal wires arranged to coincide with the lateral location of the feedpoints of the two wire antennas. The segmentation has been selected to have the segment lengths as 0.2 at 30MHz. Thus the patterns would be expected to degrade at the higher frequencies. The M.A.D. boom on the tail is represented by a single wire.

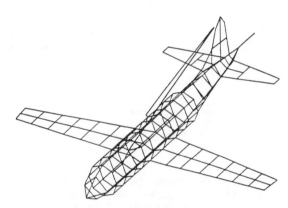

Figure 4.1. CP-140 complex model 3D,
thin wire representation

In this first model, radii have been selected to produce the equivalent surface area in the computer model. In other related studies of model bandwidth[17] it was found that the best choice lies somewhere between a radius for equal area and that for equal perimeter. Segment radii are tapered on the wing and stabilizer surfaces as well as the non-uniform portion of the fuselage in order to minimize radius discontinuities.

For validation purposes, it had been possible to find and obtain a radiation pattern data set from 1/24-th scale measurements that had been made on a reliable outdoor aircraft pattern range. Individual pattern comparisons are made by overlaying computed and measured patterns or by displaying volumetric pattern representations side-by-side. What is expected is that the lobes and nulls be reproduced and also that the relative levels of the two components of polarization be similar.

In comparisons of this type, the analyst is overwhelmed by the mass of pattern information that must be evaluated in order to properly sample the performance of the model on a volumetric basis over the frequency band. Unfortunately, at this time, there is no

possibility to produce current distribution measurements which would be directly comparable to the currents obtained from the matrix solution. In a previous work[18] single number parameters were discussed which facilitate these comparisons. Thus terms such as radiation pattern efficiency and percent E-theta are used. The first is the ratio of the amount of power in a 'useful' sector(+/-30 degrees from the horizon), and the second is the ratio of power in the vertical component to the total radiated power. These parameters are directly related to communications system performance. Percent E-theta values were also available from the experimental data, and these also are compared.

4.1 Radiation Pattern Results

In this project, computations were carried out at 2MHz intervals over the 2-30Mhz band and at other frequencies for special purposes. Of necessity, only typical results are shown as well as those representing the worst agreement with measured results.

Fig. 4.2 shows the principal plane patterns at 2,6,20 and 24 MHz for the port antenna. Note that the scale is linear and that all patterns are plotted with respect to a normalized isotropic reference level so that the relative magnitudes of E-theta and E-phi can also be compared. The execution time for a single frequency was approximately 750 seconds on a CYBER 825.

Fig. 4.3 shows the corresponding results for the starboard antenna. The results at 24 MHz represent the poorest agreement with measured results that was attained.

With complex models it is difficult to do model optimization in a systematic way because of the number of variables that can be perturbed and the difficulty of relating these variations to canonical examples. Some of the results presented here were published in the earlier paper[18] where comparisons with simple stick models are also discussed. Since that time, model refinement was carried out by reducing the radii by 5/12 to have closer agreement with segment guidelines at higher frequencies and tapering the segments at junctions with large steps in radii. In addition the segmentation at the feed-points of the antennas was made more dense. Fig. 4.4 represents the results with the latest model(3J-364 segments) at 24MHz for the port antenna. It is seen that the lobe structure is better defined. The other radiation patterns remained essentially the same.

Judgments of radiation patterns based on principal plane patterns alone are difficult and often inconclusive. It has become standard practice to also use volumetric pattern plots such as those shown in Fig. 4.5 for model 3J at 24MHz. Note that the peaks are fairly well reproduced. In addition to this format, the distorted sphere presentation of Fig. 4.6 now allows a quicker evaluation of shape factors and an easier correlation with spatial angles, at the expense of more processing time.

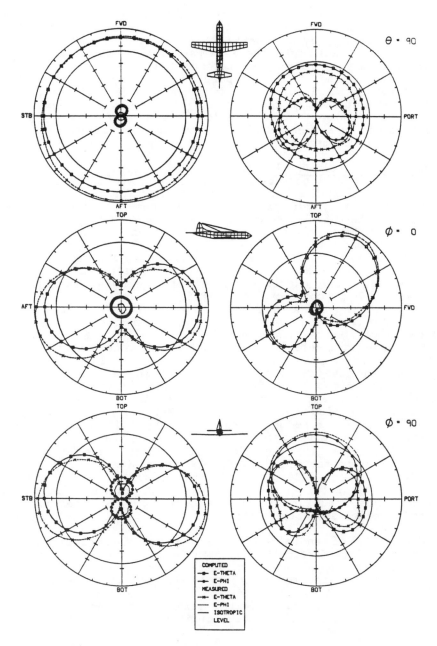

Figure 4.2.a) Comparisons of measured and computed principal plane patterns for model 3D, port antenna, at 2 and 6 MHz – Linear Scale.

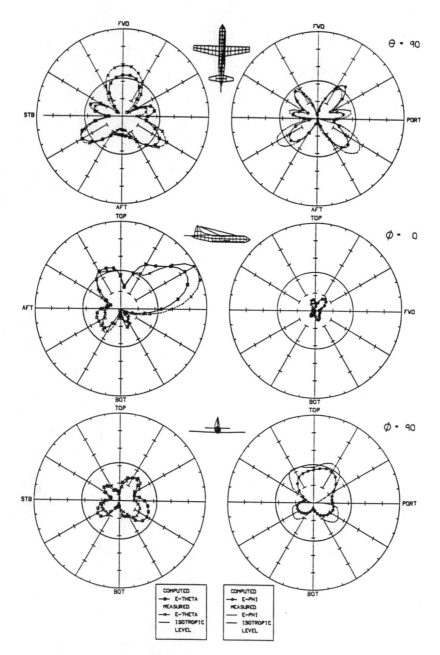

Figure 4.2.b) Comparisons of measured and computed principal plane patterns for model 3D, port antenna, at 20 MHz - Linear Scale.

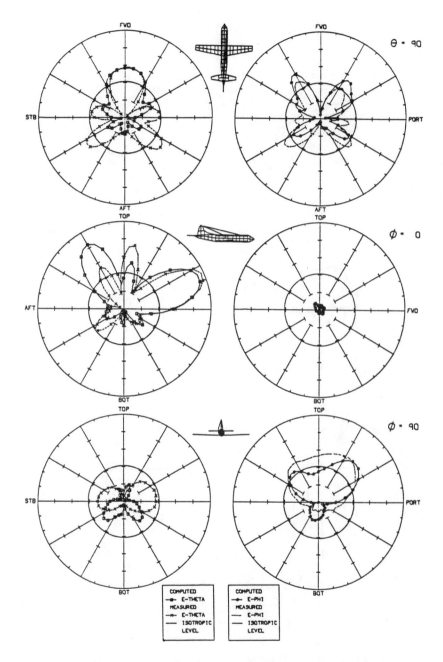

Figure 4.2.c) Comparisons of measured and computed principal plane patterns for model 3D, port antenna, at 24 MHz – Linear Scale.

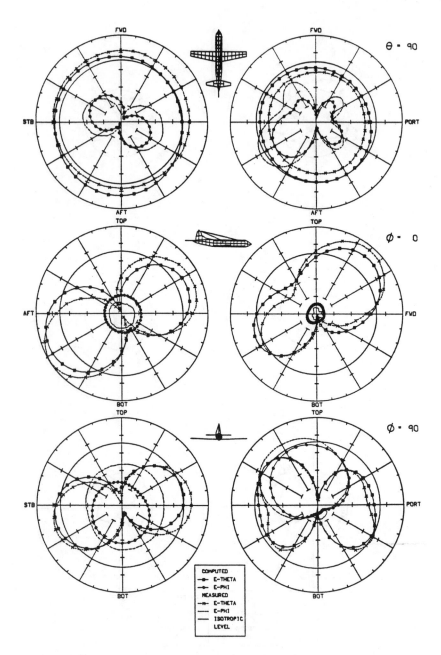

Figure 4.3.a) Comparisons of measured and computed
principal plane patterns for model 3D, starboard
antenna, at 2 and 6 MHz – Linear Scale.

272

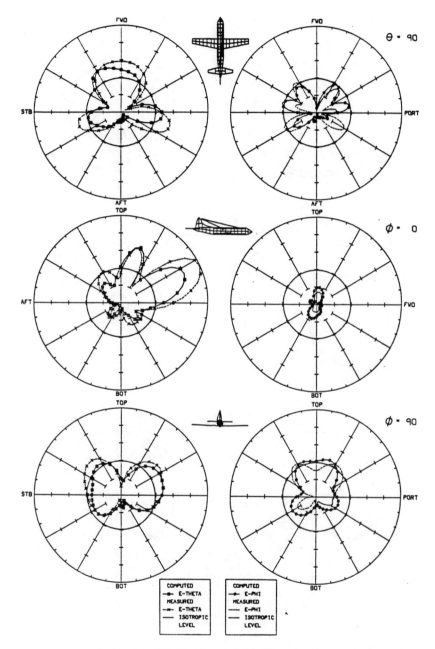

Figure 4.3.b) Comparisons of measured and computed principal plane patterns for model 3D starboard antenna, at 20 MHz - Linear Scale.

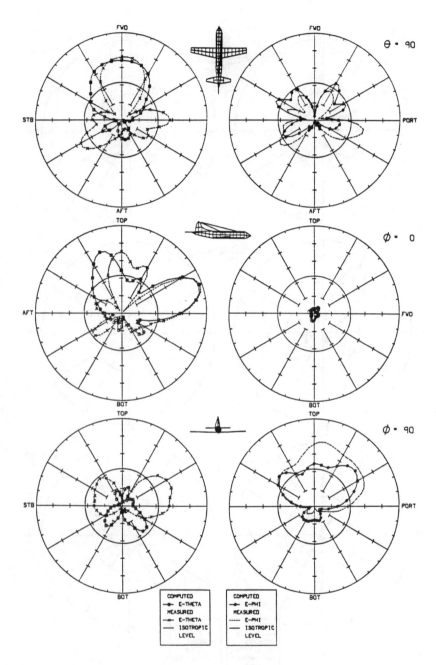

Figure 4.3.c) Comparisons of measured and computed
principal plane patterns for model 3D starboard
antenna, at 24 MHz - Linear Scale.

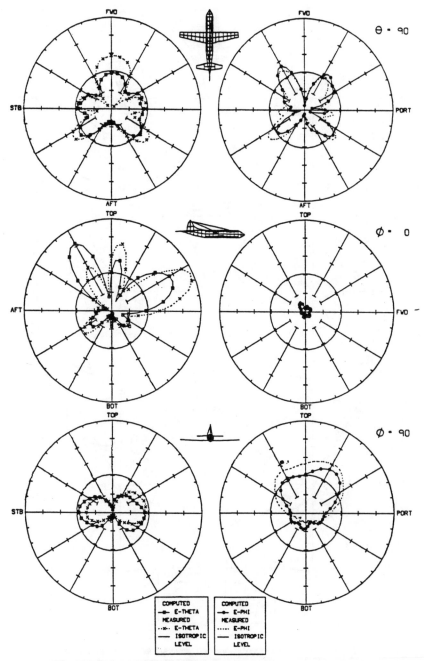

Figure 4.4. Comparisons of measured and computed principal plane patterns of model 3J, port antenna, at 24 MHz - Linear Scale.

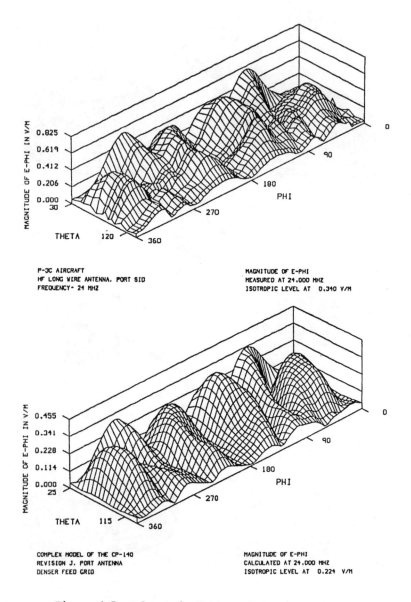

Figure 4.5 Volumetric Pattern Comparisons E_ϕ,
Model 3J - 24 MHz.

276

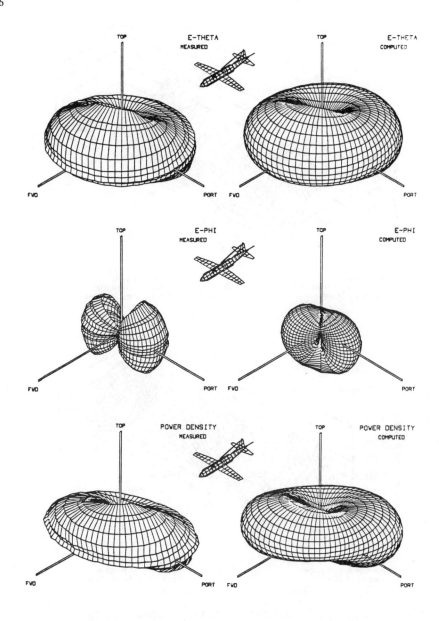

Figure 4.6. MOVIE.BYU plots of radiation patterns
for the CP-140 starboard antenna at 2 MHz, measured
and computed from model 3D - Linear Scale.

The comparison of %E-theta over the frequency band is shown in Fig. 4.7. The agreement is fairly good except at 6MHz where the radiation pattern agreement is inconsistent with this numerical value. The reason for the discrepancy is not known.

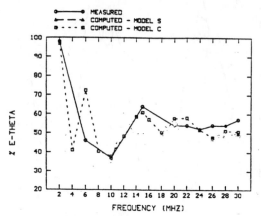

Figure 4.7. Comparisons of measured and computed %E$_\theta$ curves for model 3D port antenna.

4.2 Impedance Values

The impedance values which were computed by NEC for model 3D are shown plotted in Fig. 4.8 in comparison with measured values on the full-scale aircraft. The exact measurement conditions are not known. The impedance variations are typical of an open-circuited transmission line, but the disagreement between measured and computed values is substantial. This discrepancy gave rise to many verifications that the effective lengths of antennas on the model and on the actual aircraft were the same. No discrepancy was found. At the same time, any reasonable model refinement did not produce any radical changes in computed impedance values.

Computations of impedance have received more attention in the recent literature[19] mainly through studies of model convergence and of the method of impedance calculation itself. In this study it was important to address the essential question as to what the NEC code actually calculates for the wire-grid model and what its relationship is to the physical structure. It became apparent that the wire-grid model does not represent the shunting capacitance at the base of the feed-through insulator that is used on the aircraft. For the measurements on the actual aircraft it was also not clear whether the shunting lightning arrestor was in place. Based on these considerations, calculations at finer frequency steps were made over one frequency region and the calculated impedance values were shunted with discrete values of capacitance. Fig. 4.9 illustrates the results of these corrections. Thus a shunting capacitance of 45pF produces values in reasonable agreement with the measured values. A portion of this shunting value can be accounted for by the measured base capacitance of the feed-through insulator and the lightning arrestor.

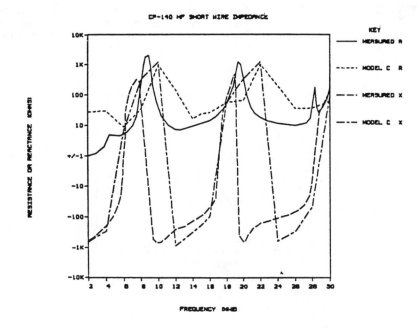

Figure 4.8. Comparisons of measured and computed impedance curves for model 3D, starboard antenna.

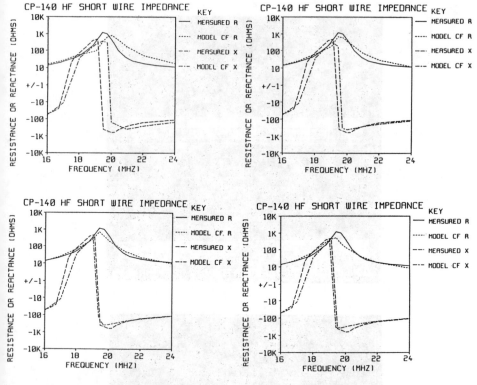

Figure 4.9. Impedance curves for starboard antenna, 16 to 24 MHz. Comparison of measurements with model 3J shunted with 30, 40, 45, and 50 pF.

Other investigators[20] have applied a similar approach to calculations of terminal impedance by moment methods.

4.3 Other Considerations

The examination of current distributions on the aircraft model is meaningful in establishing a relationship between the patterns and the currents in different portions of the aircraft. Oftentimes such analysis provides useful clues to model refinement and optimization. It is especially useful in the appreciation of the coupling between the two antennas. Although DIDEC allows the colour coding of segments in proportion to the solution currents, this process is now better done by a new computer code called SPECTRUM[21]. An example of such a display is shown in Fig. 4.10. The first illustration shows the presence of "non-physical" currents in the solution and the second the coupling from the active to the passive antenna. Note the port antenna is excited. For the sake of brevity, further discussion of these subtle but important aspects is left to the oral presentation.

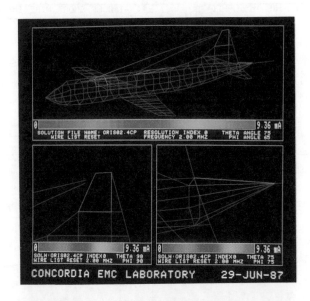

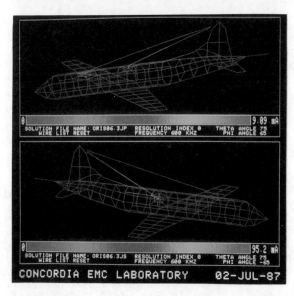

Figure 4.10.a)   Example of non-physical currents.
          b)   Antenna - Antenna Coupling at 6
MHz: port antenna excited (top frame), starboard
excited (bottom frame).

The results show that with care and experimentation, reasonable models can be designed to have considerable bandwidth for pattern computation purposes. Impedance calculations must be analyzed in relationship to the actual measured values and suitable corrections incorporated. Some details of model refinement depend very much on knowing the exact circumstances of the measurements that are being used for validation purposes. Some of these aspects are addressed further in the examples that follow.

5. FULL-SCALE VS. SCALE-MODEL MEASUREMENTS - AN EXAMPLE

The appreciation of the extent of the agreement that one can expect between full-scale, scale model and computer modelling results for complex structures is important in all applications. Special care for obtaining precise results with each of these techniques is required. Often it is not always obvious what exact feature is required in order to replicate full-scale results with scale model results and vice-versa. Examples of these considerations can be found in the recent article by Belrose[22] and in publications originating at RAE and SWRE[23]. In this paper some limited results are reported from a recent measurement on a full-scale hydro utility vehicle and a 1/7-th scale model of the same vehicle at the Canadian NRCC Ship Range. This latter range is described by Belrose[22].

The full scale vehicle and its 1/7-th scale counterpart are shown in the photograph of Figure 5.1. Note the VHF monopole on the rear bumper. Note the large turntable on which the models are mounted. The signal was received by a corner reflector and monopole in the azimuth plane at ground level and by a dipole on a traversing plastic boom for the elevation plane patterns.

Figure 5.1. Full Scale Car and 1/7-th scale model on Turntable

5.1 Pattern Variations with Position

The monole was tested at 5 distinct positions on the rear bumper. These were at the extreme driver side edge, 6 inches and 12 inches towards the center. Two additional variations of the middle positions were measured with different spacing to the trunk edge. The pattern measurements were carried out at 140MHz on the full-scale model and at 980MHz on the 1/7-th scale model.

Fig. 5.2 shows the azimuth plane patterns for the three positions of the monopole at the same distance from the trunk wall. These were taken on the full-scale vehicle. They show the extent of the variations in terms of null and lobe displacements and amplitude variations at any given heading. A decibel scale is used in these plots. It can be seen that there are several regions where the variations are of the order of 3dB with maximum variations of 10dB.

The value of these measurements is that they produce an appreciation of the differences in pattern coverage with position which is necessary in choosing the best location for the system application, and that indirectly they indicate the care which should be taken in positioning the monopole on the 1/7-th scale model.

5.2 Full-Scale vs. 1/7th Scale Results

Both full-scale and scale-model measurement results are available for the two extreme positions of the monopole on the bumper. Fig.5.3 shows the elevation plane patterns for the monopole twelve inches from the driver side. Although there is general agreement in the existence and location of the lobes, there is considerable difference in the detail structure of the patterns. Note also the differences in the pattern smoothness that are indicative of the increased scattering from the range structures at the 980MHz frequency.

The full-scale vehicle and the model differed in some small detail. For example the tail light assembly was not replicated in the 1/7-th scale model. To have the structures more similar, a metallic conducting tape was used to cover the tail lights on the conductor side of the vehicle. The curve of Fig.5.3 for this condition, shows the effect on the vertical plane pattern. It should be pointed out that the extent of the agreement would also be affected by the alignment of the vehicle for both measurements. In this case, marks on the turntable were used to re-establish the position of the model from day-to-day. The axial alignment of both vehicles was estimated to be within two degrees.

The corresponding azimuth plane patterns are shown in Fig.5.4 for both full-scale and scale-model vehicles. Although the extent of the agreement is good, there are 15dB differences in the null at 350 degrees. The repeatability of these measurements was estimated to be about 0.5dB.

Fig.5.5 shows the azimuth plane patterns for both scale-model and full-scale vehicles for the position of the monopole near the edge on the conductor side. Note the effect of placing the aluminum tape on the tail-light assembly is more pronounced in this position due to proximity of the tail light assembly.

Other measurements were taken on this project, such as the measurement of the patterns with and without the ladder on top of the vehicle. The exact location of the ladder on each of the vehicles was also important to pattern repeatability.

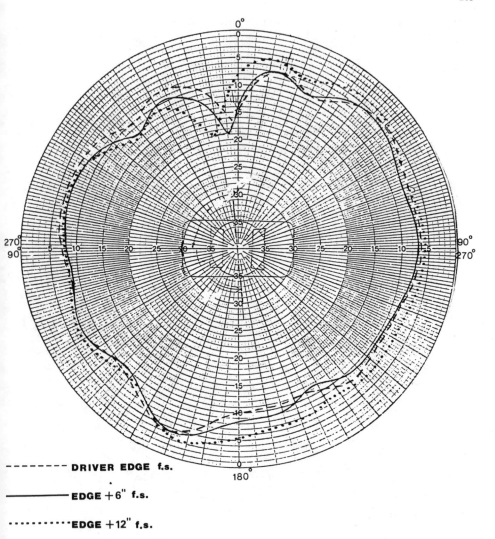

DRIVER EDGE f.s.

EDGE + 6" f.s.

EDGE + 12" f.s.

Figure 5.2.   Pattern Variation with Position on Car Bumper

284

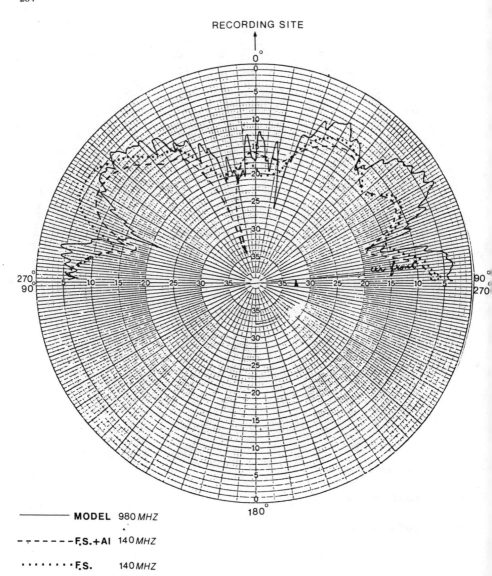

RECORDING SITE

————— **MODEL** 980 *MHZ*

– – – – – – **F.S.+AI** 140 *MHZ*

• • • • • • • • **F.S.** 140 *MHZ*

Figure 5.3. Elevation Plane Pattern Full Scale vs 1/7-Scale Model – Car & Monopole (Edge + 12")

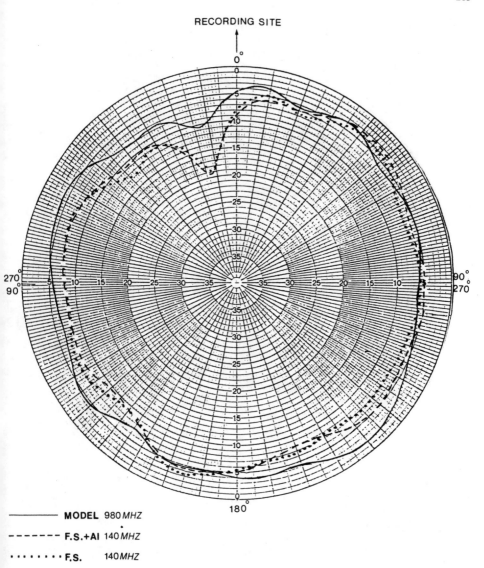

RECORDING SITE

Figure 5.4. Azimuth Plane Pattern: Full Scale vs 1/7-Scale
Model - Car & Monopole (Edge + 12")

286

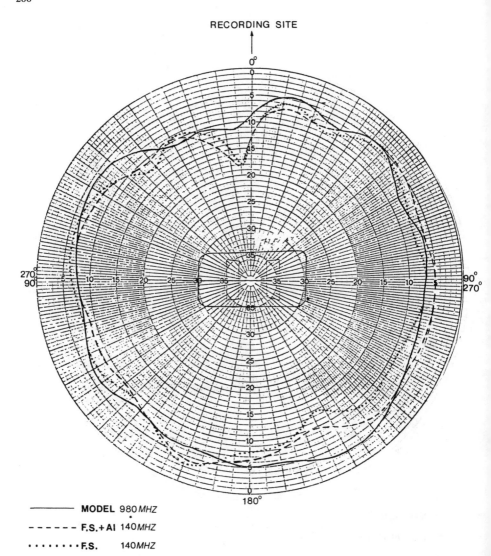

Figure 5.5. Azimuth Plane Pattern: Full-Scale vs 1/7-Scale
– Car + Monopole (Edge + 6")

5.3 Implications of the Results and Discussion

The exact correspondence between full-scale and scale-model results was not the dominant objective of this particular project. However, once the comparisons were made, the differences between measurement conditions were more carefully noted. For example, it was observed that the scale model had a round filet between the rear wall and the the fender whereas the actual car had more of a perpendicular abuttment. Once the variation of pattern with position was measured it was evident that the spacing and orientation of the monopole on the scale model would have to be carefully observed.

These comparisons emphasize that good correlation between full-scale and scale model measurements can be obtained. To achieve the best possible agreement between the two, care must be taken to identify the critical areas of shape or spacing or conductivity which are likely to have a dominant effect on the results. It will be seen that since computer models provide an appreciation of the current distribution on the complex body, they are an excellent means for obtaining the perception of critical relationships.

The results also illustrate that it best to know all the details of specific measurements when they are used for validation purposes so that one can establish the extent of any variations that are likely to exist rather than assume them to apply on an absolute level. The next section will illustrate another aspect of this relationship between measurements and computer models.

6. CASE No.2 - CH-135 TWIN HUEY HF ANTENNAS

Information on the comparative evaluation of HF antennas on aircraft and helicopters is somewhat scattered in the literature and incomplete. The systematic evaluation of aircraft HF antennas was first carried out at the Stanford Research Institute [24]. A very compact evaluation of notch, wire and cap antennas is to be found in the Transactions on Aeronautical and Navigational Electronics[25]. Valid Method of Moments modelling via NEC offers the possibility of re-examining the performance of these antennas.

In recent ACES Review Meetings[26], the modelling of HF antennas on helicopters was discussed in terms of the model creation process using DIDEC model design software and an integrated system of interactive graphics for the display and manipulation of input/output files. The present paper presents some of the results that have been obtained in the modelling of shorted transmission line(STL) and 'Zig-Zag' antennas on the CH-135/TWIN-HUEY helicopter. Figure 6.1 shows the wire-grid models and the antennas. The name of the "Zig-Zag' antenna is evident from the nature of its path around the helicopter body in order to achieve the required length for matching over the HF 2-30MHz band by commercially-available couplers.

The results of the wire-grid model analysis are presented over the entire 2-30MHz frequency range. These pattern results are also compared to initial 1/24-scale model measurements. Some serious discrepancies between measured and computed results are examined and reconciled. The process demonstrates the utility and power of a combined computation/measurement approach for complex applications.

The computer model analysis is then extended to utilize the ground imaging features within NEC to examine the radiation pattern variations with altitude. Such information is most useful in analyzing antenna pattern contributions to both short and long range communication scenarios.

There is one very important feature of computer modelling that is best illustrated from other work[18]. This is the feature called model "tuning". The following section describes this process from the example of the work on the CHSS-2/SEAKING helicopter.

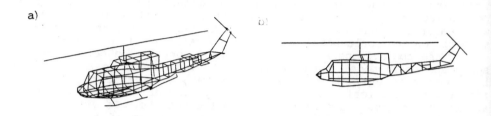

Figure 6.1. Wire-Grid Models and HF Antennas
a) Wire-Grid Model - Shorted Transmission
Line Antenna (STL)

b) Wire-Grid Model - "Zig-Zag" Antenna

6.1 "Tuning" of a Complex Model - CHSS-2/SEAKING

This example is discussed in previous papers[18,27,28] that describe the development of a complex model for the SEAKING and compare the results with scale model measurement data.

Use was made of the complex model shown in Fig.6.2 which was carefully derived at RAE Farnborough[23] for the study of HF wire antennas and loop antennas. It consists of 335 segments. All lengths are less than one tenth of a wavelength at 20MHz. A uniform radius of 0.1m is used for all wires except that the antenna wire radius is 0.01m. The results below, are for the shorted transmission line antenna.

This model was analyzed at frequencies between 2.6 and 14Mhz. The results for %-Etheta vs. frequency are shown in Fig.6.3. It can be seen that model and measurement results agree very well except at the 8.1MHz measured frequency. This had turned out to be the frequency of maximum rotor modulation effect. The corresponding azimuth radiation pattern is shown in Fig. 6.4a) where a pronounced shift in the four-lobed E-phi pattern is evident.

Two distinct investigations were carried out in order to better understand this discrepancy. In order to make sure that the solution was not contaminated by currents associated with internal resonance modes, the extended boundary condition was applied. No significant change in the solution values was obtained. Subsequently when the pattern was computed at nearby frequencies, it was found that best agreement was obtained at 7.8MHz., indicating a longer resonant path on the computer model than on the measurement model.

A corresponding adjustment in resonant path length was most readily carried out by shortening the rotor blades by 0.4m to achieve resonance at 8.1MHz. This resonance was verified by examining the current distribution along the blades and fuselage. Fig. 6.4b) shows the azimuth radiation pattern for the adjusted geometry. Note the agreement in pattern levels and orientation. The new computed %E-theta also rose to within 7% of the measured value.

This example illustrates the importance of resonant path lengths in model topology and the necessity to "tune" the model to achieve these path resonances when the model is used over bandwidths when they are likely to be excited in the application that is being analyzed.

6.2 Radiation Pattern Comparisons - Two Antennas/CH-135
Helicopter

The wire-grid model that is being used in these comparisons was developed using the considerations that had been used for aircraft antennas. Model segment lengths were selected to be approx. 0.1 in the middle of the band(15MHz). The models of Figure 6.1 have 302 and 305 segments respectively. The STL antenna is approximately 1 ft. from the fuselage and is 10 feet long. The ZIG-ZAG antenna uses the same feed point and loops around the fuselage for a total length of 46.3 feet.

Figure 6.5 shows the principal plane results for the two antennas at 2 MHz. Note that these patterns are plotted with reference to the isotropic reference level. This latter level is derived from an integration of the radiation pattern data, a procedure which is exactly the same as that used with the experimental data. Although the results are similar, there is some significant difference in the level of the E-phi component. The nature of the two antennas accounts for this but the analyst can examine the current distributions computed by NEC in order to appreciate the exact reasons for the pattern differences.

It can be seen that communication at this frequency would be very much dependent on helicopter attitude. For a full operational analysis it is necessary to examine the volumetric radiation pattern characteristics.

Again such computations were carried out at 2 MHz increments over the entire frequency band. The radiation pattern efficiency for the two antennas versus frequency is shown in Figure 6.6. Note the difference between the two antennas at the higher frequencies. Such differences would have to be examined for their operational consequences.

For ground wave propagation, %E-theta is important. Fig.6.6b) plots this parameter versus frequency. Note the ZIG-ZAG antenna has higher values at 2 MHz. Should differences such as this be considered operationally significant, the coupler efficiency must be examined in order to establish that coupler efficiency differences will not cancel any pattern distribution advantages.

Having concentrated on the parameter values, it is always important to refer back to the actual pattern shapes for an appreciation of heading and attitude differences. For example, examining the plots of Figure 6.7 for the two antennas at 26MHz, produces immediately the appreciation as to why the parameter values are so different. Thus for any special mission applications, the parameters can be weighted based on the pattern detail that is of

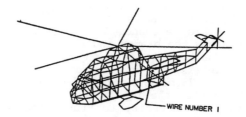

Figure 6.2.   Complex Model of CHSS-2/Sea King Helicopter

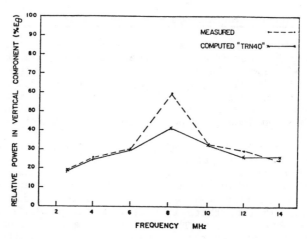

Figure 6.3.   Tranline Antenna, % E_θ vs Frequency Complex Model

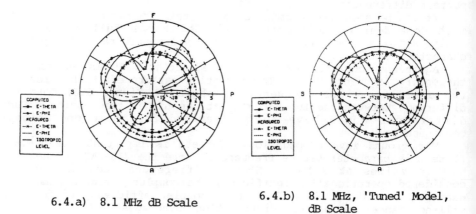

6.4.a)   8.1 MHz dB Scale

6.4.b)   8.1 MHz, 'Tuned' Model, dB Scale

Figure 6.4. Azimuth Patterns - Complex Model

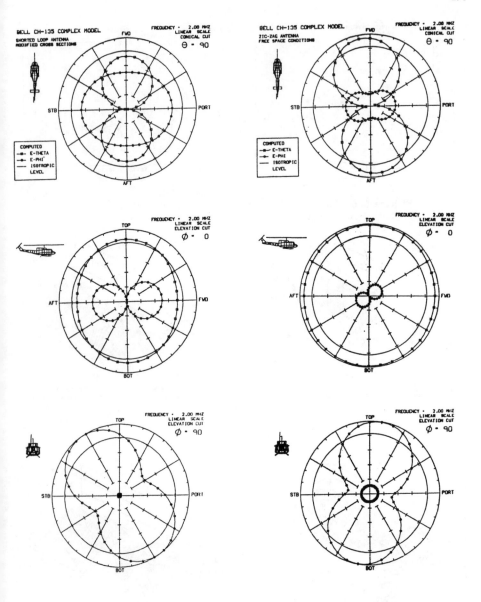

Figure 6.5. Principal Plane Patterns - 2 MHz, Linear Scale

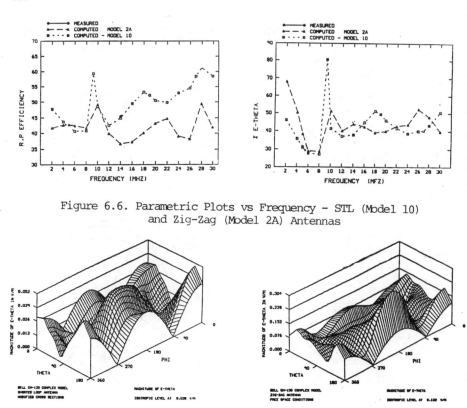

Figure 6.6. Parametric Plots vs Frequency - STL (Model 10)
and Zig-Zag (Model 2A) Antennas

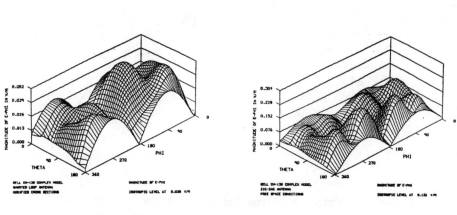

Figure 6.7. Volumetric Patterns - 26 MHz

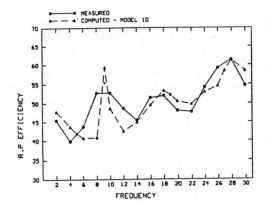

Figure 6.9. Measured and Computed η_p
vs Frequency STL Antenna

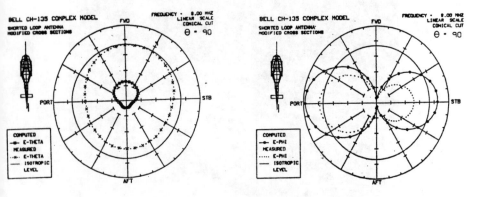

Figure 6.10. Computed vs Measured Results – 8 MHz, E_θ, E_θ, E_ϕ
$(\theta = 90°)$ – Linear Scale

294

consequence.

The ease with which the complete pattern data can be generated from valid NEC models makes this procedure a valuable adjunct to HF system performance analysis. It remains to establish that the computed patterns are valid.

Scale-model radiation pattern measurements were undertaken at the Canadian National Research Council aircraft roof range for comparison and validation purposes[29] and the initial results are summarized.

6.3 Comparison with Measured Results

A copper-plated aluminum 1/24-scale model of the CH-135/TWIN-HUEY helicopter was constructed by the National Research Council workshops. It is shown in Figure 6.8 mounted on the model support tower.

Figure 6.9 plots the radiation pattern efficiency versus frequency for computed and measured results. It can be seen that there is wide disagreement over the 6-14MHz frequency range. The radiation patterns of Fig.6.10 corroborate this disagreement and show that the disagreement is more pronounced in the E-theta polarization.

A discrepancy of this magnitude poses a most serious dilemma for computer modelling and gives rise to the plethora of doubts about model parameters, segmentation, band-width, interior body resonances, etc. The extent of the disagreement must be methodically identified. Several model stability tests were undertaken by increasing the

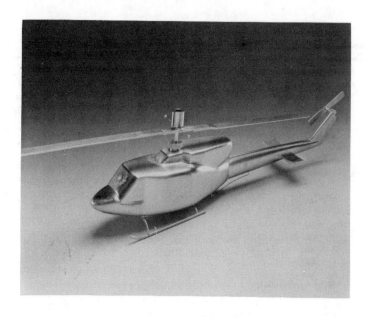

Figure 6.8. 1/24-th Scale Model of CH-135 Helicopter

density of segmentation in selected regions of the helicopter. None produced meaningful results. The field at several points in the interior of the helicopter was computed and found to be very close to zero. In addition the extended boundary condition was incorporated by introducing an interior wire across the fuselage. No significant change was computed.

Measurements at low frequencies (here 48MHz and up) usually give rise to a number of range problems involving ground and site reflections. Self-consistency tests are often an indication of this. This involves verifying that the principal plane cross-over points which involve measurements for different orientations of the measurement antenna, i.e. horizontal or vertical, agree for each of the polarizations. In this case, although some discrepancies were identified, they were insufficient to account for the extent of the disagreement. Rotor modulation studies had shown that the rotor position would have a pronounced effect on radiation patterns at frequencies near 8 MHz., however verification of rotor position during the measurement established that this was not a factor in this disagreement.

After exhausting all other possibilities and examining current distributions on the model, a vertical wire was added to the model at the model support point on the assumption that the model support rod might have been metallic. Metallic rods are often tolerated for heavy models at higher frequencies where antennas on the opposite side of the aircraft model are being measured and the rods are not likely to perturb the measurement conditions.

Figure 6.11 shows the principal plane pattern results. The agreement is very much improved. Verification of the measurement conditions confirmed that indeed the rod was metallic. It is believed that using an exact length in the computer model would have improved the agreement further. An examination of the current distributions has indicated that had the rod been used on the bottom of the model, the difference would have been smaller and the consequent more subtle differences would have been extremely difficult to diagnose.

This special example of difficulties has underlined the importance of simultaneous computational and measurement modelling in exploring details of complex body behaviour. In this project, for the sake of completenesss, the radiation pattern measurements are being repeated with a phenolic shaft.

Having established the validity of the model, it is possible to compute other operationally-significant configurations such as the variation of patterns with helicopter altitude.

6.4 Radiation Patterns vs. Altitude

Helicopters such as the CH-135 are used on many missions where the modes of operation of the HF communications system range from line-of-sight, ground wave to sky-wave including high vertical incidence ionospheric propagation. These modes can be exercised while the aircraft is on the ground, hovering at low altitude, in transit altitudes or manoevring at specific nominal altitudes. To identify the antenna contributions to these communications modes, it is necessary to know the radiation patterns under these conditions.

It has been normal practice to measure and compute radiation

patterns under free-space conditions alone. Although limited ful
scale measurements have been carried out, they seldom include th
vertical plane distribution. It would be possible to carry out suc
a measurement with scale models on a ship-range where booms allo
vertical plane probing. However there is a limitation to the rang
of altitudes that can be modelled.

The modelling of the ground plane in NEC allows the examinatio
of the radiation patterns versus aircraft altitude. Suc
computations were carried for the STL antenna at 5 MHz with NEC fo
heigths of 1 metre,15,30,100 and 300 metres. The results ar
presented in two different formats, each of which brings out specifi
features of the resultant patterns.

Figures 6.12 and 6.13 show the vertical plane patterns. Here th
formation of the lobe structure is clearly evident and instructive
It now allows the immediate identification of signal level at variou
elevation angles for ionospheric propagation purposes.

The three dimensional plots, shown in Figure 6.14 allow th
complete conical set to be viewed and provide the missing informatio
with regard to variations with elevation angle at various azimut
angles. The nature of the pattern displays make the results easy t
interpret by communications operators.

6.5 Concluding Remarks

The NEC computer models allow a complete evaluation to be made o
the performance of two different antenna types on the CH-13
helicopter. These models have been compared to initial measurement
with 1/24-scale model measurements and the discrepancies have bee
reconciled in terms of the specific measurement conditions. The us
of the models to examine the change in radiation patterns wit
altitude permits a realistic analysis of HF system performance unde
a variety of flight profiles. Corresponding patterns unde
conditions of different ground conductivities can be readil
computed.

The final case study summarizes some early experience with th
modelling of ship antennas.

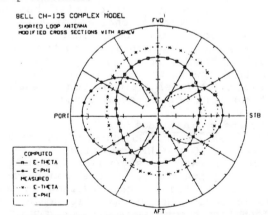

Figure 6.11. Computed vs Measured Results
8 MHz, $\theta = 90°$, Adjusted Model, Linear Scale

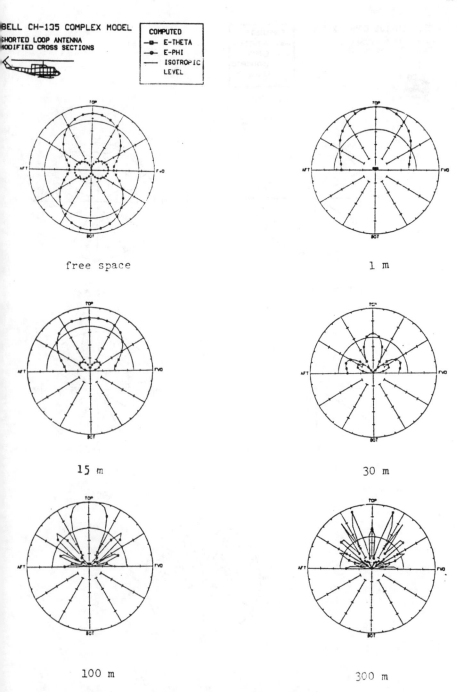

Figure 6.12. Bell CH-135 Complex Model - Shorted Loop Antenna
Pattern Variation with Altitude, $\phi = 0$, 5 MHz - Linear Scale

298

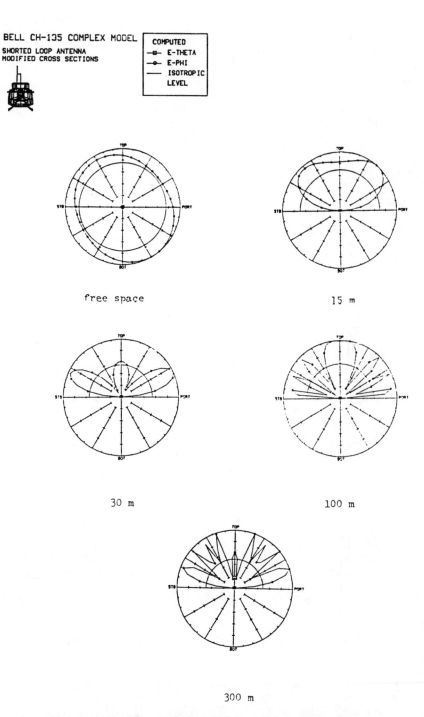

Figure 6.13. Bell CH-135 Complex Model - Shorted Loop Antenna
Pattern Variation with Altitude, $\phi = 90°$, 5 MHz - Linear Scale

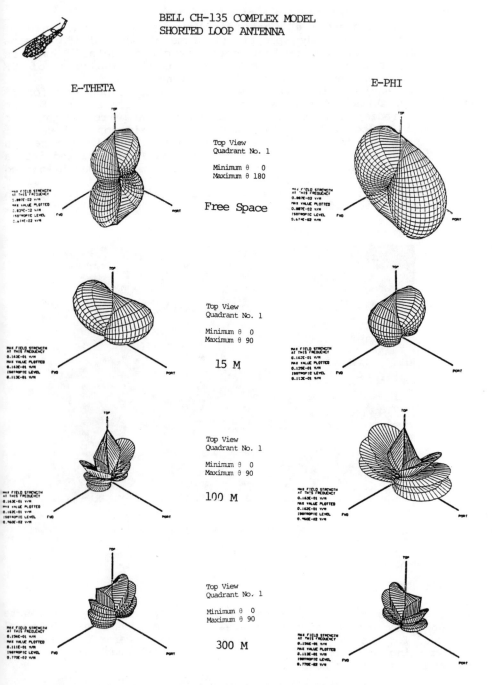

BELL CH-135 COMPLEX MODEL
SHORTED LOOP ANTENNA

E-THETA

E-PHI

Top View
Quadrant No. 1

Minimum θ   0
Maximum θ 180

Free Space

Top View
Quadrant No. 1

Minimum θ   0
Maximum θ 90

15 M

Top View
Quadrant No. 1

Minimum θ   0
Maximum θ 90

100 M

Top View
Quadrant No. 1

Minimum θ   0
Maximum θ 90

300 M

Figure 6.14. Three-Dimensional Patterns (TDPAT) vs Altitude
5 MHz – Linear Scale

7. CASE No.3 - HF ANTENNAS ON A PATROL FRIGATE

The EMC Laboratory at Concordia University has made use of moment methods to study specific problems of UHF and IFF antenna pattern distortion by localized ship structural details such as masts[21] and radar scanners and for the analysis of HF communications antennas such as whips and fans.

Comprehensive data sets of measured radiation patterns for HF antennas on ships are not easy to find in the literature. Some samples of radiation patterns of HF antennas on a patrol frigate are provided by Belrose[22]. After sustained checking of archives, it became possible to obtain the latter data as a digitized set in punched paper tape format. This file was read and translated into a format suitable for the pattern display software that has been used in the two other case studies.

A photograph of the ship scale model is shown in Figure 7.1. An initial wire-grid model was developed for use in the 2-10Mhz range.

Figure 7.1. 1/48-th Scale Model of Ship

It is shown in Fig.7.4. It consists of 472 segments whose radii were chosen to produce the equivalent area of the ship sections. The model was used for the computation of the radiation patterns of one of the transmitting whip antennas first at 2,5 and 10MHz and later for small selected frequencies nearby.

7.1 Comparison of Computed and Measured Radiation Patterns

The computed and measured principal plane radiation patterns are shown in Fig.7.2. Although there is general agreement in the pattern

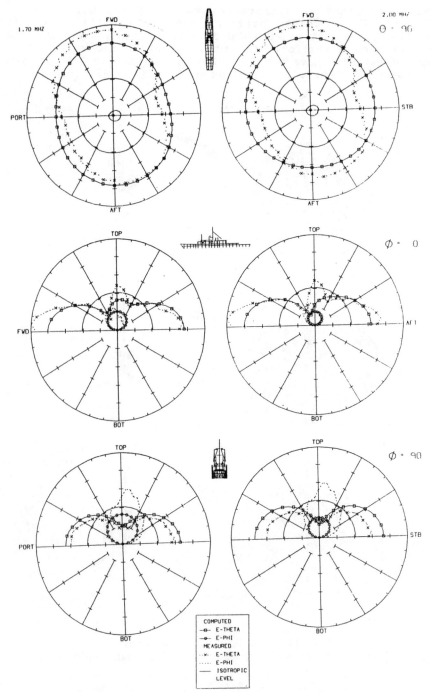

Figure 7.2. Principal Plane Patterns:
Measured (2 MHz) vs Computed (2 MHz & 1.7 MHz) — Model 5A

shapes, substantial differences exist in the elevation plane
patterns. The computed patterns show a slightly assymmetrical
elevation plane pattern with a deep null overhead, while the measured
patterns show a vertical lobe with a peak in the region where the
computed patterns have a null. The counterpart volumetric patterns
are shown in Fig.7.3. When examining such differences and the

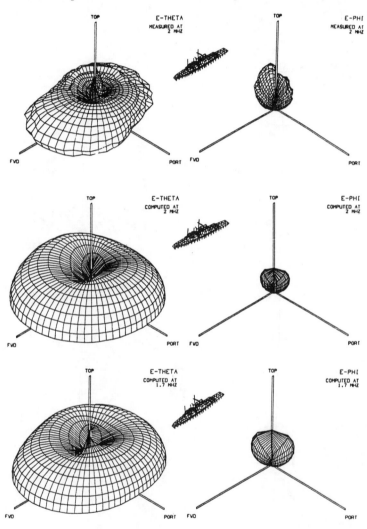

Figure 7.3. Volumetric Patterns:
Measured (2 MHz) vs Computed (2 MHz & 1.7 MHz) - Model 5A

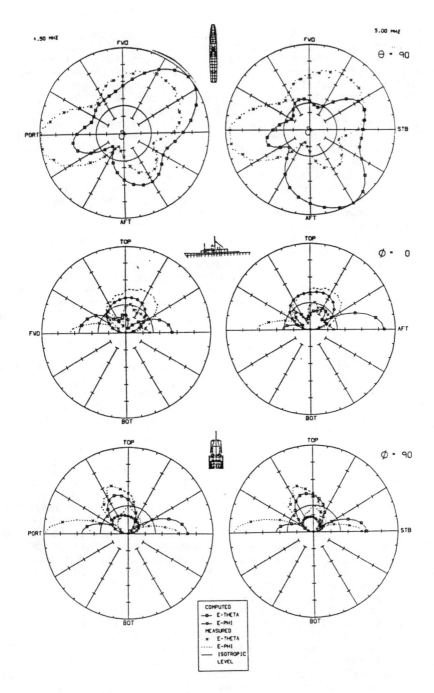

Figure 7.4. Principal Plane Patterns:
Measured (5 MHz) vs Computed (5.0 & 4.5 MHz) - Model 5A

details of the model, it is not readily obvious what can be done to resolve the differences. Since these measurements were carried out two decades ago, there is no possibility for review of measurement details or of re-measurement in order to corroborate that the measured pattern artifacts in the vertical plane are genuine.

In this case, the frequency was perturbed at 0.1MHz intervals in order to identify whether some critical path conditions might exist. It was found that at 1.7MHz the pattern agreement was superior insofar as the patterns begin to display the vertical plane artifact that is part of the measured results. The 1.7MHz results are shown in both figures.

Corresponding results for 5MHz are shown in Figures 7.4 and 7.5. This latter figure shows the results at 4.5 and 4.8MHz to illustrate that in this case the results did not have the strong frequency dependence shown in the helicopter case study.

Computations were also done at 9.8 and at 10MHz. These are shown in Figures 7.6 and 7.7. It is difficult to conclude whether the frequency change produces a meaningful difference.

The model was changed to produce a denser segmentation near the base of the whip antennas. This resulted in 509 segments. No substantial difference in radiation pattern results was obtained. Also reducing the radius of the wire segments in the same manner as in the aircraft case produced only minor changes in the computed patterns.

7.2 General Commentary

There are many parameters to vary in complex models. It becomes a very expensive exercise to undertake extensive experimentation without a clear indication of major trends towards convergence to the measured values. At the same time, there is no systematic analysis of the coupling of antennas to ship structures which could be used to guide the model optimization process, similar to that produced by Granger and Bolljahn[24] for aircraft.

To some extent, the computer model results themselves can be exploited in the abscence of other real-world information. The computed current distribution can offer some insights into the reasons for the dominant features of the radiation pattern. For example, the SPECTRUM display of Figure 7.8 shows considerable current on the center mast structure. It would be logical to assume that the path between the whip antenna and this structure could be critical and might need adjustment. However, a separate diagnosis of radiation patterns of selected segments of the ship structure shows that at 1.7MHz it is the smaller currents on the top of the hull and deck that produce the specific elevation pattern features that are not being properly matched.

Experimentation with more complex models of approximately 1500 segments is being continued in order to better understand the total behaviour of these large, complex and angular structures.

It would be most appropriate if a library of other relevant measurement and modelling data could be created and shared by the NATO technical community in order to achieve a more profound understanding of these challenging cases.

305

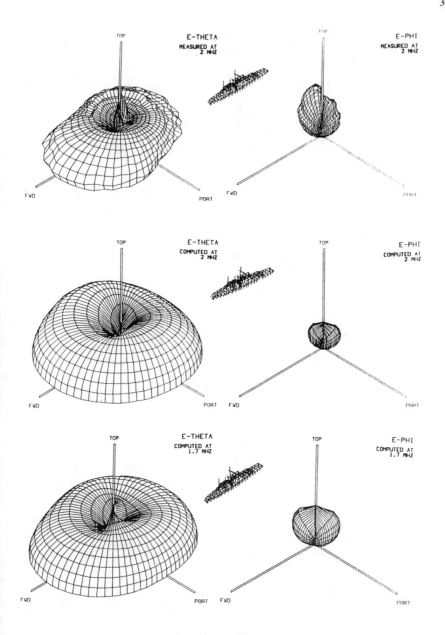

Figure 7.5. Volumetric Patterns:
Measured (5 MHz) vs Computed (4.5, 4.8 & 5.0 MHz)- Model 5A

306

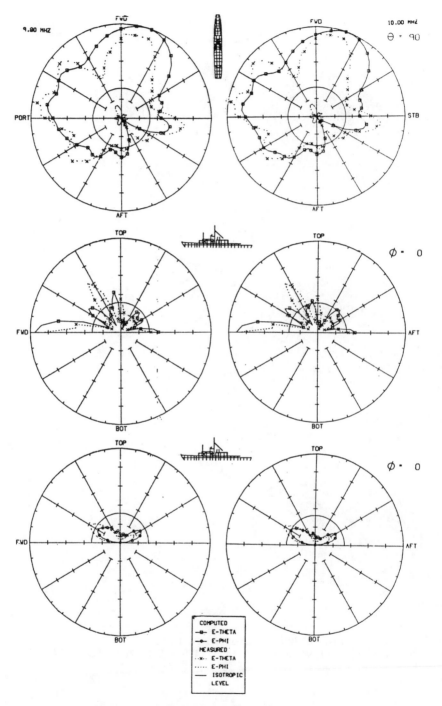

Figure 7.6. Principal Plane Patterns:
Measured (10 MHz) vs Computed (10.0 and 9.8 MHz) – Model 5A

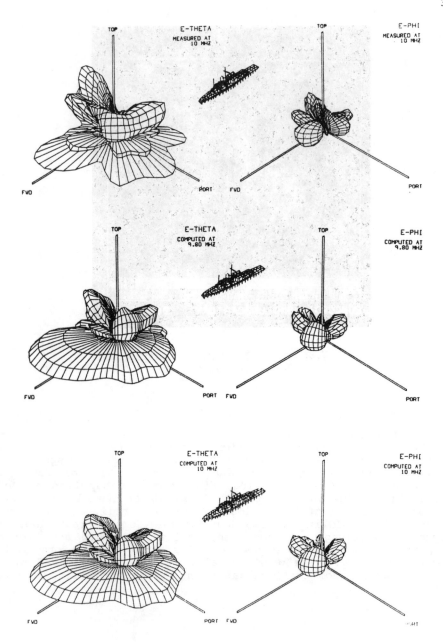

Figure 7.7. Volumetric Patterns:
Measured (10 MHz) vs Computed (10.0 & 9.8 MHz) - Model 5A

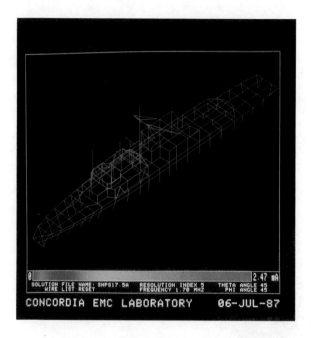

Figure 7.8. Colour-Coded Current Distribution
on Ship Model 5A, 1.7 MHz

8. ACKNOWLEDGMENTS

The work for the case studies was primarily supported by the Defence Research Establishment Ottawa (D.R.E.O.) under the sponsorship of Drs. P. Bhartia and David Liang and Mr. R. Kirk of DND. The vehicle and helicopter modelling information was supplied by John Hazell, Dr. S. Mishra and Randolph Balaberda of the Electromagnetic Engineering Section of the National Research Council. Dr. E. Cerny, David Gaudine, Colin Larose, Lan Nguyen, Sergio Prestipino, Duy Nguyen, Vito Salvagio, Victor Phan, Benita Rozensweig and Michael Wong of the EMC Laboratory were enthusiastic contributors to the research activity. The work was also partially supported by an operating grant from the National Sciences and Engineering Research Council.

9. REFERENCES

1. J. K. Skwirzynski,(Ed.), "Theoretical Methods for Determining the Interaction of Electromagnetic Waves with Structures", Proceeding of 1979 NATO ASI, Norwich U.K., Sijthoff & Noordhoff 1981.
2. Symposium Digest, 1987 IEEE Antennas and Propagation Society International Symposium and URSI Radio Science Meeting, Blacksburg, VA, June 15-19, 1987.
3. J. Moore and R. Pizer, "Moment Methods in Electromagnetics- Techniques and Applications", Research Studies Press Ltd.(Wiley), 1984.
4. G.J. Burke, A.J. Poggio, J.C. Logan and J.W. Rockway, "NEC - Numerical Electromagnetics Code for Antennas and Scattering", 1979 IEEE International Symposium on Antennas and Propagation Digest, IEEE Publication No. 79CH1456-3AP, Seattle, Washington, June, 1979.
5. Tapan K. Sarkar, " The Conjugate Gradient Method as Applied to Electromagnetic Field Problems", NATO Advanced Study Institute: "Electromagnetic Modelling for Analysis and Synthesis Problems", Tuscany, Italy, August 10-21, 1987.
6. Edmund K. Miller, " Model-Based Parameter-Estimation Applications in Electromagnetics", NATO Advanced Study Institute: "Electromagnetic Modelling for Analysis and Synthesis Problems", Tuscany, Italy, August 10-21, 1987.
7. The Applied Computational Electromagnetics Society, Dr. Richard K. Adler, Secretary, Naval Postgraduate School, Code 62AB, Monterey, CA, 93943 USA.
8. P. Bhartia, S. J. Kubina, E. Cerny and D. Gaudine, "Generation of Computer Models for the EMC Analysis and Design of Large Systems". Proceedings of the 1984 International Symposium on Electromagnetic Compatibility, Tokyo, Japan, October 1984.
9. G.J. Burke and A.J. Poggio, "Numerical Electromagnetics Code - Method of Moments," Technical Document No. 116, prepared for the Naval Electronic Systems Command, (ELEX3041), Jan. 2, 1980.
10. R. Mittra, editor, "Numerical and Asymptotic Techniques in Electromagnetics", Springer-Verlag, 1975.
11. T.T. Wu and R.W.P. King, "The Tapered Antenna and its Application to the Junction Problem for Thin Wires", IEEE Trans. on Antennas and Propagation, Vol. AP-24, No. 1, pp. 42-45, January, 1976.
12. J.N. Brittingham, E.K. Miller, and J.T. Okada, "SOMINT :An Improved Model for Studying Conducting Objects near Lossy Half Spaces," Lawrence Livermore Laboratory Report No. UCRL-52423, Feb. 24, 1978.
13. G.J. Burke, E.K. Miller, J.N. Brittingham, D.L. Lager,R.J. Lytle, and J.T. Okada, "Computer Modelling of Antennas Near the Ground," Lawrence Livermore Laboratory Report No. UCID-18626, May 13, 1980.
14. James R. Warner, "DIGRAF-Device Independent Graphics from FORTRAN" User's Guide, University of Colorado at Boulder, January 1979.
15. James R. Warner, "Principles of Device-Independent Computer Graphics Software", IEEE Transactions on Computer Graphics and Applications, Vol.1, No.4, October 1981.
16. James C. Logan, "Interactive Graphics Utility for Army Nec Automation", Software Note, Applied Computational Electromagnetics Society Newsletter, Vol.1, No.2, November 1986.

17. C.W. Trueman and S. J. Kubina, "AM Re-Radiation Project - Final Report", Technical Note EMC-80-03, Concordia University, EMC Laboratory, March 1980.
18. S. J. Kubina, "Numerical Modelling Methods for Predicting Antenna Performance on Aircraft", AGARD Lecture Series No.131, The Performance of Antennas in their Operational Environment, October 1983.
19. Ben Halpern and Raj Mittra, " On the Validity of Terminal Properties of Thin-Wire Antennas Computed Using the Numerical Electromagnetics Code(NEC)", Program Abstracts 1986 National Radio Science Meeting, Philadephia, Pa.
20. B. D. Popovic, M.B. Dragovic and A. R. Djordjevic, "Analysis and Synthesis of Wire Antennas", Research Studies Presss(Wiley), 1982.
21. S. J. Kubina and Colin Larose, "A NEC Topside Antenna Case Study with DIDEC and SPECTRUM: Model Generation and Current Display Codes", ACES Newsletter, Vol.1, No.2, November 1986.
22. J. S. Belrose, "Scale Modelling and Full Scale Techniques with Particular Reference to Antennas in their Operational Environments", AGARD Lecture Series No. 131, The Performance of Antennas ion their Operational Environment, October 1983.
23. M. J. Sidford, "Influence of the Airframe on the Performance of Antennas for Avionics", AGARD Lecture Series No. 131, The Performance of Antennas ion their Operational Environment, October 1983.
24. J. V. N. Granger and J. T. Bolljahn,"Aircraft Antennas", Proc. I.R.E., Vol.43, No.5, pp. 533-550; May 1955.
25. J.Y. Wong, "Radiation Pattern Efficiencies of Some Suppressed HF Antennas", Transactions on Aeronautical and Navigational Electronics, Vol. ANE-5, No.1, March 1958.
26. S. J. Kubina, "Helicopters, HF Antennas: Models, Measurements and All That", Proceedings of the ACES Conference, Monterey CA, Apr. 1987.
27. Yousef Bahsoun, "Evaluation of an HF Helicopter Antenna-Measurements and Numerical Techniques", M. Eng. Thesis, Faculty of Engineering, Concordia University, Montreal, Que., Canada, March 1982.
28. Y. A. Bahsoun, S. J. Kubina and C. W. Trueman, "Modelling HF Loop Antennas on the CHSS-2 'Sea King' Helicopter", Symposium Digest, 1982 AP-S Antennas and Propagation Symposium, University of New Mexico, May 1982.
29. R. Balaberda and J. Hazell, "Radiation Pattern Measurements of a TRANLINE Antenna on a Scale Model of the CH-135 Bell Helicopter", National Research Council of Canada Report ERB-991, July 1986.

BASIC PRINCIPLES OF RADAR POLARIMETRY AND ITS APPLICATIONS TO TARGET
RECOGNITION PROBLEMS WITH ASSESSMENTS OF THE HISTORICAL DEVELOPMENT
AND OF THE CURRENT STATE-OF-THE ART

WOLFGANG-M. BOERNER AND WEI-LING YAN
University of Illinois at Chicago
UIC-EECS/CL, 840 W. Taylor St.
SEL-4210, M/C 154
Chicago, IL 60680-4348

ABSTRACT. This introductory paper provides a succinct overview of the historical developments in radar polarimetry, highlighting important contributions. A historical events table and a list of pertinent references are
provided together with a critical review on the validity of various polarization dependent approaches to radar target phenomenology. Based on such a
critical assessment, the appropriate formalism for handling basic polarimetric radar concepts is introduced, providing clarification and justification. In a next step, we provide the scattering matrix formulations together with the transformation laws for measurements obtained with a coherent
polarization (scattering matrix) radar using different sets of antenna polarization bases such as linear horizontal/vertical (HV), circular left/
right (LR) or any other orthogonal elliptic (AB) transmit/receive system
leading directly to polarimetric radar target/clutter phenomenology. Introducing the polarization sphere, we will illustrate Kennaugh's target
characteristic operator theory and its extension to the partially coherent
clutter case using simple canonical target/clutter examples.

1. INTRODUCTION

Radar polarimetry has become an indispensible tool in modern electromagnetic sensor technology both in the civil and the military sectors and also in
remote sensing and radar meteorology [6,7]. From the outset, we wish to
emphasize that by incorporating coherent polarimetric phase information into radar signal and image processing, one can anticipate a breakthrough
which is at least comparable to that brought about by the advent of holography and computer assisted tomography and its applications to Synthetic
Aperture Radar (SAR) and Inverse Synthetic Aperture Radar (ISAR). Although
considerable R&D efforts have already been expanded during the past three
decades, there still exist many "grey areas" in both theory and techniques
of radar polarimetry, which is at best illustrated by the following recent
true event:
  "...Circularly polarized waves have either a right-handed polarization or a left-handed polarization, which is defined by conven-
  tion. The TELSTAR satellite sent out circularly polarized microwaves. When it first passed over the Atlantic, both the British
  station at Goonhilly and the French station at Pleumeur Bodou
  tried to receive its signals. The French succeeded, because their
  definition of the sense of polarization agreed with the American
  definition. The British station was set up to receive the wrong
  polarization because their definition of the sense of polarization
  was contrary to our definition..."
from J.R. Pierce, "Almost Everything About Waves," MIT Press, Cambridge,
MA, pp. 130-131, 1974 [38].
  In an effort to clear up some of these still remaining misconceptions,

311

B. de Neumann (ed.), Electromagnetic Modelling and Measurements for Analysis and Synthesis Problems, 311-363.
© 1991 Kluwer Academic Publishers. Printed in the Netherlands.

it is useful to first present a brief assessment of the intrinsic facility of polarization visualization by insects, fish and mammals, and then to provide a historical development of the theory of polarization and of the fundamentals of radar polarimetry. Based on such a critical historical assessment, the appropriate formalism for handling basic polarimetric radar concepts may then be introduced providing clarifications and justifications.

1.1 Visual Observations Of Polarized Light By Insects, Fish And Mammals

Insects (bees, ants, hornets, wasps, water fleas, fruit flies), cray fish, eels, various bottom feeding fish, migrating birds, and also some mammals are able to distinguish between polarized light and unpolarized light as easily as we can distinguish colors (Können [37]). When making use of this ability to orientate themselves using polarimetric parameter manipulations, these creatures sometimes perceive the orientation and ellipticity of light at very low (10%) degrees of polarization, and use these inputs for controlling dynamic flight motion even under the severest storm conditions. On the other hand, man is almost "polarization-blind" and, generally, has to use a polarizing filter to determine the polarization of light. Nonetheless, when light has an extremely high degree of polarization (e.g., >60%), most humans and especially those living in regions with almost permanent foggy conditions, can still perceive polarization with the naked eye when properly trained to observe the so-called "Haidinger brush" which was discovered by Haidinger as early as 1844. It turns out that in a plane emitting polarized light, we may see a tiny yellowish figure appear, which we do not perceive in unpolarized light. The orientation of this so-called "Haidinger brush" depends on the direction of vibration of light and co-rotates if this plane is rotating. If one wishes to visualize the outward appearance of the 'brush,' the best way is to compare it with "Brewster's brush" which one can observe by looking through certain minerals (pleo-chroic minerals: glaucophane, cordierite, etc.) and we refer to the excellent recent books by Können (1982, 1985) [37] on this subject and to his recent invited address during the recent NATO-ARW- DIMRP '88 (1989/90) [7].

We also noted that the cornea of the eye of different species of birds, fish, eels, and mammals are not polarization isotropic and may even change its polarimetric properties drastically during the species' life cycle. For example, the color sensitivity of the eel's eye is optimal towards the red during infancy and shifts to the blue-green before departing on its cross-oceanic journey and, at the same time, the cornea, passing purely linear (spinning) polarization states during infancy, changes to passing purely circularly polarized light during maturity. On the other hand, the cornea of grass feeding animals like cattle, horse, sheep, goats, hare, etc., mainly pass horizontal linearly polarized light with strong suppression of vertically polarized light. Furthermore, various sections of the cornea, for example, of migrating bird's, possess different polarimetric properties over different view angle domains, this is also true of many fish and of human beings.

In 1949, Karl von Frisch discovered [37] that many insects, such as bees, possess the ability to orient themselves using the distribution of polarization of the daylight to find their way, as well as to land on moving platforms (leaves, twigs, stems, blades of grass, etc.) under the most adverse weather conditions although direct access to sunlight is obscured.

It is not only the cornea of the eyes that possess polarimetric anisotropic properties, but also the body surface of insects, beetles, fish, moths, butterflies, etc., possess polarimetric anisotropic properties. For example, some beetles and moths are able to camouflage themselves so well against the background that they seem completely invisible when viewed with the eye of other species. In most cases, this camouflaging operation is a

strongly polarimetric phenomenon and with the use of a naturally built-in polarization filter, optimized not only at circular or linear polorizations, complete contrast enhancement against the background is obtained. This indicates that not only is the body surface polarimetrically agile and adaptive but also that the cornea of the eye of these creatures is too and that there may exist an interrelated adaptive control system.

We have introduced this section on the visual observation of polarized light in nature because from extended physiologic behavioral studies of the pertinent species of insects, fish, birds and mammals, we may be able to discern new and improved design and vector (polarization) signal processing approaches for future polarimetric radar and lidar systems, particularly in obtaining a better understanding on how and why polarization anisotropy is used in nature.

1.2 Historical Development

The history of polarimetric radar can be traced from the first recorded analytical descriptions of polarized light [23,52,54]. In the following, we shall first discuss the discovery of the properties of polarized light in nature (Können, 1988 [37]) and then of the discoveries of the phenomenon of polar electromagnetic energy spanning the entire electromagnetic spectrum, such as, the m-to-sub-mm-wavelength regions, leading to the emergence of the new discipline of polarimetric radar and polarimetric lidar technology.

1.2.1 Historical Discovery Of Polarization. The discovery of the phenomenon of polar electromagnetic energy dates back as early as 1669, when the first known quantitative work on the subject was published by Erasmus Bartolinus. It contained his observations of objects viewed through a crystal doubled and of an incident light ray splitting into ordinary and extraordinary rays (Können [37]).

At this moment, let us recall that the Vikings, using the "findlings stone," produced from some dichroic mineral like cordeirite, were able to navigate in the absence of direct sunlight. Thus, the Vikings during the eighth (8th) century when navigating in the Baltic Sea prior to traversing the Atlantic, may have been the first human beings to utilize polarimetric effects.

Bartolinus was followed by Christian Huygens, who contributed most significantly to the field of optics by proposing the wave nature of light and discovering polarized light (1677). E. Louis Malus proved Newton's suggestions that polarization is an intrinsic property of light and not something added by a crystal (1808). The next significant contribution to this field was added by Augustine Fresnel (1788-1827), who proposed that light could be considered as a transverse wave. His reflection formulas deduced from experimentation are still in use today and have been rederived using rigorous electromagnetic theory by Paul Drude in the late nineteenth century (1889). The last of the early pioneers was Sir David Brewster, who by extending the work of Malus, discovered the relationship between the polarizing angle and the relative refractive power of dielectric materials (1816). The transition to the formulation of a rigorous electromagnetic theory was paved by Michael Faraday with his postulations of the physical laws of electromagnetism in 1832, which was soon followed by his discovery in 1845 of the rotation of the polarization plane in magnetic fields. In 1852, George Gabriel Stokes laid the foundations of mathematical theories to describe polarized, unpolarized and partially polarized streams of light by introducing his four parameters now known as "Stokes parameters."

In 1873, James Clerk Maxwell succeeded in providing a rigorous formulation of Faraday's postulates which led to the formulations of the diffraction theories by Helmholtz (1881) and Kirchhoff (1883) and a strict mathematical treatment was then provided by Arnold Sommerfeld (1896) and

earlier by William O. Strutt (Lord Rayleigh) (1881). A significant contribution to the understanding of polarized light was made in 1892 by Henri Poincaré, who showed that all possible states of polarization could be represented by points on the Riemann (1872) sphere, the latitude and longitude of each point defining the eccentricity and inclination of the polarization orientation angle of the polarization ellipse. Based on these discoveries, in 1886 Heinrich Hertz demonstrated the application of the electromagnetic theory as it applies to lower frequencies such as radio waves, which marks the advent of modern applications of electromagnetic waves leading to radio wave communication, object detection and ranging. A good paper was first written by Marconi (1922). The work of most of these scientists are available in a translated form in the literature collected by William Swindell [56] and are also portrayed in Born and Wolf[12], and Skolnik [54].

1.2.2 Historical Development Of Radar Polarimetry. The use of radio-to-microwaves frequencies for aircraft detection and the design of the first radars were accomplished concurrently in Europe and America by the 1930's and further advanced before and during World War II (1937-1941) [24,49,50, 54].

Very important early basics on the properties of partially polarized waves were discovered by Norbert Wiener (1927-1929) on harmonic analysis in quantum mechanics showing that the coherency matrix is a linear combination of the Pauli spin matrices with the Stokes parameters. These studies by Wiener and Pauli had direct influence on the later work of R. Clark Jones (1941-1944) under the guidance of Professor Hans Mueller at the Massachusetts Institute of Technology, Cambridge, MA [16]. Earlier on, during 1924-29, Wolfgang Pauli introduced the concept of the spinor in quantum mechanics which has proven to be an ideally successful tool in the proper description of polarimetric radar problems. Then extensive polarimetric wave propagation analyses were carried out at MIT leading to the so-called "Jones calculus" (Azzam and Bashara, 1977) [2], which now has also become of great value to forward scattering radar analyses, for example, in polarimetric radar meteorology.

One of first extensive studies on radar polarization was initiated by George Sinclair [53] in 1946 at the Ohio State University, Antenna (later called ElectroScience) Laboratory. He showed that a radar target acts as a polarization transformer and he expressed the properties of a coherent radar target by the 2x2 coherent scattering matrix [S]. These studies were further pursued by Victor Rumsey (1949-1951) and particularly by Edward Morton Kennaugh (1948-1954, 1952). Other basic studies were conducted by Booker (1950), Deschamps (1951), Kales (1951), Bohnert (1951), and Gent (1954); and here we refer particularly to the series of papers in the Proc. IRE, May 1951, in which the paper by Deschamps (1948- 1951) [20] is still of use. Based on these studies, Kennaugh (1952) introduced a new approach to radar theory [36] and developed the "Optimal Target Polarization Concept" for the reciprocal, monostatic relative phase case which was already of particular interest in meteorological radar studies (circular polarization rain clutter rejection or cancellation). There were a number of other isolated studies initiated, for example, the GIT-Project A235 (July 1955) for the purpose of using polarization to distinguish between targets and clutter, which was well reviewed and summarized by Root [49]. The decade of the fifties closed without any real recognition for the need of decoy discrimination, and theory and techniques of polarimetric radar still remained highly underdeveloped. However, an extensive amount of measurements on the relative phase scattering matrix were made in the late fifties by J. Richard Huynen (May 1960) at the Lockheed Aircraft Corporation, Palo Alto, Ca., who exploited Kennaugh's optimal target polarization concept and developed, during the sixties, his approach to radar target phenomenology

which culminated in his dissertation (1970). We also note here that Copeland (1960), under the guidance of Kennaugh, developed the first practical scheme for classification and identification of radar targets which was later on utilized by Huynen (1960-1970), based purely on radar polarimetric concepts for symmetrical, reciprocal targets [30].

Except for the pioneering polarimetric radar developments by J. Eaves and co-workers of GIT-EES, who developed the concept of intrapulse polarization agile pulse switching and Andre J. Poelman, who introduced the polarimetric multi-notch filter [46], no other relevant studies on advancing radar polarimetry were carried out in the West during the sixties and early seventies. Whereas, we note that in the Russian literature (Kanareykin et al, 1965, 1968; Kozlov, 1979; Shupyatsky and Morgunov, 1968; Varshanchuk and Kobak, 1971; Zhivotovski, 1973, 1978; Bogoradsky et al, 1980; Rodimov and Popovski,1984) the potential applicability of radar polarimetry to target /clutter analysis was recognized early on [11,25,34,35,42,57], and immediately after the appearance of Huynen's dissertation [30], polarimetric radar investigations were pursued with new vigor.

No real progress was made in advancing the fundamentals of radar polarimetry until the early eighties when a renewed effort was made in 1978 at UIC-EECS/CL (Chicago, IL), to assess, most critically, the previous works of Kennaugh and Huynen [5] which were essentially unknown or forgotten by most workers in the field, and to generalize the target characteristic operator concept to the general bistatic case giving special consideration to polarization scattering matrix measurements in any orthogonal elliptic basis.

2. CHRONOLOGICAL TABLE OF THE HISTORY OF THE DISCOVERY OF POLARIZATION LEADING TO RADAR POLARIMETRY

We consider it useful for the interested reader to reproduce and expand on the historical tables introduced by Gehrels [23] and the more recently updated one by Können [37] with deletions and additions to meet historically important events leading to the development of radar polarimetry with which we will complete this paper.

2.1 History Of The Discovery Of Polarization

About 1000	The Vikings discovered the dichroic properties of crystals like cordeirite. With these crystals they observed the polarization of the sky light when under foggiest conditions and were thus able to navigate in the absence of the sun.
1669	Erasmus Bartolinus from Denmark discovered the double refraction of calcite crystals.
1690	Huygens discovered the polarization of the doubly-refracted rays of calcite, without, however, being able to explain the phenomenon.
1808	Malus discovered the polarization of reflected light by using a calcite crystal as a filter. This filter apparently looses its double refraction when the entering light is polarized and the crystal is held in the correct position. Afterwards, Malus formulated his law, which gives the relationship between the position of a polarizing filter and the quantity of transmitted light when the entering light is totally (linearly) polarized.
1809	Arago rediscovered the polarization of the blue sky. In 1811 he discovered the optical activity of quartz, and in 1812 he constructed a filter out of a pile of glass sheets. In 1819 he found the polarization of comet tails and in 1825 the (weak) overall polarization of 22 degree haloes. In 1824 he found the polarization of the glow emitted by hot, incandescent metals. He was also the first to

record the polarization of the moon.

1811   Biot discovered the polarization of the rainbow. In 1815 he established the optical activity of fluids such as turpentine, and in 1818 he studied the optical activity of gaseous turpentine in a gas column of 15 m length. Unfortunately, this apparatus exploded before he could finish his measurements. In 1815 Biot also discovered the strong dichroism of tourmaline.

1812   Brewster discovered the law, which was named after him, that indicates the relationship between the index of refraction and the angle of incidence at which light is totally converted by reflection into linearly polarized light. In 1818 he discovered "Brewster's brush" in pleochroic crystals.

1816   Fresnel gave a theoretical explanation of the existence of polarization.

1828   Nicol invented his prism, which can be considered to be the first easily usable polarizing filter.

1832   Faraday postulated the fundamental laws of electromagnetism.

1844   Haidinger discovered that the human eye has the ability to distinguish between unpolarized and polarized light, because, in the latter case, a yellowish figure appears on the retina (the "Haidinger's brush").

1845   Faraday discovered the rotation of the polarization plane in magnetic fields.

1852   Stokes introduced the composition and resolution of streams of polarized, unpolarized and partially polarized light in terms of four parameters known as those of Stokes.

1872   Riemann introduced the transformation of mapping the sphere's surface onto the polar map.

1873   Maxwell succeeded in providing a rigorous formulation of the electromagnetic field equations derived from Faraday's postulates.

1874   Wright discovered the polarization of zodiacal light.

1878   Helmholtz introduced the vector decomposition into lamellar and rotational vector fields.

1882–3 Kirchhoff introduced the physical optics formulation of diffraction.

1884   Kiessling recorded that the glory is polarized.

1889   Cornu discovered that artificial haloes in sodium nitrate crystals are highly polarized because of the double refraction of the crystals.

1890   Lord Rayleigh explained the reason for the blue color of the sky.

1892   Poincare introduced the polarization sphere using the Riemann transformation and Stokes parameters.

1896   Sommerfeld introduced a rigorous theory of diffraction.

1905   Umov described the relationship between the degree of polarization of light reflected from rough surfaces and the albedo of the surface.

1911   Michelson discovered that certain beetles have a gloss which is circularly polarized.

1912   Sommerfeld introduced the Green's function formulation into diffraction theory.

1926   Stern and Gerlach discovered the electron spin.

1926   Landau introduced the density matrix into quantum mechanics which is mathematically identical to the coherency matrix of polarization introduced by E. Wolf in 1959.

1926–9 Pauli introduced the spinor formalism into quantum mechanics.

1928   Land constructed his first polarizing filter. Further developments of this filter made it possible to study effects of polarization with a simple and efficient sheet filter. Such filters are also used in sunglasses, etc., to reduce the intensity of glare. Compared with the Nicol and other crystal filters used up to that

point, the development of this kind of sheet filter meant great progress.
1929   Wiener introduced the generalized harmonic analysis.
1932   Dirac introduced the (4×4) matrices, which were named after him, and
were mathematically identical with Mueller polarization operators.
1933   Max Born published the first version of OPTIK.
1939   Le Grand and Kalle reported that scattered light underwater is polarized.
1940   Bricard discovered that supernumerary fog-bows shift when one looks
at them through a linear filter which is then rotated.
1941   Jones introduced a new calculus for the treatment of optical systems
based on Hans Mueller's polarimetric studies.
1947   Van de Hulst gave the first feasible explanation of the glory and
explained its polarization directions.

2.2 Emergence Of Polarimetric Radar Studies [6,7,24,49,50]

1948   Sichak and Milazzo described antennas for circular polarization.
1949   Yeh provided the first correct definition of polarization antenna
power relations.
1949   Sinclair introduced the polarization scattering (RCS) matrix into
radar.
1949   Kennaugh provided the first correct interpretation on circular polarization rain cancellation.
1949   Max von Frisch discovered that bees are more capable than man of
distinguishing polarized from unpolarized light and use this ability
to orientate themselves.
1951   Booker, Rumsey, Deschamps, Kales, and Bohnert introduced techniques
for handling elliptically polarized waves with special reference to
antennas, Proc. IRE, Vol. 35 (May 1951).
1952   Kennaugh formulated his radar target characteristic operator theory
based on his optimal polarization null concept.
1953   Deschamps introduced a hyperbolic protractor for microwave impedance
measurements and other (polarization) purposes.
1954   Gent provided a polarimetric theory for radar target reflection.
1955   Shurcliff discovered that the human eye is also capable of distinguishing circularly polarized from unpolarized light.
1956   Graves introduced a radar polarization power scattering matrix.
1958   Duncan discovered the polarization of the aurora.
1959   Born and Wolf published the first edition of "Principles of Optics"
based on Born's Optik (1933).
1959   E. Wolf introduced a rigorous derivation of the coherency matrix in
the description of polarized fields.
1960   Copeland classified (symmetrical) radar targets by polarization properties.
1960   Crispin et al described the measurements and the use of the radar
scattering matrix.
1963   Beckmann and Spizzichino published "The scattering of electromagnetic waves from rough surfaces."
1965   Lowenschuss described radar scattering matrix applications.
1965   Bickel analyzed some invariant properties of the polarization scattering matrix.
1965   IEEE Special Issue on "Radar Reflectivity," Aug. 1965.
1966   Fung introduced his first theory on scattering and depolarization of
electromagnetic waves from a rough surface.
1966   Kanareykin, Pavlov, and Potekhin published the polarization of radar
signals (Russian).
1967   IEEE Proc., Special Issue on Partial Coherency, Jan. 1967.
1967   Hagfors described a study of the depolarization of lunar radar

echoes.

1967 Bolinder provided a geometrical analysis of partially polarized waves.

1968 Shupyatskiy and Morgunov applied polarization methods to radar studies of clouds and precipitation.

1968 Beckmann published "The depolarization of electromagnetic waves."

1968 Kanareykin, Potekhin, and Shiskin published "Maritime Polarimetry."

1968-71 McCormick and Hendry of NRC, OTTAWA/Canada, and Mueller, Seliga and Bringer of CHILL/UIUC, Urbana, IL, initiated their studies on polarization radar measurements of precipitation scattering using coherent dual polarization CW instrumentation radars with (L,R) and (H,V) antenna polarization bases, respectively.

1968 Crispin and Siegel published "Methods of Radar Cross-Section Analysis," containing A.L. Maffet, "Scattering Matrices."

1970 Huynen defended his doctoral dissertation on a "Phenomenological Theory of Radar Targets."

1970 Thiel defended his dissertation on the presentation and transformation theory of quasi-monochromatic scattering in radio astronomy.

1971 Hendry and McCormick analyzed polarization scatter properties of precipitation.

1971 Poelman provided performance evaluations of two polarization radar systems possessing linear (H,V) or circular (L,R) polarization antenna facility.

1973 Gehrels edited the first general source book on polarimetry (in English).

1974 Gorshkov published "Ellipsometry" (in Russian).

1975 Long published the first edition of "Radar Reflections from Land and Sea."

1975 Oguchi provided an in-depth analysis on "Rain Depolarization at cm/mm Wavelengths."

1975 Poelman reported on using orthogonally polarized returns to detect target echoes in Gaussian noise.

1975 Aboul-Atta and Boerner introduced a set of vectorial impedance boundary conditions for the natural dependence of harmonic polarimetric fields.

1976 Seliga and Bringi suggested the potential use of radar differential reflectivity with measurements (Z'_{DR}) at orthogonal polarizations for portraying precipitation parameters.

1976 Appling and Eaves introduced the concept and design of the intrapulse polarization-agile radar known as IPPAR.

1977 Metcalf assessed polarization diversity radar and ladar technology in meteorological research.

1978 Ishimaru published "Wave Propagation and Scattering in Random Media."

1978 Boerner reported on polarization utilization in electromagnetic imaging.

1978 Silverman reported on chaff dispersion and polarization.

1979 Schneider and Williams provided an analysis on circular polarization in radar.

1979 IEEE Standard No. 1409-1979, "Test Procedures for Antennas."

1979 Ioannidis and Hammer analyzed the optimum antenna polarization for target discrimination in clutter.

1979 Kozlov investigated the radar contrast of two objects.

1981 Root organized the First Polarimetric Radar Technology Workshop at Redstone Arsenal, Alabama.

1981 Meischner and Schroth proposed the design of an advanced polarization radar system for meteorological studies at DFVLR which was condensed for use during 1987.

1982 Können published "Polarized Light in Nature" (in Dutch; English

translation - 1985).

1983 Root organized the Second Polarimetric Radar Technology Workshop at Redstone Arsenal, Alabama.

1984 Held, Brown et al of CAL–TEC/JPL succeeded in collecting the first set of complete relative phase polarimetric SAR data over the San Francisco Bay area with the updated JPL C–Band POL-SAR System.

1985 Boerner et al, edited the Proc. of a NATO-ARW on "Inverse Methods in Electromagnetic Imaging," Bad Windsheim, FRG, 1983, with special emphasis on radar polarization.

1985 Collins published "Antennas and Radio Wave Propagation."

1985 Giuli analyzed polarization diversity methods in radar.

1986 Kostinski and Boerner critically assessed the fundamentals of radar polarimetry.

1987 Barnes, et al introduced consistent calibration methods for POL-RAD/SAR system.

1988 Root and Matkin organized the Third Polarimetric Radar Technology Workshop at Redstone Arsenal, Alabama.

1988 Boerner organized the NATO-ARW on Direct and Inverse Methods in Radar Polarimetry, Bad Windsheim, FRG.

3. THE REPRESENTATIONS OF THE POLARIZATION STATE

Polarization is a property of single-frequency electromagnetic radiation describing the shape and orientation of the locus of the extremity of the field vectors as a function of time [31]. In general, the electric vector of a harmonic plane wave traces an ellipse in the transverse plane with time [45]. In this section, we introduce several representations and descriptive parameters, commonly used to describe wave polarization, namely, the size A of the ellipse, the ellipticity angle ε, the tilt (orientation) angle τ, the polarization transformation ratio ρ, the relative phase $\phi = \phi_y - \phi_x$ (phase difference between the two orthogonal components of the electric field of the wave), and the Stokes parameters g_i, and we illustrate how these are related to each other. The Poincare sphere is also introduced and utilized to summarize this section.

3.1 The General Representation

A single monochromatic, uniform, TEM (transverse electromagnetic) traveling plane wave can be decomposed into three orthogonal components, which are linear in simple harmonic oscillations. In the right-handed cartesian coordinate system, the electric vector of this plane wave may be expressed in terms of the x, y, and z components. If the wave is traveling in the positive z direction, then the real instantaneous electric field is written by

$$\vec{\varepsilon}(z,t) = \begin{bmatrix} \varepsilon_x(z,t) \\ \varepsilon_y(z,t) \\ \varepsilon_z(z,t) \end{bmatrix} = \begin{bmatrix} |E_x|\cos(\omega t - kz + \phi_x) \\ |E_y|\cos(\omega t - kz + \phi_y) \\ 0 \end{bmatrix} . \qquad (3.1)$$

The non-zero components are $\varepsilon_x(z,t)$ and $\varepsilon_y(z,t)$, where $|E_x|$ and $|E_y|$ represent the magnitudes of these components, and ϕ_x and ϕ_y represent the respective phases of these oscillations. It is easy to show [45] that at a fixed value of z, on the transverse plane, the electric vector ε of a harmonic plane wave rotates as a function of time, with the tip of the vector

320

describing an ellipse. For simplicity, consider the plane of z = 0, Eq. 3.1 becomes

$$\vec{\varepsilon}(t) = \begin{bmatrix} \varepsilon_x(t) \\ \varepsilon_y(t) \end{bmatrix} = \begin{bmatrix} |E_x|\cos(\omega t+\phi_x) \\ |E_y|\cos(\omega t+\phi_y) \end{bmatrix} . \quad (3.2)$$

These two independent, linear, simple harmonic vibrations $\varepsilon_x(t)$ and $\varepsilon_y(t)$, which are along two orthogonal directions x and y, compose an elliptical vibration as shown in Fig. 3.1.

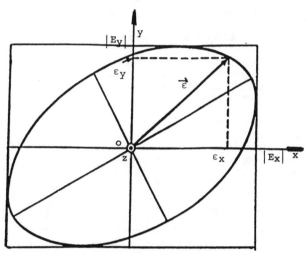

Fig.3.1 Superposition of ε_x and ε_y yields
the Polarization Ellipse

The locus of the extremity of the electric field vectors is, in general, an ellipse that may degenerate into a segment of a straight line or into a circle. Correspondingly, the polarization is called elliptical, linear, or circular [31]. The shape and the orientation of the polarization ellipse depend on the relative phase $\phi = \phi_y - \phi_x$. Fig. 3.2 displays different kinds of polarization ellipses for different relative phases ϕ. When ϕ_x and ϕ_y differ by an integer number of 2π, the two components are in phase; the wave is said to be linearly polarized and the straight line locus traverses the first and third quadrants (Fig. 3-2(a)). When ϕ_x and ϕ_y differ by an odd integer multiple of π, the components are 180° out of phase, and the straight line locus traverses the second and fourth quadrants (Fig. 3-2(e)). When ϕ_y leads ϕ_x by an angle ϕ which is less than π, the two components form an elliptic locus which traverses counterclockwise when looking down along +z-direction, and it defines left-handed polarization as shown in Figs. 3-2(b), (c) and (d). When ϕ_y lags ϕ_x by an angle ϕ less than π, it defines right-handed polarization as shown in Figs. 3-2(f), (g)

and (h).

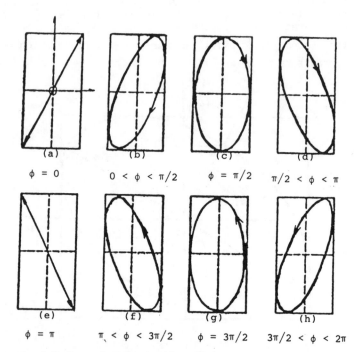

$\phi = 0$ $0 < \phi < \pi/2$ $\phi = \pi/2$ $\pi/2 < \phi < \pi$

$\phi = \pi$ $\pi < \phi < 3\pi/2$ $\phi = 3\pi/2$ $3\pi/2 < \phi < 2\pi$

Fig.3.2 General Polarization States:
    (a) and (e) are linear polarizations;
    (b),(c) and (d) are left-handed polarizations;
    (f),(g) and (h) are right-handed polarizations.

3.2 The Polarization Ellipse

Because the locus of the extremity of the field vectors is, in general, an ellipse, we can use the geometric parameters of the ellipse, τ, ε, and A, to represent the polarization state of the plane wave [30]. These geometric parameters are shown in Fig. 3.3 and are defined as follows: The orientation (or tilt) angle τ is the angle between the major axis of the ellipse and the positive x-axis. The ellipticity angle ε is defined by

$$\varepsilon = \tan^{-1}(b/a) \tag{3.3}$$

where b is the length of the semi-minor axis of the ellipse and a is the length of its semi-major axis. The size of the ellipse is

$$A = \sqrt{a^2 + b^2} \quad . \tag{3.4}$$

The polarization of the plane wave can be determined in terms of the geometric parameters [30],

$$\vec{\varepsilon}(A, \phi_x, \tau, \varepsilon) = A \begin{bmatrix} \cos\tau & -\sin\tau \\ \sin\tau & \cos\tau \end{bmatrix} \begin{bmatrix} \cos\varepsilon \\ i\sin\varepsilon \end{bmatrix} \cos\phi_x \quad . \tag{3.5}$$

In the next step, we will find the relationship between the parameters in

322

the two equivalent methods of specifying the elliptical polarization state
of a monochromatic wave. The first method describes the elliptic vibration
of the electric vector as a superposition in terms of $|E_x|$, $|E_y|$, ϕ_x, and
ϕ_y. And the second method specifies the elliptic polarization by the three
geometric parameters A, ε, and τ.

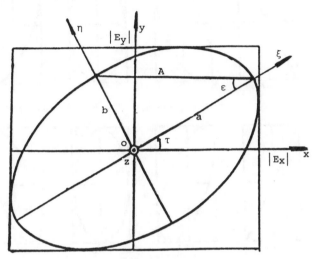

Fig. 3.3 The Three Parameters that Define the
Ellipse of Polarization are the Orientation (or Tilt)
Angle τ, the Ellipticity Angle ε and the Size A.

In Fig. 3.3, we selected the new axes ξ and η along the major and minor
axes, respectively. In this new coordinate system, the polarization ellipse becomes a regular ellipse, i.e., $\phi_\eta - \phi_\xi = \pi/2$, thus Eq. 3.1 becomes

$$\begin{bmatrix} \varepsilon_\xi \\ \varepsilon_\eta \end{bmatrix} = \begin{bmatrix} a\cos(\omega t - kz + \phi_\xi) \\ b\cos(\omega t - kz + \phi_\eta) \end{bmatrix} = \begin{bmatrix} a\cos(\omega t - kz + \phi_\xi) \\ -b\sin(\omega t - kz + \phi_\xi) \end{bmatrix} . \tag{3.6}$$

The new axes ξ and η are obtained by rotating the old axes x and y through
the angle τ. The electric field in the x–y coordinate system can be transformed into that in the ξ–η coordinate system as follows,

$$\begin{bmatrix} \varepsilon_\xi \\ \varepsilon_\eta \end{bmatrix} = \begin{bmatrix} \cos\tau & \sin\tau \\ -\sin\tau & \cos\tau \end{bmatrix} \begin{bmatrix} \varepsilon_x \\ \varepsilon_y \end{bmatrix} . \tag{3.7}$$

Substituting Eq. 3.1 and Eq. 3.6 into Eq. 3.7, yields

$$|E_x|\cos(\omega t - kz + \phi_x)\cos\tau + |E_y|\cos(\omega t - kz + \phi_y)\sin\tau = a\cos(\omega t - kz + \phi_\xi)$$

$$-|E_x|\cos(\omega t - kz + \phi_x)\sin\tau + |E_y|\cos(\omega t - kz + \phi_y)\cos\tau = -b\sin(\omega t - kz + \phi_\xi) .$$

Expanding cos and sin terms, and equating the coefficients of $\cos(\omega t - kz)$
and $\sin(\omega t - kz)$, yield

$$|E_x|\cos\phi_x\cos\tau + |E_y|\cos\phi_y\sin\tau = a\cos\phi_\xi \tag{3.8}$$

$$|E_x|\sin\phi_x\cos\tau + |E_y|\sin\phi_y\sin\tau = a\sin\phi_\xi$$

$$-|E_x|\cos\phi_x\sin\tau + |E_y|\cos\phi_y\cos\tau = -b\sin\phi_\xi$$

$$|E_x|\sin\phi_x\sin\tau - |E_y|\sin\phi_y\cos\tau = -b\cos\phi_\xi \quad .$$

Through simple algebra, the following relationship can be obtained:

$$a^2 + b^2 = |E_x|^2 + |E_y|^2 \tag{3.9}$$

$$\frac{b}{a} = \frac{|E_x|\cos\phi_x\sin\tau - |E_y|\cos\phi_y\cos\tau}{|E_x|\sin\phi_x\cos\tau + |E_y|\sin\phi_y\sin\tau}$$

and

$$\frac{b}{a} = \frac{-|E_x|\sin\phi_x\sin\tau + |E_y|\sin\phi_y\cos\tau}{|E_x|\cos\phi_x\cos\tau + |E_y|\cos\phi_y\sin\tau} \quad .$$

After equating the above two equations, we find

$$(|E_x|^2 - |E_y|^2)\frac{1}{2}\sin2\tau = |E_x||E_y|\cos(\phi_y-\phi_x)\cos2\tau \quad ,$$

so that with $\phi = \phi_y - \phi_x$,

$$\tan2\tau = \frac{2|E_x||E_y|\cos\phi}{|E_x|^2 - |E_y|^2} \quad . \tag{3.10}$$

Also from Eq. 3.8, we find

$$ab\sin^2\phi_\xi + ab\cos^2\phi_\xi = |E_x||E_y|\sin\phi_y\cos\phi_x - |E_x||E_y|\sin\phi_x\cos\phi_y \quad ,$$

thus

$$ab = |E_x||E_y|\sin(\phi_y-\phi_x) = |E_x||E_y|\sin\phi \quad . \tag{3.11}$$

According to the definition of the ellipticity angle ε of Eq. 3.3, we obtain

$$\sin2\varepsilon = \frac{2\tan\varepsilon}{1 + \tan^2\varepsilon} = \frac{2(b/a)}{1 + (b^2/a^2)} = \frac{2|E_x||E_y|\sin\phi}{|E_x|^2 + |E_y|^2} \quad . \tag{3.12}$$

Eq. 3.10 and Eq. 3.12 give us the relationship between ε, τ, $|E_x|$, $|E_y|$, and ϕ. The polarization states in terms of ε and τ are given in Table 3.1.

3.3 The Jones Vector [2,15]

In the previous sections we described the electric field using the real measurable quantity ε. Now, we introduce the complex electric field vector $\vec{E}$. For a monochromatic field whose variation in space is also sinusoidal, the instantaneous vector $\vec{\varepsilon}$ takes the form [2,15]

$$\vec{\varepsilon}(r,t) = \mathrm{Re}(\vec{E}(\vec{r})e^{j\omega t}) \tag{3.13}$$

where the complex vector $\vec{E}$ depends on the position vector $\vec{r}$ and is given by

$$\vec{E}(\vec{r}) = \vec{E}e^{-j\vec{k}\cdot\vec{r}} \quad . \tag{3.14}$$

$\vec{E}$ is a complex-constant amplitude vector and is given by

$$\vec{E} = \begin{bmatrix} E_x \\ E_y \\ E_z \end{bmatrix} = \begin{bmatrix} |E_x| e^{j\phi_x} \\ |E_y| e^{j\phi_y} \\ |E_z| e^{j\phi_z} \end{bmatrix} . \tag{3.15}$$

For a plane wave which propagates along the +z-axis, the complex electric field is given by

$$\vec{E}(z,t) = \begin{bmatrix} E_x(z,t) \\ E_y(z,t) \end{bmatrix} = \begin{bmatrix} |E_x| e^{j(\omega t - kz + \phi_x)} \\ |E_y| e^{j(\omega t - kz + \phi_y)} \end{bmatrix} . \tag{3.16}$$

It is the same as Eq. 3.1, where $E_x(z,t)$ and $E_y(z,t)$ are the complex time invariant terms associated with each real, time-varying electric field component.

In a time harmonic (monochromatic) TEM wave field, each cartesian component of the electric vector varies sinusoidally with time, at all points in space. We can use the phasor notation to suppress the temporal term, thus Eq. 3.16 can be replaced by

$$\vec{E}(z) = \begin{bmatrix} |E_x| e^{j\phi_x} \\ |E_y| e^{j\phi_y} \end{bmatrix} e^{-jkz} .$$

Also, the spatial term can be dropped because the electric field has the same phase at all points on a z = constant plane. If we consider the plane at z = 0, the above equation becomes

$$\vec{E}(0) = \begin{bmatrix} |E_x| e^{j\phi_x} \\ |E_y| e^{j\phi_y} \end{bmatrix} . \tag{3.17}$$

This vector is called the "Jones Vector" of the wave. If we extract and disregard ϕ_x, the absolute phase of E_x, then Eq. 3.17 becomes

$$\vec{E} = \begin{bmatrix} |E_x| \\ |E_y| e^{j\phi} \end{bmatrix} \tag{3.18}$$

where $\phi = \phi_y - \phi_x$ denotes the relative phase.

3.4 The Polarization Ratio ρ

Any wave can be resolved into two orthogonal components (they may be two orthogonal linear, elliptical, or circularly polarized components) [31] in the plane transverse to the propagation direction. So, we can choose any two orthogonal polarized components as the polarization basis. For an arbitrary polarization basis (AB) with unit vectors $\hat{A}$ and $\hat{B}$, one may define the polarization state (spinor: Boehm 1951 [38]) by

$$\vec{E}(AB) = E_A\hat{A} + E_B\hat{B} \tag{3.19}$$

where the two components, E_A and E_B, are complex numbers. The polarization ratio ρ_{AB} in an arbitrary basis (AB) is also a complex number, and it may be defined as

$$\rho_{AB} = \frac{E_B}{E_A} = \frac{|E_B|}{|E_A|}e^{j(\phi_B-\phi_A)} = |\rho_{AB}|e^{j\phi_{AB}} \tag{3.20}$$

where $|\rho_{AB}|$ is the ratio of magnitude of the two orthogonal components of the field E_A and E_B, and ϕ_{AB} is the phase difference between E_A and E_B. The complex polarization ratio ρ_{AB} depends on the polarization basis (AB) and can be used to specify the polarization of an electromagnetic wave.

$$\vec{E}(AB) = \begin{bmatrix} E_A \\ E_B \end{bmatrix} = |E_A|e^{j\phi_A}\begin{bmatrix} 1 \\ \rho_{AB} \end{bmatrix}$$

$$= |E_A|e^{j\phi_A}\sqrt{\frac{1 + \dfrac{E_B E_B^*}{E_A E_A^*}}{1 + \dfrac{E_B E_B^*}{E_A E_A^*}}}\begin{bmatrix} 1 \\ \rho_{AB} \end{bmatrix}$$

$$= |E|e^{j\phi_A}\frac{1}{\sqrt{1 + \rho_{AB}\rho_{AB}^*}}\begin{bmatrix} 1 \\ \rho_{AB} \end{bmatrix} \tag{3.21}$$

where $|E| = \sqrt{E_A E_A^* + E_B E_B^*}$ is the amplitude of the wave $\vec{E}(AB)$. If we choose $|E| = 1$ and disregard the absolute phase ϕ_A, the above representation becomes

$$\vec{E}(AB) = \frac{1}{\sqrt{1 + \rho_{AB}\rho_{AB}^*}} \begin{bmatrix} 1 \\ \rho_{AB} \end{bmatrix} . \tag{3.22}$$

The above representation of the polarization state using the polarization ratio ρ_{AB} is very useful. For example, we want to represent a left-handed circular (LHC) polarization state and a right-handed circular (RHC) polarization state in a linear basis (HV) using the polarization ratio. For a left-handed circular polarization, $|E_H| = |E_V|$, $\phi_{HV} = \phi_V - \phi_H = \pi/2$, and according to Eq. 3.20, the polarization ratio ρ_{HV} of the LHC is j. One uses Eq. 3.22 with $\rho_{AB} = j$ and becomes for the left-handed circular polarization

$$\vec{E}(HV) = \frac{1}{\sqrt{1 + j(-j)}} \begin{bmatrix} 1 \\ j \end{bmatrix} = \frac{1}{\sqrt{2}} \begin{bmatrix} 1 \\ j \end{bmatrix} . \tag{3.23}$$

Similarly, the polarization ratio ρ_{HV} of a right-handed circular polarization in the (HV) basis is $-j$ because the relative phase ϕ_{HV} is $-\pi/2$. The representation of RHC in the linear (HV) basis is

$$\vec{E}(HV) = \frac{1}{\sqrt{2}} \begin{bmatrix} 1 \\ -j \end{bmatrix} .$$

Note: The polarization ratio ρ_{AB} is important in polarimetry. However, the value of the polarization ratio ρ defined in a certain polarization basis is different from that defined in the other polarization basis even if the physical polarization state is the same (see Table 3.1).

3.4.1 The Polarization Ratio ρ_{HV} in the Linear Basis(HV).

In the linear (HV) basis with unit vectors $\hat{H}$ and $\hat{V}$, a polarization state may be expressed as

$$\vec{E}(HV) = E_H\hat{H} + E_V\hat{V} .$$

The polarization ratio ρ_{HV}, according to definition of Eq. 3.20, can be described as

$$\rho_{HV} = \frac{E_V}{E_H} = \frac{|E_V|}{|E_H|} e^{j(\phi_V - \phi_H)} = \tan\gamma e^{j\phi_{HV}} . \tag{3.24}$$

where the angle γ is defined in Fig. 3.4, only in the (HV) basis, and

$$|E_H| = \sqrt{E_H^2 + E_V^2} \cos\gamma \tag{3.25}$$

$$|E_V| = \sqrt{E_H^2 + E_V^2} \, \sin\gamma \quad .$$

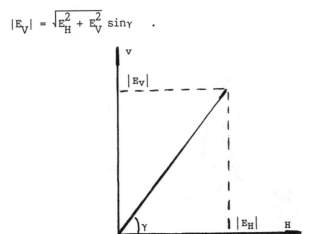

Figure 3.4 The Parameter γ

In a cartesian coordinate system, the +x-axis is commonly chosen as the horizontal basis and the +y-axis as the vertical basis. Substituting Eq. 3.25 into Eq. 3.1, we find

$$\varepsilon(z,t) = \begin{bmatrix} \sqrt{E_x^2 + E_y^2} \, \cos\gamma\cos(\omega t - kz + \phi_x) \\[2ex] \sqrt{E_x^2 + E_y^2} \, \sin\gamma\cos(\omega t - kz + \phi_y) \end{bmatrix}$$

$$= \sqrt{E_x^2 + E_y^2} \, \text{Re} \left\{ \begin{bmatrix} \cos\gamma \\ \sin\gamma e^{j\phi} \end{bmatrix} e^{j(\omega t - kz + \phi_x)} \right\}$$

where $\phi = \phi_y - \phi_x$ is the relative phase. In the literature, ϕ_x is commonly denoted by α, thus

$$\varepsilon(z,t) = \sqrt{E_x^2 + E_y^2} \, \text{Re} \left\{ \begin{bmatrix} \cos\gamma \\ \sin\gamma e^{j\phi} \end{bmatrix} e^{j(\omega t - kz + \alpha)} \right\} \quad . \qquad (3.26)$$

The expression in the square bracket is a spinor which is independent of the time-space dependence of the traveling wave. The spinor parameters (γ, ϕ) are easy to be found on the Poincaré sphere and can be used to represent the polarization state of a plane wave. In Fig. 3.5, the polarization state, described by the point P on the Poincaré sphere, can be expressed in terms of these two angles, where

2γ = the angle subtended by the great circle drawn from the point P on the equator.

ϕ = the angle between the great circle and the equator.

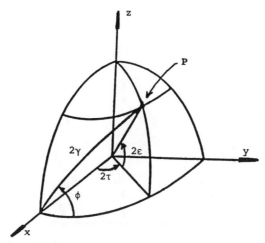

Figure 3.5 Poincaré Sphere Showing
Relation of Angles γ, τ, ε, and ϕ

After insertion of Eq. 3.25 into Eq. 3.12, the spinor parameters, γ and ϕ, can be related to the elliptic parameter, ε and τ, in the following manner,

$$\sin 2\varepsilon = 2\cos\gamma\sin\gamma\sin\phi = \sin 2\gamma\sin\phi \quad . \tag{3.27}$$

Dividing the numerator and the denominator of Eq. 3.10 by $|E_x|^2$, we find

$$\tan 2\tau = \frac{2\tan\gamma}{1 - \tan^2\gamma}\cos\phi = \tan 2\gamma\cos\phi \quad . \tag{3.28}$$

Eq. 3.27 and Eq. 3.28 describe the elllipticity angle ε and the tilt angle τ in terms of the variables γ and ϕ. Also, we can derive an inverse pair relationship which describes the γ and ϕ in terms of ε and τ. Eq. 3.27 and Eq. 3.28 provide $\sin\phi$ and $\cos\phi$ expressions, thus

$$\frac{\sin^2 2\varepsilon}{\sin^2 2\gamma} + \frac{\tan^2 2\tau}{\tan^2 2\gamma} = 1 \quad ,$$

by multiplying both sides by $\sin^2 2\gamma\cos^2 2\tau$ and moving $\sin^2 2\tau\cos^2 2\gamma$ to the right side, and after some calculation, we obtain

$$\cos 2\gamma = \cos 2\tau\cos 2\varepsilon \quad . \tag{3.29}$$

The ratio of Eq. 3.27 and Eq. 3.28 yields

$$\tan\phi = \frac{\sin 2\varepsilon}{\tan 2\tau\cos 2\gamma} = \frac{\tan 2\varepsilon}{\sin 2\tau} \quad . \tag{3.30}$$

It is convenient to describe the polarization state by either of the two sets of angles (γ,ϕ) or (ε,τ) which describe a point on the Poincaré sphere. The complex polarization ratio ρ_{HV} can be used to specify the po-

arization of an electromagnetic wave expressed in an (HV) basis. Some common polarization states expressed in terms of ρ_{HV} are listed in Table .1.

.4.2 Polarization Ratio ρ_{LR} in the Cicular Basis(LR). In the circular basis (LR), we have two unit vectors $\hat{L}$ and $\hat{R}$. $\hat{L}$ stands for left–handed circular unit vector and $\hat{R}$ is the right–handed circular unit vector. Any polarization state of a plane wave can be expressed by

$$\vec{E}(LR) = E_L\hat{L} + E_R\hat{R}$$

where E_L and E_R is the L component and the R component, respectively.

A unit amplitude left–handed circular polarization has only the L component in the circular basis (LR). It can be expressed by

$$\vec{E}(LR) = 1 * \hat{L} + 0 * \hat{R} = \begin{bmatrix} 1 \\ 0 \end{bmatrix} .$$

The above representation of a unit LHC polarization in the circular basis LR) is different from that in the linear basis (HV) of Eq. 3.23. Simi-_arly, a unit amplitude right–handed circular polarization has only the R component in the circular basis (LR).

$$\vec{E}(LR) = 0 * \hat{L} + 1 * \hat{R} = \begin{bmatrix} 0 \\ 1 \end{bmatrix} .$$

Also, this representation of RHC polarization in the circular basis (LR) is different from that in the linear (HV) basis.

The polarization ratio ρ_{LR}, according to the definition of the polarization ratio, Eq. 3.20, is

$$\rho_{LR} = \frac{E_L}{E_R} = \frac{|E_L|}{|E_R|} e^{j(\phi_L - \phi_R)} = |\rho_{LR}|e^{j\phi_{LR}} = \tan\beta e^{j\phi_{LR}} \tag{3.31}$$

where $|\rho_{LR}|$ is the ratio of magnitudes of the two orthogonal components E_L and E_R, and ϕ_{LR} the phase difference. The angles β and ϕ_{LR} are also easy to be found on the Poincaré sphere like the angles γ and ϕ_{HV} (see Fig. 3.8 of Section 3.6). Some common polarization states expressed in terms of ρ_{LR} are listed in Table 3.1.

Table 3.1: Examples of polarization states expressed in terms of geometric parameters (ε, τ), polarization ratio ρ, and normalized Jones vector $\vec{E}$.

Polarization	ε	τ	(HV)basis		(LR)basis	
			ρ_{HV}	$\vec{E}$	ρ_{LR}	$\vec{E}$
Linear Horizontal	0	0	0	$\begin{bmatrix} 1 \\ 0 \end{bmatrix}$	1	$\frac{1}{\sqrt{2}}\begin{bmatrix} 1 \\ 1 \end{bmatrix}$

Linear Vertical	0	$\pm\dfrac{\pi}{2}$	∞	$\begin{bmatrix}0\\1\end{bmatrix}$	-1	$\dfrac{1}{\sqrt{2}}\begin{bmatrix}-j\\j\end{bmatrix}$
45° Linear	0	$\dfrac{\pi}{4}$	1	$\dfrac{1}{\sqrt{2}}\begin{bmatrix}1\\1\end{bmatrix}$	j	$\dfrac{1}{2}\begin{bmatrix}1-j\\1+j\end{bmatrix}$
135° Linear	0	$-\dfrac{\pi}{4}$	-1	$\dfrac{1}{\sqrt{2}}\begin{bmatrix}-1\\1\end{bmatrix}$	$-j$	$\dfrac{1}{2}\begin{bmatrix}-1-j\\-1+j\end{bmatrix}$
Left-Handed Circular	$\dfrac{\pi}{4}$		j	$\dfrac{1}{\sqrt{2}}\begin{bmatrix}1\\j\end{bmatrix}$	0	$\begin{bmatrix}1\\0\end{bmatrix}$
Right-Handed Circular	$-\dfrac{\pi}{4}$		$-j$	$\dfrac{1}{\sqrt{2}}\begin{bmatrix}1\\-j\end{bmatrix}$	∞	$\begin{bmatrix}0\\1\end{bmatrix}$

3.5 The Stokes Parameters [12]

The previous sections deal with completely polarized waves, i.e., waves for which $|E_A|$, $|E_B|$, and ϕ_{AB} are constants (or at least slowly varying function of time). If we need to deal with partial polarization, it is convenient to use the Stokes parameters, introduced by Sir George Stokes in 1852[43].

3.5.1 The Stokes Vector for the Completely Polarized Wave.
First, let us consider the simple case. For a monochromatic wave, in terms of the linear (HV) basis, the four Stokes parameters are

$$g_0 = |E_H|^2 + |E_V|^2 \tag{3.32}$$
$$g_1 = |E_H|^2 - |E_V|^2$$
$$g_2 = 2|E_H||E_V|\cos\phi$$
$$g_3 = 2|E_H||E_V|\sin\phi$$

where $|E_H|$, $|E_V|$, and ϕ are the magnitudes and the phase difference between the two orthogonal components E_H and E_V, respectively. For a completely polarized wave, there are only three independent parameters, which are related in the following manner,

$$g_0^2 = g_1^2 + g_2^2 + g_3^2 \ . \tag{3.33}$$

The Stokes parameters are sufficient to characterize the magnitude, phase, and polarization of a wave. It is possible to show that the Stokes parameter g_0 is always equal to the total power (density) of the wave, g_1 is equal to the power in the linear horizontal or vertical polarized components, g_2 is equal to the power in the linearly polarized components at tilt angles $\tau = 45°$ or $135°$, and g_3 is equal to the power in the left-handed or right-handed circular polarized components [43]. If any of the parameters

g_1, g_2, or g_3 has a non-zero value, it indicates the presence of a polarized component in the plane wave. The Stokes parameters are not only related to $|E_H|$, $|E_V|$, and ϕ through Eq. 3.32, but also related to the geometric parameters A, ε, and τ, used previously for describing the polarization ellipse. From Eq. 3.4, Eq. 3.9, and Eq. 3.32, we obtain

$$A^2 = |E_x|^2 + |E_y|^2 = g_0 \quad . \tag{3.34}$$

From Eq. 3.12 and Eq. 3.32, we obtain

$$\sin 2\varepsilon = \frac{g_3}{g_0} \quad ,$$

or $\quad g_3 = g_0 \sin 2\varepsilon = A^2 \sin 2\varepsilon \quad . \tag{3.35}$

From Eq. 3.10 and Eq. 3.32, we obtain

$$\tan 2\tau = \frac{g_2}{g_1} \quad ,$$

or $\quad g_2 = g_1 \tan 2\tau \quad . \tag{3.36}$

By inserting Eq. 3.35 and Eq. 3.36 into Eq. 3.33, we find

$$g_1 = g_0 \cos 2\varepsilon \cos 2\tau = A^2 \cos 2\varepsilon \cos 2\tau \quad . \tag{3.37}$$

Substitution of this equation into Eq. 3.36, in turn, gives

$$g_2 = g_0 \cos 2\varepsilon \sin 2\tau = A^2 \cos 2\varepsilon \sin 2\tau \quad . \tag{3.38}$$

Eq. 3.34, Eq. 3.37, Eq. 3.38, and Eq. 3.35 form a set of expressions which describe the Stokes parameters in terms of the geometric parameters A, ε, and τ. Now, we can group the Stokes parameters in a 4 x 1 column vector called the "Stokes Vector" (in terms of the two set of parameters given below in Eq. 3.39),

$$\vec{g} = \begin{bmatrix} g_0 \\ g_1 \\ g_2 \\ g_3 \end{bmatrix} = \begin{bmatrix} |E_H|^2 + |E_V|^2 \\ |E_H|^2 - |E_V|^2 \\ 2|E_H||E_V|\cos\phi \\ 2|E_H||E_V|\sin\phi \end{bmatrix} = \begin{bmatrix} A^2 \\ A^2\cos 2\varepsilon\cos 2\tau \\ A^2\cos 2\varepsilon\sin 2\tau \\ A^2\sin 2\varepsilon \end{bmatrix} \quad . \tag{3.39}$$

3.5.2 The Stokes Vector for the Partially Polarized Wave.

The partially polarized wave can be defined quantitatively by introducing a coherency matrix [J]. The polarization coherency matrix [J] is evaluated by direct product of the electric field of the wave and its conjugate, then by taking time average.

$$[J] = \langle \vec{E} \otimes \vec{E}^* \rangle = \begin{bmatrix} \langle E_H E_H^* \rangle & \langle E_H E_V^* \rangle \\ \langle E_V E_H^* \rangle & \langle E_V E_V^* \rangle \end{bmatrix} = \begin{bmatrix} J_{HH} & J_{HV} \\ J_{VH} & J_{VV} \end{bmatrix} \tag{3.40}$$

where $\langle\cdots\rangle = \lim_{T\to\infty} [\ \dfrac{1}{2T} \displaystyle\int_{-T}^{T} \langle\cdots\rangle \ dt\ ].$

We can associate the Stokes Vector $\vec{g}$ with the coherency matrix $[J]$. From Eq. 3.32, we have

$$g_0 = |E_H|^2 + |E_V|^2 = \langle E_H E_H^* \rangle + \langle E_V E_V^* \rangle = J_{HH} + J_{VV} \tag{3.41}$$

$$g_1 = |E_H|^2 - |E_V|^2 = \langle E_H E_H^* \rangle - \langle E_V E_V^* \rangle = J_{HH} - J_{VV}$$

$$g_2 = 2|E_H||E_V|\cos\phi = |E_H||E_V|(e^{i\phi}+e^{-j\phi}) = \langle E_H E_V^* \rangle + \langle E_V E_H^* \rangle = J_{HV} + J_{VH}$$

$$g_3 = 2|E_H||E_V|\sin\phi = -j|E_H||E_V|(e^{i\phi}-e^{-j\phi}) = -j\langle E_H E_V^* \rangle + j\langle E_V E_H^* \rangle = -jJ_{HV}+jJ_{VH}$$

A partially polarized wave may be regarded as the sum of a completely unpolarized wave and a completely polarized wave. Thus, we may write g for a partially polarized wave as follows,

$$
\begin{bmatrix} g_0 \\ g_1 \\ g_2 \\ g_3 \end{bmatrix}
= A^2 \begin{bmatrix} 1-p \\ 0 \\ 0 \\ 0 \end{bmatrix}
+ A^2 \begin{bmatrix} p \\ P\cos2\varepsilon\cos2\tau \\ P\cos2\varepsilon\sin2\tau \\ P\sin2\varepsilon \end{bmatrix}
\tag{3.42}
$$

where p is the degree of polarization which is defined as the ratio of unpolarized power to the total power and can be written as

$$p = \dfrac{\sqrt{g_1^2+g_2^2+g_3^2}}{g_0} . \tag{3.43}$$

Table 3.2 Jones Vector $\vec{E}$, Coherency Vector $\vec{J}$, and Stokes Vector $\vec{g}$ for Some States of Polarization.

Polarization	$\vec{E}$	$\vec{J}$	$\vec{g}$
Linear Horizontal	[1 0]	[1 0 0 0]	[1 1 0 0]
Linear Vertical	[0 1]	[0 0 0 1]	[1 -1 0 0]
45° Linear	[1 1]	[1 1 1 1]	[1 0 1 0]
135° Linear	[1 -1]	[1 -1 -1 1]	[1 0 -1 0]

Left-handed Circular	[1 j]	[1 -j j 1]	[1 0 0 -1]
Right-handed Circular	[1 -j]	[1 j -j 1]	[1 0 0 1]

Using Eq.3.41 and Eq.3.43, we can obtain a expression of the degree of polarization p in terms of the coherency matrix [J] as follow

$$p = \sqrt{1 - \frac{4 \left(\det [J] \right)}{(J_{HH} + J_{VV})^2}} \qquad (3.44)$$

Table 3.2 gives the Jones Vector $\vec{E}$, the coherency vector $\vec{J}$ (to be introduced in section 4.4), and the Stokes Vector $\vec{g}$ for special cases of purely
monochromatic wave fields in specific states of polarization.

3.6 The Poincare Sphere

The Poincaré sphere is a useful graphical aid for the visualization of polarization effects. There is a one-to-one correspondence between all possible polarizations and points on the Poincare sphere. Since any wave can
be described by the Stokes Vector $\vec{g}$, from Eq. 3.39, we may say that g_1, g_2,
and g_3 are the Cartesian Coordinate components of vector $\vec{g}$.

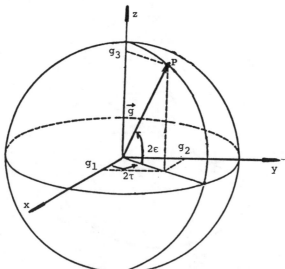

Fig. 3.6 Representation of a Polarization State on
the Poincaré Sphere with Definition of Angle τ and ε

In the Poincaré sphere representation of wave polarization, the polarization state is described by a point P on the sphere where the three
cartesian coordinate components are g_1, g_2 and g_3. So, for any state of a

completely polarized wave which corresponds to one point $P(g_1, g_2, g_3)$ on the sphere of radius g_0, and vice versa. From Figure 3.6, it can be seen that the longitude and latitude of the point P are related to the geometric parameters of the polarization ellipses as follow,

longitude = 2τ

latitude = 2ε

In addition, the magnitude $|\rho_{HV}| = \tan\gamma$, and the phase ϕ, of polarization ratio ρ_{HV} can also be shown on the Poincaré sphere directly as shown in Figure 3.7. From Eq. 3.29 and Eq. 3.39, we find that

$$\frac{g_1}{g_0} = \cos2\varepsilon\cos2\tau = \cos2\gamma \quad .$$

So, the angle 2γ is related to the direction cosine of the Stokes vector $\vec{g}$ with the x-axis, i.e., the angle between $\vec{g}$ and the x-axis. Drawing a projecting line from point P to the YOZ plane, the point P' is the intersecting point; and connecting the point P' and the origin, we have the angle $\angle g_2 op'$. Consider the angle $\angle g_2 OP'$ on the YOZ plane, we find that

$$\tan(\angle g_2 OP') = g_3/g_2 \quad .$$

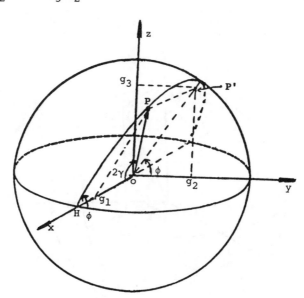

Figure 3.7 The Poincaré Sphere with Parameters γ and ϕ

it is easy to show that the angle $\angle g_2 OP'$ is equal to the relative phase ϕ because Eq. 3.32 gives us $\tan\phi = g_3/g_2$. Also, we know that points P and P' are on the plane of the great circle, and that the angle between the great

circle and the equator is equal to the angle $\angle g_2 OP'$, i.e., equal to the relative phase ϕ.

Also, it can be shown that a polarization state can be represented in different polarization bases. Any polarization basis consists of two unit vectors which are located at two corresponding opposite points on the Poincaré sphere. Figure 3.8 shows how the polarization state P on the Poincaré sphere can be represented in three polarization bases, (HV), (45°135°), and (LR). The complex polarization ratios are given by

$$\rho_{HV} = |\rho_{HV}|e^{j\phi_{HV}} = \tan\gamma \; e^{j\phi_{HV}} \qquad (3.45)$$

$$\rho_{45°135°} = |\rho_{45°135°}|e^{j\phi_{45°135°}} = \tan\alpha \; e^{j\phi_{45°135°}}$$

$$\rho_{LR} = |\rho_{LR}|e^{j\phi_{LR}} = \tan\beta \; e^{j\phi_{LR}}$$

where $\tan\gamma$, $\tan\alpha$, and $\tan\beta$ are the ratio of the magnitudes of the corresponding orthogonal components, and ϕ_{HV}, $\phi_{45°135°}$, and ϕ_{LR} are the phase differences between the corresponding orthogonal components.

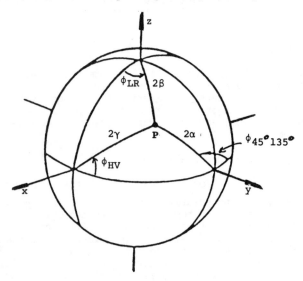

Figure 3.8 The Polarization State P
in the Different Polarization Bases

4. RADAR SCATTERING MATRICES [S] AND [M]

In radar polarimetry, one is concerned with the polarization dependence of a radar scattering process as seen at the receiving antenna terminal at a fixed point in space. Let us use $\vec{E}_T$ and $\vec{E}_R$ to represent transmitting and receiving monochromatic TEM traveling plane wave, respectively. In addition, we can use them to describe the polarization of transmitter and receiver. Then we use antenna height $\vec{h}$ to describe the wave which the re-

ceiving antenna would radiate if it acted as a transmitter [17].

When an electromagnetic plane wave illuminates a target, the polarization of the scattered wave $\vec{E}_R(AB)$, in general, is different from that of the incident wave $\vec{E}_T(AB)$ expressed in the general polarization basis (AB). This change in polarization states represents the characteristic properties of the target expressed by the coherent scattering matrix [S(AB)] also known as the Sinclair matrix (1950) [53]. At the target an incident wave $\vec{E}_T(AB)$ gives rise to a scattered wave $\vec{E}_R(AB)$ which at the receiver is related via the scattering matrix [S(AB)] for the polarization basis (AB), as presented in Mott [45]

$$\vec{E}_R(AB) = \begin{bmatrix} E_{R_A} \\ E_{R_B} \end{bmatrix} = \begin{bmatrix} S_{AA} & S_{AB} \\ S_{BA} & S_{BB} \end{bmatrix} \begin{bmatrix} E_{T_A} \\ E_{T_B} \end{bmatrix} = [S(AB)]\vec{E}_T(AB) \qquad (4.1)$$

using normalization with respect to range and the far-field assumption throughout the remainder of this paper.

Whereas, Eq. 4.1 represents the "radar wave" operator equation which behaves very similar to the transmission matrix in Jones' calculus (Azzam and Bashara 1977 [2]) for the forward scatter case in optics, in radar polarimetry we also need to introduce the voltage equation (Collin 1985 [17])

$$V = \vec{h}^T \vec{E}_R = \vec{h}^T [S] \vec{E}_T \qquad (4.2)$$

which mathematically behaves like a bilinear form when transformation from one basis to another occurs (Kostinski and Boerner 1986 [38]).

The range normalized scattering matrix [S] is used to describe the properties of the radar target at a given direction (or fixed direction of illumination and reception) and the radar frequency, independent of the measurement system.

4.1 Coordinate System and Polarization Bases

The value of the elements S_{ij} of the scattering matrix [S] depend on the chosen coordinate system and polarization basis. It is covenient to choose a fixed cartesian coordinate system (x,y,z) with the origin at the center inside of the scattering target as shown in Fig. 4.1, where the (x,y,z) direction are chosen to coincide with the invariant symmetry axes of the target. However, there are many choices of polarization bases because any two orthogonal unit vectors in the transverse plane can be chosen for a polarization basis. We would like to define a polarization basis uniquely determined by a general coordinate system [14,18,27] described in Fig. 4.1. At any point $T(r_1, \theta_1, \phi_1)$ in the space, we consider a distinct orthogonal basis formed by three spherical unit vectors $-\hat{r}_1$, $-\hat{\theta}_1$, and $\hat{\phi}_1$. They are defined as

$$-\hat{r}_1 = -\sin\theta_1\cos\phi_1\hat{x} - \sin\theta_1\sin\phi_1\hat{y} - \cos\theta_1\hat{z}$$

$$-\hat{\theta}_1 = -\cos\theta_1\cos\phi_1\hat{x} - \cos\theta_1\sin\phi_1\hat{y} + \sin\theta_1\hat{z}$$

$$\hat{\phi}_1 = -\sin\phi_1\hat{x} + \cos\phi_1\hat{y}$$

These unit vectors define a right-handed vector triplet, i.e.,

$$(-\hat{r}_1) \times (-\hat{\theta}_1) = \hat{\phi}_1 \qquad (4.3)$$
$$(-\hat{\theta}_1) \times \hat{\phi}_1 = -\hat{r}_1$$
$$\hat{\phi}_1 \times (-\hat{r}_1) = -\hat{\theta}_1$$

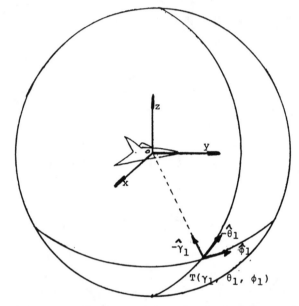

Fig. 4.1 Coordinate System and Polarization Basis
for the Monostatic Case

For an electromagnetic plane wave traveling through the point T, or an antenna located at the point T, we can choose the direction of the unit vector $-\hat{r}_1$ as the plane wave traveling direction $k_1 = -r_1$, and then use the other two spherical unit vectors $-\hat{\theta}_1$ and $\hat{\phi}_1$, to form a polarization basis for this plane wave traveling in the $-\hat{r}_1$ direction.

For the monostatic (backscattering) case, the transmitting antenna T and the receiving antenna R are located at the same position $T(r_1, \theta_1, \phi_1)$. The polarization bases for transmission and reception are chosen to be the same, i.e., $-\theta_1$ and ϕ_1. The polarization states of the transmitter and the receiver can be written as

$$\vec{E}_T = E_{T_{\theta_1}}(-\hat{\theta}_1) + E_{T_{\phi_1}}(\hat{\phi}_1) \qquad (4.4)$$

$$\vec{E}_R = E_{R_{\theta_1}}(-\hat{\theta}_1) + E_{R_{\phi_1}}(\hat{\phi}_1)$$

$$\vec{h} = h_{\theta_1}(-\hat{\theta}_1) + h_{\phi_1}(\hat{\phi}_1)$$

For the bistatic case, the transmitting antenna T and the receiving antenna

R are placed at separate locations $T(r_1, \theta_1, \phi_1)$ and $R(r_2, \theta_2, \phi_2)$, as shown in Fig. 4.2, respectively. The two sets of spherical unit vectors $(-\hat{\theta}_1, \hat{\phi}_1)$ and $(-\hat{\theta}_2, \hat{\phi}_2)$ form the transmitter polarization basis and the receiver polarization basis, respectively. Note,

$$-\hat{r}_2 = - \sin\theta_2\cos\phi_2\hat{x} - \sin\theta_2\sin\phi_2\hat{y} - \cos\theta_2\hat{z}$$
$$-\hat{\theta}_2 = - \cos\theta_2\cos\phi_2\hat{x} - \cos\theta_2\sin\phi_2\hat{y} + \sin\theta_2\hat{z}$$
$$\hat{\phi}_2 = - \sin\phi_2\hat{x} + \cos\phi_2\hat{y}$$

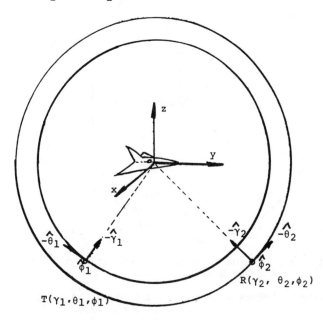

Fig. 4.2 Coordinate System and Polarization
Basis for the Bistatic Case

Thus the polarization states can be written as follows,

$$\vec{E}_T = E_{T_{\theta_1}}(-\hat{\theta}_1) + E_{T_{\phi_1}}(\hat{\phi}_1) \qquad (4.5)$$
$$\vec{E}_R = E_{R_{\theta_2}}(-\hat{\theta}_2) + E_{R_{\phi_2}}(\hat{\phi}_2)$$
$$\vec{h} = h_{\theta_2}(-\hat{\theta}_2) + h_{\phi_2}(\hat{\phi}_2)$$

where E_{T_θ}, E_{R_θ} and h_θ are the θ components of E_T, E_R and h, and E_{T_ϕ}, E_{R_ϕ} and h_ϕ are the ϕ components of E_T, E_R and h.

4.2 The Scattering Matrix for the Bistatic Case

From Eq. 4.5, we know the polarization of incident wave is

$$\vec{E}_T = E_{T_{\theta_1}}(-\hat{\theta}_1) + E_{T_{\phi_1}}(\hat{\phi}_1) = E_{T_A}\hat{A} + E_{T_B}\hat{B}$$

and the polarization of the scattered wave is

$$\vec{E}_R = E_{R_{\theta_2}}(-\hat{\theta}_2) + E_{R_{\phi_2}}(\hat{\phi}_2) = E_{R_A}\hat{A} + E_{R_B}\hat{B}$$

Now the bistatic scattering process according to Eq. 4.1, can be written as

$$
\begin{bmatrix} E_{R_{\theta_2}} \\ E_{R_{\phi_2}} \end{bmatrix}
=
\begin{bmatrix} S_{\theta_2\theta_1} & S_{\theta_2\phi_1} \\ S_{\phi_2\theta_1} & S_{\phi_2\phi_1} \end{bmatrix}
\begin{bmatrix} E_{T_{\theta_1}} \\ E_{T_{\phi_1}} \end{bmatrix}
\tag{4.6}
$$

here $S_{\theta_2\phi_1}$ is a complex number, which is proportional to the received $-\hat{\theta}_2$ component of the scattered wave, while the target is illuminated by $\hat{\phi}_1$ polarized transmitter only, i.e., $E_{R_{\theta_2}} = S_{\theta_2\phi_1}E_{T_{\phi_1}}$ with $E_{T_{\theta_1}} = 0$. In general, $-\hat{\theta}_2$ and $\hat{\phi}_1$ are not orthogonal, thus, in the strict sense, the incident wave polarized in $\hat{\phi}_1$ direction and the scattered wave polarized in $-\hat{\theta}_2$ direction do not constitute a Cross-polarized channel. But, in a broad sense, we may treat them as a cross-polarization channel. Similarly, $\hat{\phi}_1$ and $\hat{\phi}_2$ are not parallel, and in the strict sense, they do not form a pair of Co-polarized channels.

In general, the scattering matrix [S] is not symmetric. Thus, a total of eight parameters which depend on $\hat{\theta}_1$, $\hat{\theta}_2$, $\hat{\phi}_1$, and $\hat{\phi}_2$ are required to give the complete bistatic scattering matrix, however, if only the relative phases are of interest [14,18,27], the total number of parameters would be seven.

4.3 The Scattering Matrix in the Backscattering Case

In the monostatic (backscattering) case, most imaging radars use the same antenna for transmission and reception of the signals located at position $T(r_1,\theta_1,\phi_1)$. The incident wave $\vec{E}^I$ for this target is the electromagnetic traveling plane wave produced by the transmitting antenna $\vec{E}_T$. The scattered wave $\vec{E}^S$ from this target is the electromagnetic traveling plane wave received by the receiving antenna $\vec{E}_R$.

From Eq. 4.4, we know that $\vec{E}_T$ and $\vec{E}_R$ have the same polarization basis $(-\hat{\theta}_1$ and $\hat{\phi}_1)$. For convenience, let us call unit vector $-\hat{\theta}_1$ as horizontal unit vector $\hat{H}$ and unit vector $\hat{\phi}_1$ as vertical unit sector $\hat{V}$. In the basis (HV), one may define a complex 2x2 scattering matrix [S(HV)] according to Eq. 4.1 and Eq. 4.4 as

$$
\begin{bmatrix} E_H^S \\ E_V^S \end{bmatrix} = \begin{bmatrix} S_{HH} & S_{HV} \\ S_{VH} & S_{VV} \end{bmatrix} \begin{bmatrix} E_H^I \\ E_V^I \end{bmatrix}
\tag{4.7}
$$

Here, S_{HV} is also a complex number which is proportional to the received horizontally polarized component of the scattered signal (in amplitude and phase) when the target is only illuminated by a vertically polarized transmitter, i.e., $E_H^S = S_{HV} E_V^I$ with $E_H^I = 0$, and similarly for the other elements. From Eq. 4.6, we can see that the elements S_{HH} and S_{VV} produce the power return in the copolarized channel and the elements S_{HV} and S_{VH} produce the power return in the cross-polarized channel.

If the role of the transmitting and the receiving antenna are interchanged, the reciprocity theorem requires that the scattering matrix be symmetric, provided that the propagation medium is reciprocal, that is

$$
S_{HV} = S_{VH}
\tag{4.8}
$$

There are only five independent parameters in the scattering matrix [S] if only the relative phases are of interest.

4.4 The Mueller Matrix [M] for the Backscattering Case

If scatterer properties change with time, or a number of independent incoherent scatterers interest, the Mueller Matrix [M] must be used. The Mueller Matrix [M] can be defined by

$$
\vec{g}^S = [M]\vec{g}^I
\tag{4.9}
$$

Here, $\vec{g}^S$ is the scattered Stokes Vector and $\vec{g}^I$ is the incident Stokes Vector. The Mueller Matrix [M(HV)] is a 4x4 real matrix. It can be related to the scattering matrix [S(HV)] for the coherent case. In order to obtain the relationship, we introduce a coherency vector $\vec{J}$ which is formed by the elements of the coherency matrix [J] as

$$
\vec{J} = \vec{E} \otimes \vec{E}^* = \begin{bmatrix} E_H E_H^* \\ E_H E_V^* \\ E_V E_H^* \\ E_V E_V^* \end{bmatrix} = \begin{bmatrix} J_{HH} \\ J_{HV} \\ J_{VH} \\ J_{VV} \end{bmatrix}
\tag{4.10}
$$

According to Eq.3.41, it is

$$
\begin{bmatrix} g_0 \\ g_1 \\ g_2 \\ g_3 \end{bmatrix} = \begin{bmatrix} J_{HH} + J_{VV} \\ J_{HH} - J_{VV} \\ J_{HV} + J_{VH} \\ -jJ_{HV} + jJ_{VH} \end{bmatrix}
$$

If we define a matrix $[A]$ as

$$[A] = \begin{bmatrix} 1 & 0 & 0 & 1 \\ 1 & 0 & 0 & -1 \\ 0 & 1 & 1 & 0 \\ 0 & -j & j & 0 \end{bmatrix} \tag{4.11}$$

then

$$\vec{g} = [A]\vec{J} \tag{4.12}$$

Now, by using the relationship between the incident and the scattered electric field of the wave, from Eq. 4.10 and Eq. 4.1

$$\vec{J}^S = \vec{E}^S \otimes \vec{E}^{S*} = ([S]\vec{E}^I) \otimes ([S]^* \vec{E}^{I*})$$

$$= \begin{bmatrix} S_{HH}E_H^I + S_{HV}E_V^I \\ S_{VH}E_H^I + S_{VV}E_V^I \end{bmatrix} \otimes \begin{bmatrix} S_{HH}^* E_H^{I*} + S_{HV}^* E_V^{I*} \\ S_{VH}^* E_H^{I*} + S_{VV}^* E_V^{I*} \end{bmatrix}$$

$$= \left(\begin{bmatrix} S_{HH} & S_{HV} \\ S_{VH} & S_{VV} \end{bmatrix} \otimes \begin{bmatrix} S_{HH}^* & S_{HV}^* \\ S_{VH}^* & S_{VV}^* \end{bmatrix} \right) \begin{bmatrix} J_{HH}^I \\ J_{HV}^I \\ J_{VH}^I \\ J_{VV}^I \end{bmatrix}$$

$$= ([S] \otimes [S]^*)\vec{J}^I$$

So the scattered Stokes Vector $\vec{g}^S$ can be written as

$$\vec{g}^S = [A]\vec{J}^S = [A]([S] \otimes [S]^*)\vec{J}^I = [A]([S] \otimes [S]^*)[A]^{-1}\vec{g}^I$$

Comparing the above equation with Eq. 4.9, we obtain the relationship between the Mueller matrix $[M]$ and the scattering matrix $[S]$.

$$[M] = [A]([S] \otimes [S]^*)[A]^{-1} \tag{4.13}$$

The elements of $[M]$ can be given in terms of the elements of the scattering matrix $[S]$ by

$$M_{11} = \tfrac{1}{2}(S_{HH}S_{HH}^* + S_{VH}S_{VH}^* + S_{HV}S_{HV}^* + S_{VV}S_{VV}^*) = \tfrac{1}{2}(|S_{HH}|^2 + |S_{VH}|^2 + |S_{HV}|^2 + |S_{VV}|^2)$$

$$M_{12} = \tfrac{1}{2}(S_{HH}S_{HH}^* + S_{VH}S_{VH}^* - S_{HV}S_{HV}^* - S_{VV}S_{VV}^*) = \tfrac{1}{2}(|S_{HH}|^2 + |S_{VH}|^2 - |S_{HV}|^2 - |S_{VV}|^2)$$

$$M_{13} = \tfrac{1}{2}(S_{HH}S_{HV}^* + S_{VH}S_{VV}^* + S_{HV}S_{HH}^* + S_{VV}S_{VH}^*) = \mathrm{Re}(S_{HH}S_{HV}^*) + \mathrm{Re}(S_{VH}S_{VV}^*)$$

$$M_{14} = (j/2)(S_{HH}S_{HV}^* + S_{VH}S_{VV}^* - S_{HV}S_{HH}^* - S_{VV}S_{VH}^*) = \mathrm{Im}(S_{HH}S_{HV}^*) + \mathrm{Im}(S_{VH}S_{VV}^*)$$

$$M_{21} = \tfrac{1}{2}(S_{HH}S_{HH}^* - S_{VH}S_{VH}^* + S_{HV}S_{HV}^* - S_{VV}S_{VV}^*) = \tfrac{1}{2}(|S_{HH}|^2 - |S_{VH}|^2 + |S_{HV}|^2 - |S_{VV}|^2)$$

$$M_{22} = \frac{1}{2}(S_{HH}S_{HH}^* - S_{VH}S_{VH}^* - S_{HV}S_{HV}^* + S_{VV}S_{VV}^*) = \frac{1}{2}(|S_{HH}|^2 - |S_{VH}|^2 - |S_{HV}|^2 + |S_{VV}|^2)$$

$$M_{23} = \frac{1}{2}(S_{HH}S_{HV}^* - S_{VH}S_{VV}^* + S_{HV}S_{HH}^* - S_{VV}S_{VH}^*) = \mathrm{Re}(S_{HH}S_{HV}^*) - \mathrm{Re}(S_{VH}S_{VV}^*)$$

$$M_{24} = (j/2)(S_{HH}S_{HV}^* - S_{VH}S_{VV}^* - S_{HV}S_{HH}^* + S_{VV}S_{VH}^*) = \mathrm{Im}(S_{HH}S_{HV}^*) + \mathrm{Im}(S_{VV}S_{VH}^*)$$

$$M_{31} = \frac{1}{2}(S_{HH}S_{VH}^* + S_{VH}S_{HH}^* + S_{HV}S_{VV}^* + S_{VV}S_{HV}^*) = \mathrm{Re}(S_{HH}S_{VH}^*) + \mathrm{Re}(S_{HV}S_{VV}^*)$$

$$M_{32} = \frac{1}{2}(S_{HH}S_{VH}^* + S_{VH}S_{HH}^* - S_{HV}S_{VV}^* - S_{VV}S_{HV}^*) = \mathrm{Re}(S_{HH}S_{VH}^*) - \mathrm{Re}(S_{HV}S_{VV}^*)$$

$$M_{33} = \frac{1}{2}(S_{HH}S_{VV}^* + S_{VH}S_{HV}^* + S_{HV}S_{VH}^* + S_{VV}S_{HH}^*) = \mathrm{Re}(S_{HH}S_{VV}^*) + \mathrm{Re}(S_{HV}S_{VH}^*)$$

$$M_{34} = (j/2)(S_{HH}S_{VV}^* + S_{VH}S_{HV}^* - S_{HV}S_{VH}^* - S_{VV}S_{HH}^*) = \mathrm{Im}(S_{HH}S_{VV}^*) + \mathrm{Im}(S_{VH}S_{HV}^*)$$

$$M_{41} = (j/2)(S_{VH}S_{HH}^* - S_{HH}S_{VH}^* + S_{VV}S_{HV}^* - S_{HV}S_{VV}^*) = \mathrm{Im}(S_{VH}S_{HH}^*) + \mathrm{Im}(S_{VV}S_{HV}^*)$$

$$M_{42} = (j/2)(S_{VH}S_{HH}^* - S_{HH}S_{VH}^* - S_{VV}S_{HV}^* + S_{HV}S_{VV}^*) = \mathrm{Im}(S_{VH}S_{HH}^*) + \mathrm{Im}(S_{HV}S_{VV}^*)$$

$$M_{43} = (j/2)(S_{VH}S_{HV}^* - S_{HH}S_{VV}^* + S_{VV}S_{HH}^* - S_{HV}S_{VH}^*) = \mathrm{Im}(S_{VH}S_{HV}^*) + \mathrm{Im}(S_{VV}S_{HH}^*)$$

$$M_{44} = -\frac{1}{2}(S_{VH}S_{HV}^* - S_{HH}S_{VV}^* - S_{VV}S_{HH}^* + S_{HV}S_{VH}^*) = \mathrm{Re}(S_{HH}S_{VV}^*) - \mathrm{Re}(S_{HV}S_{VH}^*)$$

5. CHANGE OF POLARIZATION BASIS

The change of polarization basis plays an important role in applications of radar polarimetry. All measurable quantities such as voltage, energy density, etc., must remain invariant and orthonomality of any two vectors must be preserved under the change of basis transformation.

5.1 The Basis Transformation Matrix [T] and the Vector Transformation Matrix [U]

In general, consider that there are two systems, let the old system be

$$C = (\hat{e}_1, \hat{e}_2, \hat{e}_3, \ldots, \hat{e}_n) \tag{5.1}$$

and the new system be

$$C' = (\hat{e}_1', \hat{e}_2', \hat{e}_3', \ldots, \hat{e}_n') \tag{5.2}$$

where $\hat{e}_1$, $\hat{e}_2$, ..., and $\hat{e}_n$ are the unit vectors of system C, and $\hat{e}_1'$, $\hat{e}_2'$, ..., and $\hat{e}_n'$ are the unit vectors of system C'. We can express the unit vectors of the new system C' in terms of the unit vectors of the old system C

$$C' = \begin{bmatrix} \hat{e}_1' \\ \hat{e}_2' \\ \cdot \\ \cdot \\ \hat{e}_n' \end{bmatrix} = [T] \begin{bmatrix} \hat{e}_1 \\ \hat{e}_2 \\ \cdot \\ \cdot \\ \hat{e}_n \end{bmatrix} = [T]C \tag{5.3}$$

Eq. 5.3 is the transformation from the old basis to the new basis, $[T]$ is the basis transformation matrix

$$[T] = (\hat{e}_1'|_C \quad \hat{e}_2'|_C \quad \cdots \quad \hat{e}_n'|_C)^T \ , \tag{5.4}$$

where $\hat{e}_i'|_C$ means the ith unit vector of the new basis presented in the old system C. It can be shown that the vector transformation matrix $[U]$ is the transpose matrix of the bases transformation matrix $[T]$, i.e.,

$$[U] = [T]^T \tag{5.5}$$

and

$$\vec{x}|_C = [U]\vec{x}|_{C'} \tag{5.6}$$

where $[U]$ is the vector transformation matrix. It is a unitary matrix, and transforms a vector from the new basis to the old basis.

Finally, from Eq. 5.5 and Eq. 5.4, we obtain

$$[U] = (\hat{e}'_1|_C \quad \hat{e}'_2|_C \quad \cdots \quad \hat{e}'_n|_C) \tag{5.7}$$

which indicates that we can use the new basis unit vectors in the old basis as the column to form the unitary vector transformation matrix $[U]$.

5.2 Unitary Transformation Matrix in Terms of the Polarization Ratio

Now consider the polarization basis which is the two dimensional system. We have two bases C and C', the old basis C is formed by two orthogonal unit vectors $\hat{x}$ and $\hat{y}$, and the new basis C' is formed by another two orthogonal unit vectors α and β; thus, a vector $\vec{E}$ can be expressed in both bases. In the old basis

$$\vec{E}|_C = \begin{bmatrix} E_x \\ E_y \end{bmatrix} = E_x\hat{x} + E_y\hat{y} \tag{5.8}$$

In the new basis

$$\vec{E}|_{C'} = \begin{bmatrix} E_\alpha \\ E_\beta \end{bmatrix} = E_\alpha\hat{\alpha} + E_\beta\hat{\beta}$$

From Eq. 5.7, the unitary transformation matrix $[U]$ is

$$[U] = (\hat{\alpha}|_C \quad \hat{\beta}|_C) \ , \tag{5.9}$$

where $\hat{\alpha}|_C$ and $\hat{\beta}|_C$ are the column matrices which are the unit vectors of the new basis C' and expressed in terms of the old system C. From Eq. 5.6, the vector transformation is

$$\vec{E}|_C = [U]\vec{E}|_{C'}$$

that is

$$\begin{bmatrix} E_x \\ E_y \end{bmatrix} = [U] \begin{bmatrix} E_\alpha \\ E_\beta \end{bmatrix} \tag{5.10}$$

It is easy to express the unitary transformation matrix [U] in terms of the polarization ratio ρ. From Eq. 3.22, the unit vectors $\hat{\alpha}$ and $\hat{\beta}$ of the new system C' can be expressed in the old system C in terms of the polarization ratio ρ.

$$\hat{\alpha}\big|_C = \hat{\alpha}(xy) = \begin{bmatrix} \alpha_x \\ \alpha_y \end{bmatrix} = \frac{1}{\sqrt{1 + \rho\rho^*}} \begin{bmatrix} 1 \\ \rho \end{bmatrix} \tag{5.11}$$

$$\hat{\beta}\big|_C = \hat{\beta}(xy) = \begin{bmatrix} \beta_x \\ \beta_y \end{bmatrix} = \frac{1}{\sqrt{1 + \rho'\rho'^*}} \begin{bmatrix} 1 \\ \rho' \end{bmatrix}$$

where

$$\rho = \frac{\alpha_y}{\alpha_x} = \frac{|\alpha_y|}{|\alpha_x|} e^{j\phi} \quad \text{and} \quad \phi = \phi_{\alpha_y} - \phi_{\alpha_x}$$

$$\rho' = \frac{\beta_y}{\beta_x} = \frac{|\beta_y|}{|\beta_x|} e^{j\phi'} \quad \text{and} \quad \phi' = \phi_{\beta_y} - \phi_{\beta_x}$$

To find the relationship between ρ and ρ', and because $\hat{\alpha}$ and $\hat{\beta}$ are ortho-normal, we find

$$||\hat{\alpha}|| = ||\hat{\beta}|| = 1 \tag{5.12}$$

and

$$\hat{\alpha} \cdot \hat{\beta} = 0 \tag{5.13}$$

The dot product of $\hat{\alpha}$ with $\hat{\beta}$, yields

$$\hat{\alpha} \cdot \hat{\beta} = \left(\frac{1}{\sqrt{1 + \rho\rho^*}} \begin{bmatrix} 1 & \rho^* \end{bmatrix} \right) \left(\frac{1}{\sqrt{1 + \rho'\rho'^*}} \begin{bmatrix} 1 \\ \rho' \end{bmatrix} \right)$$

$$= \frac{1}{\sqrt{1 + \rho\rho^*} \sqrt{1 + \rho'\rho'^*}} (1 + \rho^*\rho')$$

then comparing the result with Eq. 5.13, yields $1 + \rho^*\rho' = 0$, that means $\rho' = -1/\rho^*$, substituting this into Eq. 5.11, we find

$$\hat{\beta}\big|_C = \frac{1}{\sqrt{1 + (-1/\rho^*)(-1/\rho)}} \begin{bmatrix} 1 \\ -1/\rho^* \end{bmatrix}$$

$$= \frac{|\rho|}{\sqrt{1 + \rho\rho^*}} \; (-1/\rho^*) \begin{bmatrix} -\rho^* \\ 1 \end{bmatrix}$$

$$= \frac{|\rho|}{\sqrt{1 + \rho\rho^*}} \; \frac{-1}{|\rho|e^{-j\phi}} \begin{bmatrix} -\rho^* \\ 1 \end{bmatrix}$$

$$= \frac{e^{j\phi}}{\sqrt{1 + \rho\rho^*}} \begin{bmatrix} +\rho^* \\ -1 \end{bmatrix} \tag{5.14}$$

Using Eq. 5.11 and Eq. 5.14 to form unitary transformation matrix [U] according to Eq. 5.9, yields

$$[U] = (\hat{\alpha}|_C \quad \hat{\beta}|_C) = \frac{1}{\sqrt{1 + \rho\rho^*}} \begin{bmatrix} 1 & \rho^* e^{j\phi} \\ \rho & -e^{j\phi} \end{bmatrix} \cdot \tag{5.15}$$

5.3 Transformation of a Polarization State from Linear to Circular Basis

Assuming an old basis C is the linear basis (HV), $\hat{H}$ is horizontal unit vector, $\hat{V}$ is the vertical unit vector, and a new basis C' is the circular basis (LR), $\hat{L}$ is the left-handed circular unit vector, $\hat{R}$ is the right-handed circular unit vector, from Eq. 3.23, expressed in the (HV) basis, $\hat{L}$ and $\hat{R}$ can be written as

$$\hat{L}|_C = \hat{L}(HV) = \frac{1}{\sqrt{2}} \begin{bmatrix} 1 \\ j \end{bmatrix} \tag{5.16}$$

$$\hat{R}|_C = \hat{R}(HV) = \frac{1}{\sqrt{2}} \begin{bmatrix} 1 \\ -j \end{bmatrix}$$

According to Eq. 5.7, the unitary transformation matrix [U] then becomes

$$[U] = (\hat{L}|_C \quad \hat{R}|_C) = \frac{1}{\sqrt{2}} \begin{bmatrix} 1 & 1 \\ j & -j \end{bmatrix} \tag{5.17}$$

If a polarization state is $\vec{E}(HV)$ in the linear basis C, and is $\vec{E}'(LR)$ in the circular basis C', then the polarization state transformation from Eq. 5.6, is

$$\vec{E}(HV) = [U]\vec{E}'(LR) \tag{5.18}$$

or

$$\vec{E}'(LR) = [U]^{-1}\vec{E}(HV)$$

Because [U] is a unitary matrix, we find

$$[U]^{-1} = [U]^{*T} = \frac{1}{\sqrt{2}} \begin{bmatrix} 1 & -j \\ 1 & j \end{bmatrix} \tag{5.19}$$

For example, consider a left-handed circular polarization state is
$\vec{E}(HV) = \dfrac{1}{\sqrt{2}} \begin{bmatrix} 1 \\ j \end{bmatrix}$ in the linear basis (HV) and is $\vec{E}'(LR) = \begin{bmatrix} 1 \\ 0 \end{bmatrix}$ in the
circular basis C'. Now we apply Eq. 5.18 and Eq. 5.17 to transform this polarization state from the circular basis to the linear basis.

$$\vec{E}(HV) = \frac{1}{\sqrt{2}} \begin{bmatrix} 1 & 1 \\ j & -j \end{bmatrix} \begin{bmatrix} 1 \\ 0 \end{bmatrix} = \frac{1}{\sqrt{2}} \begin{bmatrix} 1 \\ j \end{bmatrix}$$

Also, we can use Eq. 5.18 and Eq. 5.19 to transfer this polarization state from a linear basis to the circular basis.

$$\vec{E}'(LR) = \left(\frac{1}{\sqrt{2}} \begin{bmatrix} 1 & -j \\ 1 & j \end{bmatrix} \right)\left(\frac{1}{\sqrt{2}} \begin{bmatrix} 1 \\ j \end{bmatrix} \right) = \frac{1}{2} \begin{bmatrix} 2 \\ 0 \end{bmatrix} = \begin{bmatrix} 1 \\ 0 \end{bmatrix}$$

Representing [U] by the polarization ratio ρ, from Eq. 5.11, we obtain

$$\hat{L}\big|_C = \hat{\alpha}\big|_C = \frac{1}{\sqrt{1 + \rho\rho^*}} \begin{bmatrix} 1 \\ \rho \end{bmatrix} = \frac{1}{\sqrt{2}} \begin{bmatrix} 1 \\ j \end{bmatrix}$$

which gives $\rho = j$. From Eq. 5.14, we obtain

$$\hat{R}\big|_C = \hat{\beta}\big|_C = \frac{e^{j\phi}}{\sqrt{1 + \rho\rho^*}} \begin{bmatrix} \rho^* \\ -1 \end{bmatrix} = \frac{1}{\sqrt{2}} \begin{bmatrix} 1 \\ -j \end{bmatrix}$$

and we find that $e^{j\phi} = j$ and that $\phi = \pi/2$. Substituting above values into Eq. 5.15, we find

$$[U] = \frac{1}{\sqrt{1 + \rho\rho^*}} \begin{bmatrix} 1 & +\rho^* e^{+j\pi/2} \\ \rho & -e^{+j\pi/2} \end{bmatrix} = \frac{1}{\sqrt{1 + \rho\rho^*}} \begin{bmatrix} 1 & j\rho^* \\ \rho & -j \end{bmatrix} \tag{5.20}$$

5.4 Transformation of the Scattering Matrix [S] from (HV) Basis to a New Basis (AB) for the Monostatic Case

The received voltage according to Eq. 4.2, is

$$V = \vec{h}^T [S]\vec{E}_T = \vec{h} \cdot [S]\vec{E}_T = \vec{E}_R \cdot [S]\vec{E}_T \tag{5.21}$$

In the linear (HV) basis, the voltage is

$$V(HV) = E_R(HV) \cdot [S(HV)]E_T(HV) \tag{5.22}$$

In the new basis (AB), the voltage is

$$V'(AB) = E'_R(AB) \cdot [S'(AB)]E'_T(AB) \tag{5.23}$$

Appling unitary transformation matrix $[U]$ in Eq. 5.22, yields

$$V(HV) = ([U]E'_R(AB)) \cdot [S(HV)]([U]E'_T(AB))$$

$$= E'_R(AB)^T[U]^T[S(HV)][U]E'_T(AB)$$

$$= E'_R(AB) \cdot ([U]^T[S(HV)][U])E'_T(AB)$$

Comparing this result with Eq. 5.23, and because the received voltage equation remains unchanged under a transformation of the basis, the transformation of the scattering matrix $[S(AB)]$ becomes

$$[S'(AB)] = [U]^T[S(HV)][U] = \begin{bmatrix} S'_{AA} & S'_{AB} \\ S'_{AB} & S'_{BB} \end{bmatrix} \tag{5.24}$$

Substituting Eq. 5.20 into Eq. 5.24, yields

$$[S'(AB)] = \frac{1}{1 + \rho\rho^*} \begin{bmatrix} 1 & \rho \\ j\rho^* & -j \end{bmatrix} \begin{bmatrix} S_{HH} & S_{HV} \\ S_{HV} & S_{VV} \end{bmatrix} \begin{bmatrix} 1 & j\rho^* \\ \rho & -j \end{bmatrix}$$

We can get the following elements of $[S'(AB)]$ in terms of the elements of $[S(HV)]$,

$$S'_{AA} = \frac{1}{1 + \rho\rho^*} (S_{HH} + 2\rho S_{HV} + \rho^2 S_{VV}) \tag{5.25}$$

$$S'_{AB} = \frac{1}{1 + \rho\rho^*} (\rho S_{VV} + (1-\rho\rho^*)S_{HV} - \rho^* S_{HH})$$

$$S'_{BB} = \frac{-1}{1 + \rho\rho^*} (\rho^{*2} S_{HH} - 2\rho^* S_{HV} + S_{VV})$$

In general, it can be shown that

$$Span\{[S'(AB)]\} = |S'_{AA}|^2 + |S'_{AB}|^2 + |S'_{BA}|^2 + |S'_{BB}|^2 \tag{5.26}$$

$$= |S_{HH}|^2 + |S_{HV}|^2 + |S_{VH}|^2 + |S_{VV}|^2$$

$$= Span\{[S(HV)]\}$$

Thus, the $Span\{[S]\}$ is invariant under the change of basis transformation.

6. OPTIMAL POLARIZATION STATES

The Optimal polarization problem is to find such polarization states of the transmitted and received waves for a target of known scattering matrix $[S]$, that the voltage developed across the receiving antenna terminals is maximized (or minimized). Mathematically, it is to find $\vec{E}_T$ and $\vec{h}$ so that the received voltage in the voltage Eq. 5.21 is maximized (or minimized) for a

given [S].

6.1 Optimal Polarization for the Completely Polarized Wave

First, let us consider the completely polarized wave case, i.e., both of the transmitted and the received waves completely polarized.

6.1.1 Optimal Polarizations for the Monostatic Case.

There are two methods to solve this problem.

Method 1. Graves (1956) [26] defined the so-called "power scattering matrix [G]" as

$$[G] = [S]^+[S] \tag{6.1}$$

As usual, the $[\ ]^+ = [*]^T$ denotes Hermitian conjugate. This problem becomes an eigenvalue problem.

$$[G]\vec{E}_{T_n} = \mu_n \vec{E}_{T_n} \tag{6.2}$$

where $n = 1,2,3,\ldots$. The eigenvectors $\vec{E}_{T_n}$ of [G] are "optimal" polarization states which yield the optimal backscattered power at the transmitter location T for the given target of known scattering matrix [S]. [G] is Hermitian for any [S]. The eigenvalues μ_n of [G] are real which yield the power returned from the target at position T. It can be obtained by premultiplying $\vec{E}_T$ on both sides of Eq. 6.2.

$$\vec{E}_{T_n}^+([S]^+S)\vec{E}_{T_n} = ([S]\vec{E}_{T_n})^+([S]\vec{E}_{T_n}) = |\vec{E}_{R_n}|^2 = \mu_n$$

Method 2. Kennaugh's "Pseudo-Eigenvalue Equation"

Kennaugh (1952) [36] originated the concept of optimal polarization states in the monostatic case and demonstrated that the optimal polarization states for a given target satisfy the following nonlinear equation,

$$[S]X_n = \mu_n X_n^* \tag{6.3}$$

where $n = 1,2,3,\ldots$. The eigenvalues μ_n are complex numbers. X_n are optimal in the sense that the absolute values of the complex scalar obtained by the Euclidean inner product as the $|C_n| = |\langle X_n \cdot [S]X_n \rangle| = |\mu_n|$ are optimal. Kennaugh denoted the C_n to be the backscatter voltages.

6.1.2 The Three-step Optimization Approach for the General Case.

This method enables one to treat symmetric, asymmetric, monostatic and bistatic cases in an identical manner [36].

Step 1

The total normalized power (energy density) P_W in the scattered wave is given by $\vec{E}_R^+ \vec{E}_R$, where

$$P_W = \vec{E}_R^+ \vec{E}_R = ([S]\vec{E}_T)^+[S]\vec{E}_T = \vec{E}_T^+[S]^+[S]\vec{E}_T = \vec{E}_T^+[G]\vec{E}_T \tag{6.4}$$

We need to find such an $\vec{E}_T$ for which P_W is extremum for a given [S]. It translates into the following eigenvalue problem:

$$[G]\vec{E}_T = \lambda\vec{E}_T \tag{6.5}$$

The characteristic equation of Eq. 6.5, is

$$\{[G] - \lambda[I]\}\vec{E}_{T.OPT} = 0 \tag{6.6}$$

where [I] is the identity matrix. The explicit solution for the eigenvalues is given by a simple quadratic equation

$$\lambda^2 - (G_{11} + G_{22})\lambda + (G_{11}G_{22} - G_{12}G_{21}) = 0 \tag{6.7}$$

where

$$\lambda_{1,2} = \frac{1}{2}\left[\text{Tr}\{[G]\} \pm \sqrt{\text{Tr}^2\{[G]\} - 4\det\{[G]\}}\right]$$

The eigenvalues λ_1 and λ_2 are real because [G] is Hermitian which agrees with their physical interpretation as power. Substituting the eigenvalues (λ_1, λ_2) into Eq. 6.6 and solving for the components of $\vec{E}_{T,OPT}$, the eigenvector $\vec{E}_{T,OPT}$ is the polarization state of a transmitter so that the power in the scattered wave is maximized. If the two eigenvalues (λ_1, λ_2) are not equal, the two eigenvectors $(\vec{E}_{T,OPT})$ are orthogonal and the maximum is achieved by $\vec{E}_{T,OPT}$ corresponding to the largest eigenvalue (the minimum—for the smallest one). Furthermore, the sum of the two eigenvalues, i.e., the total energy, is an invariant, as it was shown before in Eq.5.26

$$\lambda_1 + \lambda_2 = \text{Tr}\{[G]\} = \text{Span}\{[S]\} = \text{invariant} . \tag{5.26}$$

Step 2

Compute this scattered wave by using the known scattering matrix [S] and $\vec{E}_{T,OPT}$ from Eq. 6.6

$$\vec{E}_{R,OPT} = [S]\vec{E}_{T,OPT} \tag{6.8}$$

so that the scattered polarization state $\vec{E}_{R,OPT}$, associated with the optimal transmitting polarization state $\vec{E}_{T,OPT}$, is completely specified. In general, $\vec{E}_{R,OPT} \neq \vec{E}_{T,OPT}$, because the eigenvectors of [G] are not identical to those of [S], unless [S] is a normal matrix.

Step 3

In order to ensure polarization match, i.e., to receive all of the power contained in the scattered wave, one must adjust the receiver polarization state, which results in

$$\vec{h}_{OPT} = \frac{\vec{E}^*_{R,OPT}}{||\vec{E}_{R,OPT}||} = \frac{\{[S]\vec{E}_{T,OPT}\}^*}{||[S]\vec{E}_{T,OPT}||} \tag{6.9}$$

where $|| \cdot ||$ indicates the norm. Eq. 6.9 completes the optimization process.

6.1.3 Basis Transformation Procedure. Let $\vec{E}_T$ and $\vec{E}_R$ represent the transmitting and receiving monochromatic traveling plane waves in the old basis (HV), and $\vec{E}'_T$ and $\vec{E}'_R$ in the new basis (AB), respectively. For the coherent case the total received power according to the voltage equation of Eq. 4.2 can be expressed as

$$P = |V|^2 = |\vec{E}_R^T [S] \vec{E}_T|^2 = |\vec{E}'_R{}^T [S'] \vec{E}'_T|^2 \tag{6.10}$$

so that for the monostatic backscattering case, the co-polarized power P_{co} can be calculated for $\vec{E}_R = \vec{E}_T = \vec{E}$ and $\vec{E}'_R = \vec{E}'_T = \vec{E}'$ as

$$P_{co} = |V_{co}|^2 = |\vec{E}^T [S] \vec{E}|^2 = |\vec{E}'^T [S'] \vec{E}'|^2 \tag{6.11}$$

Similarly, the cross-polarized power P_x can be calculated for $\vec{E}_R = \vec{E}_T = \vec{E}$ and $\vec{E}'_R = \vec{E}'_T = \vec{E}'$ with indicating the orthogonal vector as

$$P_x = |V_x|^2 = |\vec{E}^T [S] \vec{E}|^2 = |\vec{E}'^T [S'] \vec{E}'|^2 \tag{6.12}$$

In reference to the optimal polarization states, we refer to those polarization states which produce maximum and minimum returns in the co/cross-polarized channels. Following Kennaugh [36] and Huynen [30], it is shown in Agrawal and Boerner [1] that there exist four pairs of cross-polarization null states (X-POL NULLS), co-polarization null states (CO-POL NULLS), cross-polarization maximum states (X-POL-MAXS), and co-polarization maximum states (CO-POL-MAXS), which may be derived as follows:

X-POL Nulls and CO-POL Maxima

It can be shown that for the monostatic reciprocal case the X-POL Nulls and CO-POL Maxima are identical, which can be determined either from Eq. 6.6 and Eq. 6.8, or from Eq. 6.12 and Eq. 5.25 ($S'_{AB}=0$) so that

$$\rho_{xn1,2} = \frac{-B \pm \sqrt{B^2 - 4AC}}{2A} \tag{6.13}$$

with $A = S_{HH}^* S_{HV} + S_{HV}^* S_{VV}$, $B = |S_{HH}|^2 - |S_{VV}|^2$, $C = -A^*$. where $\rho_{xn1} \rho_{xn2}^* = -1$, satisfying the orthonormality condition. Equivalently, the scattering matrix in the new X-POL-NULL (eigenvector $\rho_{xn1,2}$) basis becomes

$$[S_d] = \begin{bmatrix} \lambda_1 & 0 \\ 0 & \lambda_2 \end{bmatrix} \tag{6.14}$$

where $\lambda_1 = |\lambda_1| e^{j\phi_1}$ and $\lambda_2 = |\lambda_2| e^{j\phi_2}$. So that the maximum power in the co-polarized channels becomes

$$P_{co_1}(\rho_{xn1}) = |\lambda_1|^2 = m^2$$

$$P_{co_2}(\rho_{xn2}) = |\lambda_2|^2$$

CO-POL Nulls and X-POL Maxima

From Eq. 6.11 and Eq. 5.25 it can be shown that the COPOL-Nulls are given from $S'_{AA} = 0$ by

$$\rho_{cn1,2} = \frac{-S_{HV} \pm \sqrt{S_{HV}^2 - S_{HH}S_{VV}}}{S_{VV}} \tag{6.15}$$

and taking $P_{co} = 0$, the COPOL-Nulls in the XPOL-Null basis are found from

$$\rho'^2\lambda_2 + \lambda_1 = 0 \quad \text{or} \quad \rho'^2 = -\left|\frac{\lambda_1}{\lambda_2}\right|e^{j(\phi_1 - \phi_2)} \tag{6.16}$$

Then, by introducing for $|\lambda_1| = m$, $|\lambda_2/\lambda_1| = \tan \gamma$, $\phi_1 = \xi + 2\nu$, $\phi_2 = \xi - 2\nu$ the scattering matrix in the X-POL Null basis can be rewritten as

$$[S_d] = e^{j\xi}\begin{bmatrix} me^{j2\nu} & 0 \\ 0 & m\tan^2\gamma\, e^{-j2\nu} \end{bmatrix} \tag{6.17}$$

which is equivalent to Huynen's diagonal matrix, where the parameter m^2 represents the co-polarized maximum power return; 4ν represents the phase between the maximum returns in the co-polarized channels (target skip angle with $-45° \leq \nu \leq 45°$), and the ratio of the maximum energy amplitudes $\tan^2\gamma$ defines the target characteristic angle γ ($0 \leq \gamma \leq 45°$), with ξ representing the absolute target phase according to Huynen[30].

It can now be shown [1] that the maximum returns in the co-polarized channel become

$$P_{co}^{max}(\rho_{xn1}) = |\lambda_1|^2 = m^2 \quad \text{and} \quad P_{co}(\rho_{xn2}) = m^2\tan^2\gamma \tag{6.18}$$

which represent the true maxima according to Eq. 6.6 and Eq. 6.8. Similarly, the maximum power obtained in the cross polarized channels can be determined [1] to become

$$P_x(\rho_{xm1,2}) = [m/(2\cos^2\gamma)]^2 \tag{6.19}$$

with the corresponding power in the co-polarized channel not being necessarily zero but

$$P_{co}(\rho_{xm1,2}) = [(m\cos 2\gamma)/2\cos^2\gamma)]^2 \tag{6.20}$$

and the cross-pol max locations ($\rho_{xm1,2} = \pm e^{j(2\gamma+\pi/2)}$ do not coincide with that of the co-pol nulls unless $\gamma = 45°$.

The Polarization Tree (Fork)

It can be shown that the pairs of co/cross-pol nulls and maxima all are lying on one main circle on the Polarization Sphere for the mono-static reciprocal case, which is the "so-called target characteristic circle". Because the pair of co-pol maxima (eigenvectors) and cross-pol nulls is identical, there are a total of six optimal polarization states and they form the "polarization tree (fork)" with one handle and six prongs as shown in

352

Fig. 6.1. It is shown that the XPOL-Nulls (COPOL Maxs) X_1, X_2 are orthogonal to each other, where at X_1 ($|\lambda_1|$ = m), the maximum power return occurs. The COPOL-Nulls (C_1, C_2) are located closer to the second XPOL-NULL (X_2), where the power sub-maximum occurs, and the diameter X_1X_2 connecting the two antipodal orthogonal polarization states X_1 and X_2 bisects the angle C_1 – C_2. The XPOL-MAXS (S_1,S_2) are also perpendicular to the diameter X_1, X_2 on the main circle as shown in Fig. 6.1.

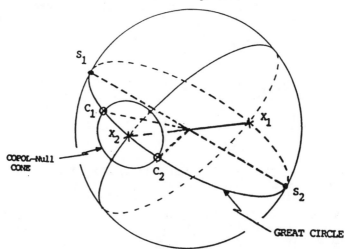

Figure 6.1   Representation of the Characteristic Polarization States on the Poincare Sphere (XPOL-Nulls: $X_{1,2}$; COPOL-Nulls: $C_{1,2}$; XPOL-Maxs: $S_{1,2}$)

Graphic Presentation for Coherent Case

Whereas for the illustration of the discret mini-max polarimetric target description the polarization fork presentation on the Poincaré sphere is very useful, it is found expedient for the continuous plots of the received co/cross-polarized powers to introduce planar spectral power maps, where for normalized norm $||\vec{E}||^2$ = 1, we obtain according to Eq. 6.11 abd Eq. 6.12

$$\text{Co-Pol Power Plot: } P_{co} = |S_{AA}(\phi,\tau)|^2 \qquad (6.21)$$

$$\text{Cross-Pol Power Plot: } P_x = |S_{AB}(\phi,\tau)|^2 \qquad (6.22)$$

as well as the relative

$$\text{Co-Pol Phase Diff.: } \Phi(\phi,\tau) = |\phi_{AA}-\phi_{BB}|^2 \qquad (6.23)$$

$$\text{Cross-Pol Phase Diff.: } X(\phi,\tau) = |\phi_{AA}-\phi_{AB}| \qquad (6.24)$$

expressed in terms of the general transceiver polarization states (ϕ,τ).

For one specific case, the scattering matrix together with the Poincaré sphere polarization tree description and the four spectral plots are pre-

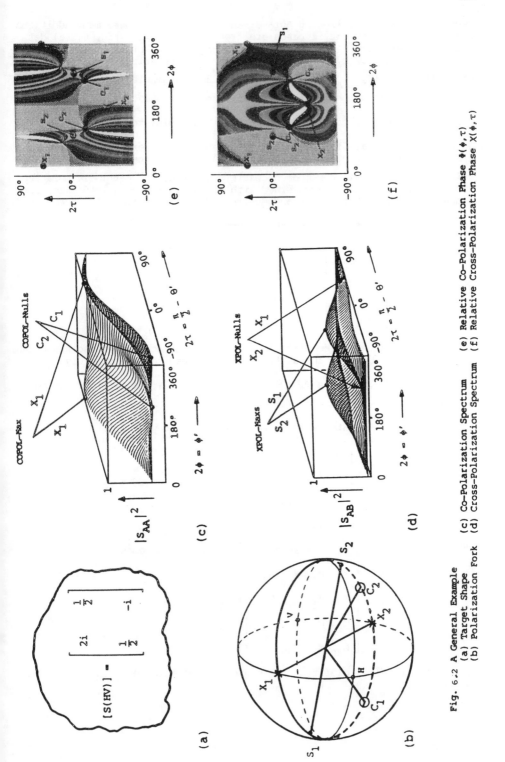

Fig. 6.2 A General Example
(a) Target Shape
(b) Polarization Fork
(c) Co-Polarization Spectrum
(d) Cross-Polarization Spectrum
(e) Relative Co-Polarization Phase $\Phi(\phi, \tau)$
(f) Relative Cross-Polarization Phase $X(\phi, \tau)$

sented in Fig. 6.2 following Agrawal and Boerner [1], where many additional specific scattering matrix cases are treated in detail. Fig. 6.2(a) describes the sysmmetric scattering matrix, Fig. 6.2(b) presents the polarization fork, Fig. 6.2(c,d) displays the co-/cross-polarization power spectral plots $|S_{AA}(\phi,\tau)|^2$, $|S_{AB}(\phi,\tau)|^2$, and Fig.6.2(e,f) displays the relative co-/cross-polarization phase plots $\Phi(\phi,\tau)$, $\chi(\phi,\tau)$, respectively[1].

6.2 Optimization Procedures for the Partially Polarized Case

We consider the case for which a time-dependent scatterer is illuminated by a monochromatic (completely polarizes) wave $\vec{E}_T$, and the reflected wave $\vec{E}_R$, which is, in general, non-monochromatic and therefore, partially polarized. From here on, we will be concerned with the partially polarized case, consequently, the Stokes vector and Mueller matrix formalism will be employed.

Let $\vec{T}$ be the Stokes vector of the transmitted wave $\vec{E}_T$,

$$\vec{T} = \begin{bmatrix} T_0 \\ T_1 \\ T_2 \\ T_3 \end{bmatrix} = \begin{bmatrix} T_0 \\ \hline \tilde{T} \end{bmatrix} \tag{6.25}$$

where ($\tilde{}$) indicates three-dimensional Stokes subspaces. Because the transmitted wave incident on the scatterer is assummed to be completely polarized, p = 1, we have $T_0 = \sqrt{T_1^2 + T_2^2 + T_3^2} = || \tilde{T} || = 1$ for the normalized input.

Let $\vec{R}$ stand for the Stokes vector of the reflected wave $\vec{E}_R$,

$$\vec{R} = \begin{bmatrix} R_0 \\ R_1 \\ R_2 \\ R_3 \end{bmatrix} = \begin{bmatrix} R_0 \\ \hline \tilde{R} \end{bmatrix} \tag{6.26}$$

Then the scattering process according to Eq. 4.9, becomes

$$\vec{R} = \begin{bmatrix} R_0 \\ R_1 \\ R_2 \\ R_3 \end{bmatrix} = \begin{bmatrix} M_{00} & M_{01} & M_{02} & M_{03} \\ M_{10} & M_{11} & M_{12} & M_{13} \\ M_{20} & M_{21} & M_{22} & M_{23} \\ M_{30} & M_{31} & M_{32} & M_{33} \end{bmatrix} \begin{bmatrix} T_0 \\ T_1 \\ T_2 \\ T_3 \end{bmatrix} \tag{6.27}$$

As it is partially polarized, the degree of polarization p for the reflected wave is

$$p = \frac{\sqrt{R_1^2 + R_2^2 + R_3^2}}{R_0} = \frac{||\tilde{R}||}{R_0} \tag{6.28}$$

According to Eq. 3.43, the reflected wave $\vec{R}$ can be decomposed into its completely polarized component R_p and unpolarized component R_U as

$$\vec{R} = R_p + R_U = \begin{bmatrix} pR_0 \\ R_1 \\ R_2 \\ R^3 \end{bmatrix} + \begin{bmatrix} (1-p)R_0 \\ 0 \\ 0 \\ 0 \end{bmatrix} \tag{6.29}$$

Since the energy density is given by the first element of the Stokes vector $\vec{R}$, we can write the following expression for the total available intensity (normalized),

$$P = pR_0 + \frac{1}{2}(1-p)R_0 = \frac{1}{2}(1+p)R_0 \tag{6.30}$$

where the first term represents adjustable intensity and the second term corresponds to noise-like intensity with 50% reception efficiency [40].

6.2.1 Determination of Optimization Criteria. There are four types of power terms that can be optimized:

R_0 Total power in the scattered wave before it reaches the receiver (R);   (6.31a)

pR_0 Completely polarized part of the power;   (6.31b)

$(1-p)R_0$ Noise or the unpolarized part: Regardless of the receiver, one half of the unpolarized part is always accepted. Thus, $R_0(1-p)/2$ is always accepted;   (6.31c)

$(1+p)R_0/2$ $pR_0 + (1-p)R_0/2$ = maximum of the total received power, i.e., the sum of the matched polarized part and one half the unpolarized part. However, if the polarized part is mismatched (cancelled with the proper receiver tuning), the total received power is minimal and equal half the unpolarized part.   (6.31d)

6.2.2 Adjustable "Completely Polarized" Intensity Optimization (6.31b)
Optimizing the "adjustable" (i.e., polarization-dependent) part, $pR_0 = \sqrt{R_1^2 + R_2^2 + R_3^2} = ||\tilde{R}||$, requires that pR_0 be optimized (maximized or minimized) as a function of T for a given [M]. In order to optimize the "adjustable" intensity term pR_0, we rewrite Eq.6.12 in the more convenient form

$$\begin{bmatrix} R_0 \\ \tilde{R} \end{bmatrix} = \begin{bmatrix} M_{00} & \tilde{n}^T \\ \tilde{m} & [M] \end{bmatrix} \begin{bmatrix} T_0 \\ \tilde{T} \end{bmatrix}. \tag{6.32a}$$

where

$$\tilde{n} = \begin{bmatrix} M_{01} \\ M_{02} \\ M_{03} \end{bmatrix}, \quad \tilde{m} = \begin{bmatrix} M_{10} \\ M_{20} \\ M_{30} \end{bmatrix}, \quad [\tilde{M}] = \begin{bmatrix} M_{11} & M_{12} & M_{13} \\ M_{21} & M_{22} & M_{23} \\ M_{31} & M_{32} & M_{33} \end{bmatrix}$$

For the normalized incident wave $\vec{T}$, which is completely polarized, i.e., $||T|| = T_0 = 1$, Eq. 6.16 is reduced to

$$R_0 = M_{00} + \tilde{n}^T \tilde{T} \tag{6.32b}$$
$$\tilde{R} = \tilde{m} + [\tilde{M}]\tilde{T}$$

From above equation we find

$$||\tilde{R}|| = \tilde{R}_T\tilde{R} = (\tilde{m} + [\tilde{M}]\tilde{T})^T(\tilde{m} + [\tilde{M}]\tilde{T})$$
$$= ||\tilde{m}||^2 + 2\tilde{m}^T[\tilde{M}]\tilde{T} + \tilde{T}^T[\tilde{G}]\tilde{T} \tag{6.33}$$

where $[\tilde{G}] = [\tilde{M}]^T[\tilde{M}]$

Let us for convenience rewrite Eq. 6.18 in the index notation as

$$F = \Sigma_{ij} \, a_{ij} \, x_i x_j + 2b_i x_i + c \tag{6.34}$$

where we define

$$F = ||\tilde{R}||^2, \quad a_{ij} = [\tilde{G}], \quad b_i = \Sigma_j m_j M_{ji}, \quad c = ||\tilde{m}||^2, \quad x_i = \tilde{T}$$

We are looking for an extremum of F as a function of the x_i terms on a unit sphere. The problem can be visualized as a search for an extremum of a distance from an origin to a displaced "ellipsoidal" suface. We search for such an extremum according to the following sequence of steps

(1) Rotate to principal axis via a rotation matrix O_{ij} resulting in

and
$$F' = \Sigma_i \lambda_i y_i^2 + 2b_i' y_i + c$$
$$b_i' \equiv \Sigma_j O_{ji} b_j \quad \text{with } i,j=1,2,3 \tag{6.35}$$

where

$$y_i = \Sigma_j O_{ij} x_j \quad \Leftrightarrow \quad x_i = \Sigma_j O_{ji} y_j . \tag{6.36}$$

The rotation matrix O_{ij} has the eigenvectors of the matrix a_{ij} as its columns [51,55]. Note that the rotation does not change the magnitude of the distances and, therefore, does not affect the sought extrema.

(2) Determining the extrema of F' via the method of Lagrange multipliers (μ), results in the inhomogeneous set of linear equations

$$(\lambda_i - \mu) \, y_i = -b_i' \tag{6.37}$$

which, after inversion, give (with yet unknown Lagrange multiplier μ and non-zero b_i-s)

$$y_i = \left[\frac{-b_i'}{\lambda_i - \mu} \right] \tag{6.38}$$

3) Substituting Eq. 6.38 in the normalization constraint $\Sigma_i |y_i|^2 = 1$ leads to a sixth order polynomial equation for μ,

$$\sum_{i=1}^{3} \left[\frac{-b_i'}{\lambda_i - \mu} \right]^2 = 1 \tag{6.39}$$

which, in general, must be solved numerically for the μ's. These roots are then substituted into Eq. 6.38 to find the y_i's and into Eq. 6.36 to find the x_i's. Finally, the intensity is computed according to Eq. 6.34 for all six roots of Eq. 6.39, and the largest (or smallest) intensity is used to choose the optimal solution.

6.2.3 Numerical Example and Graphical Presentation of Results for Partially Coherent Case. Using the numerical example given in Kostinski et al [40], the power spectral plots

$$||\tilde{R}(\phi,\tau)|| = pR_0 \tag{6.33}$$

for the adjustable coherent power term as functions of orientation angle (ϕ) and ellipticity (τ) is presented in Fig. 6.3.

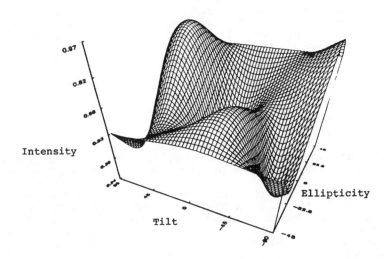

Fig. 6.3 Polarimetric Dependence of the Adjustable Intensity $||\tilde{R}(\phi,t)|| = pR_0$: The general topology of the power density (surface) plot is determined by the six stationary points corresponding to the six characteristic roots of Eq. 6.39.

From inspection of Fig. 6.3, we observe that we are obtaining six "characteristic polarization states", similar to the coherent case consisting of two power maxima (X_1,

X_2), two power minima (C_1, C_2) and two "saddle-point" extrema (S_1, S_2) which, may be related to the XPOL-Nulls (X_1, X_2), the CO-POL-Nulls (C_1, C_2) and the XPOL-Maxs (S_1, S_2) of the coherent case, respectively. For further details, we refer the reader to pertinent treatise of van Zyl [40] and those presented in [7].

7. CONCLUSIONS

A succinct overview was presented on the historical development of radar polarimetry, highlighting the important contributions including concise definitions and descriptions of the polarization state, the radar scattering matrices, and the polarimetric optimization problem. Based on such a critical assessment, the approximate formalism for handling basic polarimetric radar concepts was introduced. In a next step, we provided the scattering matrix formulations together with the transformation laws for measurements obtained with coherent dual polarization (scattering matrix) radars using different sets of antenna polarization bases such as linear horizontal/vertical (H,V) circular left/right (L,R) or any other orthogonal elliptic (A,B) transmit/recieve systems leading directly to polarimetric radar target and clutter phenomenology. Although the basic target versus clutter optimization procedures for determining the maximum and minimum transmit and receive polarization states have been given, no extensive applications will be shown here because of lack of space. Instead, we wish to refer to specific state-of-the-art review texts such as by Bechmann and Spizzichino on electromagnetic scattering from rough surfaces [3,4] and related Russian texts [11,25,35,42,48,50], by Long [44], and by Ulaby, Moore and Fung [57]. Target down- and cross-range polarimetric mapping applications are discussed in [5,6,7,14,18,19,30,36] and especially summarized in [24] and [46] using the coherent approach. For the partially polarized case [12,23] we refer to [41,29,32] which will also be covered in great detail in the forthcoming proceedings on radar polarimetry by Boerner et al [7], Root and Matkin [50]. It would be most illuminating to introduce the application of the Polarimetric Matched Filter Concept introduced in [41] to POL-SAR image analysis as will be discussed in great detail [6,7] and [49,50].

8. ACKNOWLEDGEMENTS

We wish to express our sincere gratitude to Mr. Lloyd W. Root and Mrs. Brenda L. Matkin of MI-COM, Advanced Sensors Laboratory, Redstone Arsenal, AL. for their patience and for accepting the delay of this final manuscript.

This research was supported, in part, by the U.S. Office of Naval Research, Arlington, VA, under contract US ONR N00014-C-0773; by the U.S. Naval Air Systems Command, Washington, DC, under contract US ONR N00019-81-C-0352; by the U.S. Naval Air Development Center, Warminster, PA, under contracts USONR N00019-82-C-0306 and N62269-85-C-0383; by the Army Research Office, Res. Tri. Park, NC, under contracts DAA629-80- K0027 and DAAK-84-C-0101; by the U.S. Army MICOM, Advanced Sensors Laboratory, Redstone Arsenal AL, under contracts DAA629-81-D-0101 and DAAL03-86-D-0001; and by the NATO Scientific Affairs Division, Grants SA.5-2IMOR.2038-(189)84 and SA.5-2-05RA0653(85).

We wish to thank our contract officers for the continued interest shown in our research and Ms. Julie A. Skroch for typing the final manuscript.

REFERENCES

[1]   A.P. Agrawal and W-M. Boerner, "Re-Development of Kennaugh's
      Target Characteristic Polarization State Theory Using the Polari-
      zation Transformation Ratio Formalism for the Coherent Case", IEEE
      Trans. GSRS, Vol. 27, No. 1, pp. 2-14, January 1989.

[2]   R.M.A. Azzam and N.M. Bashara, Ellipsometry and Polarized Light,
      Amsterdam/North Holland, 1977.

[3]   P. Beckmann, "The Depolarization of Electromagnetic Waves",
      Boulder, CO, The Golem Press, 1968.

[4]   P. Beckmann, and A. Spizzichino, "The Scattering of Electromagne-
      tic Waves from Rough Surfaces", New York: MacMillan, 1963.

[5]   W-M. Boerner, "Polarization Utilization in Electromagnetic Inverse
      Scattering", Chpt. 7 in "Inverse Scattering Problems in Optics",
      Vol. 2 (Ed. H.P. BALTES), Topics in Current Physics, Vol. 20,
      Springer Verlag, July 1980, pp. 237-305.

[6]   W-M. Boerner, et al. eds., "Inverse Methods in Electromagnetic
      Imaging",proc. NATO Advanced Res. Workshop on IMEI, Bad Windsheim,
      FR. Germany, sept. 18-24, 1983, NATO ASI Series, Series C, Math. &
      Phys. Sci., Vol.143, D. Reidel Publ. Co., Dordrecht, Holland,1985.

[7]   W-M. Boerner et al. (eds), "Direct and Inverse Method in Radar
      Polarimetry", Proc. NATO-ARW-DIMRP (W-M. Boerner, Director), 1988
      sept. 18-24 Bad Windsheim FRG, NATO-ASI-Series C, (Math. & Phys.
      Sci.), D.Reidel publ. Co., Dordrecht/Borton, 1989

[8]   W-M. Boerner and H.P.S. Ahluwalia, "On a Set of Continuous Wave
      Electro magnetic Inverse Boundary Conditions", Can. J. Phys.,
      50(23), pp. 3023-3061, Dec. 15, 1972 (also see: IEEE Trans.
      AP-21(5), pp. 663-672, May 1973; IEEE Trans. AP-22(5), pp.
      673-682, May 1974; Can. J. Phys. 53, pp. 1404-1407, May 1975).

[9]   W-M. Boerner, A.K. Jordan and I.W. Kay, "Introduction to the
      Special Issue on Inverse Methods in Electromagnetics", IEEE Trans.
      AP 29, March 1981, Guest Editors: W-M. Boerner, A.K. Jordan, I.W.
      Kay, pp. 185-189.

[10] W-M. Boerner, B-Y. Foo, H.J. Eom, "Interpretation of the Polari-
      metric Co-Polarization Phase Term ($\phi_{HH}-\phi_{VV}$) in High Resolution SAR
      Imaging Using the JPL CV-990 Polarimetric L-Band SAR Data, Special
      IGARSS '85 Issue of the IEEE Transactions on GeoScience & Remote
      Sensing, Vol. GE-25, No. 1, pp. 77-82, January 1987.

[11] V.V. Bogorodsky, D.B. Kanareykin and A.E. Kozlov, "Polarization of
      the Scattered Radio Radiation of the Earth Covers", Gidsometeo-
      rizdat, Leningrad, (in Russian) 1981.

[12] M. Born and E. Wolf, "Principles of Optics", 3rd Ed., Pergamon
      Press, New York, 1964.

[13] C-Y. Chan, "Studies on the Power Scattering Matrix of Radar Tar-
      gets", M.Sc. Thesis, Dept. of Electr. Engr. & Comp.Sci., Univer-
      sity of Illinois at Chicago, Chicago, IL, 1981.

[14] S.K. Chaudhuri, W-M. Boerner, "A Polarimetric Model for the Recovery of High-Frequency Scattering Centers from Bistatic-Monostatic Scattering Matrix Data", IEEE Trans. Antennas Propagat., Vol. A, p-35, No. 1, January 1987.

[15] H.C. Chen, "Theory of Electromagnetic Waves", McGraw-Hill Book Company, 1983.

[16] D. Clarke and J.F. Grainger, "Polarized Light and Optical Measurement", Pergamon Press, Oxford, 1971.

[17] R.E. Collin, "Antennas and Radio Wave Propagation", McGraw Hill, N.Y., 1985.

[18] M. Davidovitz and W-M. Boerner, "Reduction of Bistatic Scattering Matrix Measurements for Inversely Symmetric Radar Targets", IEEE Trans. Antennas Propagation, Vol. AP-31, No. 2, March 1983.

[19] M. Davidovitz and W-M. Boerner, "Extension of Kennaugh's Optimal Polarization Concept to the Asymmetric Matrix Case", IEEE Trans. A&P, Vol. AP-34(4), pp. 569-574, Apr. 1986.

[20] G.A. Deschamps, "Part 2: Geometrical Representation of the Polarization of a Plane Electromagnetic Wave", Proc. IRE, Vol. 39, May 1951, pp. 540-544.

[21] G.A. Deschamps and P.E. Mast, "Poincari Sphere Representation of Partially Polarized Fields", IEEE Trans. AP-21(4), 1973, pp. 474-478.

[22] B-Y. Foo, S.K. Chaudhuri and W-M. Boerner, "A High Frequency Inverse Scattering Model to Recover the Specular Point Curvatures from Polarimetric Scattering Data", IEEE Trans. A&P, Vol. 32, No. 11, pp. 1174-1178, Nov. 1984.

[23] T. Gehrels, Ed., "Planets, Stars and Nebulae Studied With Photopolarimetry", The University of Arizona Press, Tucson, Arizona, 1974 (Extensive Lists of Important References).

[24] D. Giuli, "Polarization Diversity in Radar", Proc. IEEE, Vol. 74(2), pp. 245-269, Feb. 1986.

[25] M.M. Gorshkov, "Ellipsometry", Sovetskoye Radio Press, (in Russian), 1974.

[26] C.D. Graves, "Radar Polarization Power Scattering Matrix", Proc. IRE, Vol. 44, Feb. 1956, pp. 248-252.

[27] G.E. Heath, "Bistatic Scattering Reflection Asymmetry, Polarization Reversal Asymmetry, and Polarization Reversal Reflection Symmetry", IEEE, Vol. AP-29, No. 3, May 1987.

[28] W.A. Hiltner, "Polarization Measurements", Actron. Technique, Chicago U. Press, 1962.

[29] J.W. Hovenier, H.C. van de Hulst and C.V.M. van der Mee, "Conditions for the Elements of the Scattering Matrix", J. Astron. and Astrophysics, Vol. 157, pp. 301-310, 1986.

30] J.R. Huynen, "Phenomenological Theory of Radar Targets", Ph.D.
     Dissertation, Technical University, Delft, The Netherlands, 1970.

31] IEEE Standard Number 149-1979: Standard Test Procedures 1973, Re-
     vision of IEEE Stds. 145-1969, Definitions of Terms for Antennas,
     Published by the Institute of Electrical and Electronics Engi-
     neers, Inc., New York, 1979.

32] G.A. Ioannidis and D.E. Hammers, "Optimum Antenna Polarization for
     Target Discrimination in Clutter", IEEE Trans. AP-27, May 1979,
     pp. 357-363.

33] R.C. Jones, "A New Calculus for the Treatment of Optical Systems",
     I. Description and Discussion, pp. 488-493, II. Proof of the Three
     General Equivalence Theorems, pp. 493-499, III. The Sohnke Theory
     of Optical Activity, pp. 500-503, J. Opt. Soc. Am., July 31 1941
     (also see: W. Swindell, "Polarized Light", Halsted Press/John
     Wiley & Sons, Stroudsburg, PA, 1975, pp. 186-240).

34] D.B. Kanareykin, N.F. Pavlov and V.A. Potekhin, "The Polarization
     of Radar Signals", Sovyet Radio, Moscow, Chap. 1-10 (in Russian),
     1966. (English Translation of Chpts. 10-12: Radar Polarization
     Effects. CCM Int. Corp., G.Collier and McMillan, 900 Third Ave,
     New York, N.Y. 10023).

35] D.B. Kanareykin, V.A. Potekhin, and M.F. Shisikin, "Maritime Po-
     larimetry", Sudostroyenie, Leningrad, 1968.

36] E.M. Kennaugh, "Polarization Properties of Radar Reflections",
     M.Sc. Thesis, Dept. of Elec. Engr., The Ohio State University,
     Columbus, OH, 43212, 1952. (also see: D.L. Moffatt and R.J. Gar-
     bacz, "Research Studies on the Polarization Properties of Radar
     Targets", by Prof. Edward M. Kennaugh, The Ohio State University,
     Electro Science Laboratory, 1420 Kinnaer Rd., Columbus, OH 43212,
     July 1984, Vols. 1 & 2 (to be re-edited in Spring 1989)).

37] G.P. Können, "Polarized Light in Nature", English Translation,
     Cambridge University Press, Cambridge, U.K., 1985.

38] A.B. Kostinski and W-M. Boerner, "On Foundations of Radar Pola-
     rimetry", IEEE A&P, Vol. AP-34, No. 12, pp. 1395-1404, also see:
     comments by H. Mieras, pp. 1470-1471, and author's reply, pp.
     1471-1473, Dec. 1986.

39] A.B. Kostinski and W-M. Boerner, "On the Polarimetric Contrast
     Optimization", IEEE Trans. A&P, Vol. 35, No. 8, pp. 988-991,
     August 1987.

40] A.B. Kostinski, B.D. James and W-M. Boerner, "On the Optimal
     Reception of Partially Polarized Waves", J. Optical Society of
     America, Part A, Optics & Image Sciences, Series 2, Vol. 5, No. 1,
     pp. 58-64, Jan. 1988.

41] A.B. Kostinski, B.D. James and W-M. Boerner, "Polarimetric Matched
     Filter for Coherent Imaging", Can J. Phys., Vol. 66, Issue 10,
     Special Issue on Coherent Imaging in Optics, pp. 871-877, Oct.
     1988.

42] A.L. Kozlov, "Radar Contrast of Two Objects", Izvestiya Vuz.,

362

Radioelekronika, Vol. 22, No. 7, July 1979, pp. 63-67.

[43] J.D. Kraus, "Electromagnetics", McGraw-Hill Book Company, 1984.

[44] M.W. Long, "Radar Reflectivity of Land and Sea", Lexington Books, D.C. Heath and Company, Lexington, MA, 1975.

[45] H. Mott, "Polarization in Antennas & Radar", John Wiley & Sons, Inc., 1986.

[46] A.J. Poelman and J.R.F. Guy, "Polarization Information Utilization in Primary Radar: An Introduction & Update to Activities at Shape Technical Center", Proc. NATO-ARW on IMEI, Bad Windsheim, FR. Germany, Sept. 18-24, 1983, Session RP.5, Section III: Paper No. III.2, pp. 521-572.

[47] H. Poincaré, "Théorie Mathématique de la Lumière", II-12, Georges Carré, Paris, 1892, pp. 282-285.

[48] A.P. Rodimov and V.V Popovski, "Statistical Theory of Polarimetric Temporal Signal and Clutter Processing in Communication (propagation paths and lines), Moscow, Vol. 21 in Series of Statistical Communications, Moscow: Radio & Comm., 1984 (in Russian).

[49] L.W. Root, Chairman & Editor (Proceedings), Second Workshop on Polarimetric Radar Technology, 1983, May 3-5, Redstone Arsenal, AL. (3 Volumes), published by: GACIAC, IIT-RI, 10 W. 35th Street, Chicago, IL 60615: GACIAC PR-83-01, August 1983 (also see: Proc. First Workshop on Polarimetric Radar Technology, 1981, June 25-26, US Army MI-COM-DRSM-REG, Redstone Arsenal, AL: GACIAC PR 81-02, Feb. 1982).

[50] L.W. Root and B.L. Matkin, Chairman/Editors, Proceedings, (Third) Polarimetric (Radar)/(Technology) Workshop, Redstone Arsenal, AL., 1988 August 16-18, GACIAC IIT-RI, 10 W. 35th St., Chicago, IL 60612, July 1989: PR-88-03.

[51] G.E. Schilov, "Linear Algebra", Dover, New York,1977.

[52] W.A. Shurcliff, "Polarized Light", Harvard Press, Cambridge, MA, 1962.

[53] G. Sinclair, "The Transmission and Reception of Elliptically Polarized Waves", Proc. IRE, Vol. 38, Feb. 1950, pp. 148-151.

[54] M.E. Skolnik, "Radar Handbook", New York, McGraw-Hill, 1970.

[55] G. Strang, "Linear Algebra and Its Applications", Academic, New York,1976.

[56] W. Swindell, "Polarized Light", Halsted Press, 1975.

[57] F.T. Ulaby, R.K. Moore and A.K. Fung, "Microwave Remote Sensing",: Vols. 1-3, Addison-Wesley, Reading, MA, 1981.

[58] G. Wanielik, Signaturuntersuchungen an einem polarimetrischen Pulsradar, (in German/English title: Signature Analyses using a Polarimetric Pulse Radar), Dr. Ing. Dissertation, Univ. Karlsruhe, Karlsruhe, FRG, Sept. 1988 (published in Fortschrittsberichte VDI,

Reihe 10: Informatik/Kommunikations–Technik, Nr. 97: ISBN 3–18–14 9710–x–1988).

[59] L.A. Zhivotovskiy, "Optimum Polarization of Radar Signals", Radio Eng. and Electronic Phys., 1814, 1973, pp. 630–632

[60] J.J. van Zyl, "On the Importance of Polarization in Radar Scattering Problems", Ph.D. Dissertation, California Institute of Technology, Pasadena, CA, January 1986.

MICROWAVE DIVERSITY IMAGING AND AUTOMATED TARGET RECOGNITION BASED ON
MODELS OF NEURAL NETWORKS

N.H. FARHAT

University of Pennsylvania
Electrical Engineering Department
Electro-Optics and Microwave-Optics Laboratory
Philadelphia, PA 19104, USA

1. INTRODUCTION

There are two distinct approaches or philosophies in radar target
identification. One is microwave image formation followed by recognition
and identification by a human observer i.e., by the eye-brain system.
Here one is concerned with concepts and methodologies for imparting to the
images formed the highest resolution possible to facilitate their analysis
and identification by human observers. Near optical resolution and
cost-effectiveness are usually the objective. The second approach is
automated recognition of the target by a machine using suitable target
signatures or representations. Here one is concerned with issues of
correct identification given partial or sketchy information irrespective
of range or aspect of the target or its location within the field of view
and with systems that can do this in robust and fault-tolerant manner. In
this approach the role of the eye-brain system in identifying the image in
the first approach is to be mimicked by a machine. The motives for
automated recognition are varied with speed and cost-effectiveness ranking
high among them.

In this paper we will discuss both approaches and show how they are
interrelated and how a thorough understanding of the microwave imaging
process is needed for the formulation of innovative methods for automated
recognition. We begin in section 2 with a review of the principles and
methodologies of microwave diversity imaging extensively studied and
developed in our laboratory where it was shown that microwave diversity
imaging provides 3-D tomographic or projective images of scattering
objects with near optical resolution employing spectral, angular, and
polarization degrees of freedom. Having adequately described this aspect
of our work in earlier publications [1]-[8] the discussion here is made
intentionally brief but with sufficient details to provide the background
for ensuing discussion of automated target identification. This is done
by bringing out those aspects of microwave diversity imaging that are
relevant to automated machine recognition. This is followed in section 3
by a discussion of a new approach to automated target identification from
incomplete or sketchy information based on models of neural networks.
The work describes a new direction in our microwave imaging and identification work pursued during the past three years. The work is motivated by
the desire to further reduce the projected cost of microwave diversity
imaging systems and by the fact that there exist important circumstances
when a real (physical) or synthetic baseline for an imaging aperture is
not available. The aim is to achieve automated recognition from partial
information when the amount of information available about the target is
so meager that formation of an image is out of the question. Our interest

365

B. de Neumann (ed.), Electromagnetic Modelling and Measurements for Analysis and Synthesis Problems, 365–389.
© 1991 Kluwer Academic Publishers. Printed in the Netherlands.

in neural signal processing or "brain-like" processing is readily appreci-
ated when one notes the associative memory attributes of the eye-brain
system, its amazing ability at supplementing or completing missing
information, and the apparent ease and speed with which it solves ill-posed
problems of the type encountered in vision, speech, and cognition in
general. Neural processing furnishes a new powerful approach to signal
processing that is both robust and fault tolerant and can be extremely
fast when implemented opto-electronically in order to fully exploit the
fit between what neural models can offer (powerful collective, nonlinear,
and iterative processing) and what optics can offer (parallelism and
massive interconnectivity) [9], [10]. The discussion in section 3
includes neural net models and opto-electronic architectures for realizing
content addressable associative memories that can be useful in radar
target recognition. Results representing the performance of software
implementation of such neural processors in the recognition of scale
models of aerospace targets employing sinogram representations are given.
The sinogram representation is chosen as an example of a feature space
that is suitable for use with neural processors. Other representations
involving low frequency polarization maps e.g., plots of the state of
polarization of the scattered field as function of frequency on an inclination angle versus ellipticity angle coordinate plane or pole-zero
representation of the scattered field can be equally utilized.

Machine recognition with artificial neural networks relies therefore on
the generation of target or object representations or feature spaces that
can lead to "distortion tolerant" recognition i.e., recognition irrespective of target range, orientation, or location within the field of view
traditionally referred to as scale, rotation, and shift invariant recognition. The generation of such representations usually involves the same
gear employed in microwave (μw) and millimeter wave (mmw) diversity
imaging where spectral, angular, and polarization degrees of freedom are
combined to realize images of scattering targets with near optical resolution. In fact the sinogram representation contains exactly the same
information contained in a μw/mmw image of the target except it is
arranged in a slightly different format that is more amenable for use in
automated recognition schemes. The work presented shows that super-
resolved recognition of complex shaped scattering objects from partial
information that can be as low as 20 to 10 percent of the sinogram
representation is possible with neural net analog processors employing
hetero-associative storage and recall where the outcome is a word label
describing the recognized object. The neural net in this sense performs
the functions of storage, processing, and recognition simultaneously.

2. MICROWAVE DIVERSITY IMAGING

In this section a brief qualitative outline of the principles, methodologies, and capabilities of microwave diversity imaging are presented.
The discussion by no means is comprehensive. It is meant to bring out the
salient features of microwave diversity imaging and emphasize those
attributes that are attractive for automated machine recognition. Thus
the references given are not exhaustive but those given contain between
them an extensive list to provide a good source of literature on the
subject.

2.1. Principles. Target shape estimation in the context of inverse
scattering from far field data is a longstanding problem with considerable
present day interest that has been studied by many (see for example

references [2], [3], [11]-[19]). It can be shown from basic electromagnetic scattering theory, assuming that physical optics and Born approximations hold, that monostatic or bistatic measurement of the far field scattered by an object as a function of illuminating frequency and object aspect can be used to access the Fourier space $\Gamma(\bar{p})$ of the object scattering function $\gamma(\bar{r})$. Here $\bar{p}$ and $\bar{r}$ are 3-D position vectors in Fourier space and object space respectively. The object scattering function γ is loosely taken to represent the 3-D geometrical distribution and strength of those scattering centers of the object that contribute to the measured field. The Fourier space data manifold $\Gamma_m(\bar{p})$ measured in practice is necessarily of a finite extent which depends on the values of $\bar{p}$ realized in the measurement. These depend in turn on geometry and on the angular and spectral windows utilized. It is possible then to retrieve a diffraction and noise limited version γ_d of the object scattering functions by 3-D Fourier inversion of Γ_m. In particular tomographic or projective reconstruction of γ_d based on the projection-slice theorem or the Radon transform have been demonstrated computationally [2], [3], [15], [16] and experimentally [5]-[6]. Image reconstruction using a filtered back-projection algorithm has also been demonstrated [20] and shown to yield images with equivalent quality to those obtained by Fourier inversion.

Accessing the Fourier space of a scatterer in practice is not direct. It requires preprocessing of the scattered far field one measures in order to remove an undesirable phase factor due to propagation between the target and the receiver and to remove the effects of clutter and measurement system response [6]-[8]. The range phase removal is synonomous with synthesizing a common phase reference or phase center on the target and hence can be viewed as a target derived reference (TDR) method [21] in which the target itself is made to furnish the reference phase for the complex field measurements at the observation point (or points). This has far reaching practical advantages that are discussed more fully elsewhere [8]. The vector nature of electromagnetic scattering can be treated by assuming that the scattering matrix which characterizes the polarization properties of the target and hence provides added information is measured at every frequency and aspect angle at which the scattering target is observed. A polarization enhanced image can in principle be obtained by incoherent superposition (addition of intensities) of the images formed from the accessed Fourier space data associated with each of the four components of the scattering matrix.

The above concepts represent the basic principles on which the methodologies of microwave diversity imaging are based.

2.2. Methodologies. In our work the Fourier space of a scattering object is accessed using an automated experimental radar scattering and microwave imaging facility (see Fig. 1). The facility is quite unique and versatile in that it allows accessing the Fourier space of scale models of targets of interest placed in an anechoic chamber over extended microwave windows (10 MHz - 26.5 GHz), for any state of polarization of the transmitting and receiving antennas, and for any aspect or target viewing angle. The instrumentation shown measures the stepped frequency response of the scatterer. Virtually any radar imaging configuration or innovative imaging concept can be readily simulated cost-effectively. Inverse synthetic aperture radar (ISAR), spot-light imaging, and array imaging can

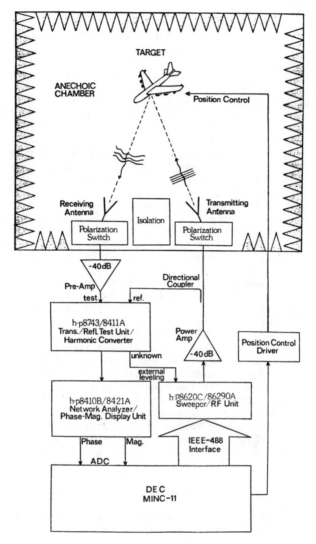

TARGET

SYSTEM CONFIGURATION

Figure 1. Experimental microwave measurement and imaging facility.

all be simulated and studied. Also any illuminating pulse can be
synthesized by controlling the amplitude and phase of the CW signals used
to illuminate and acquire the stepped frequency response of the target.
In the arrangement shown in Fig. 1 the transmitting and receiving antennas
are nearly monostatic but bistatic and also multi-static measurement can
also be performed. The state of the art of microwave instrumentation used
enable, making complex scattered field measurements with extremely high
accuracy (±0.1 dB, ±0.5 degree) over a dynamic range of 80 dB.
Better accuracy is achieved by averaging of several independent readings

t each measurement frequency. Frequency can be set automatically with an
accuracy of better than 4 Hz and stability of better than 240 Hz. Results
demonstrating the capabilities of the facility in microwave diversity
imaging of several representative targets are shown in Figs. 2 and 3. The
Fourier slices shown in Fig. 2 consist of 128 range phase corrected
frequency responses of the test object taken over an angular window
extending in azimuth from head—on to broad—side at a fixed elevation angle

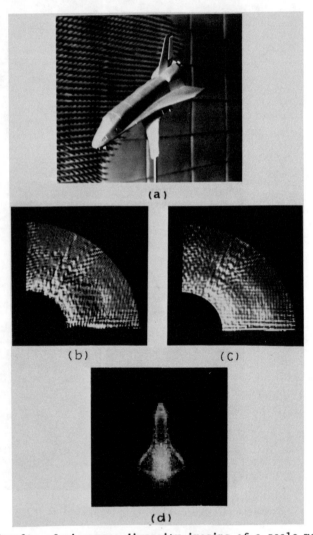

(a)

(b)   (C)

(d)

Figure 2. Results of microwave diversity imaging of a scale model of the
space shuttle. (a) Object shown mounted on azimuth positioner (turntable)
at an inclination angle $\theta = 30°$, magnitude of copolarized (b) and
cross—polarized (c) Fourier space slices taken. (d) Polarization and
symmetry enhanced projection image.

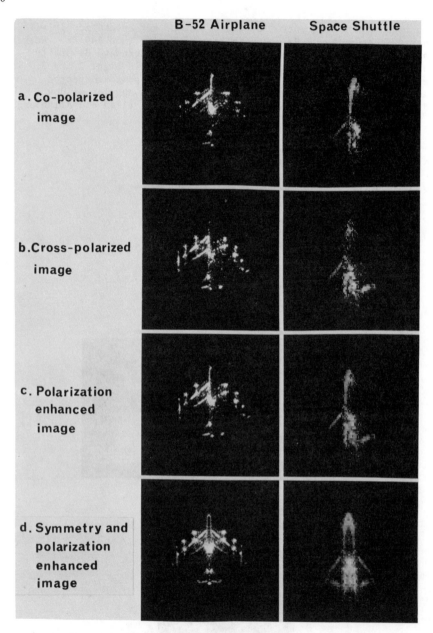

B-52 Airplane   **Space Shuttle**

a. Co-polarized image

b. Cross-polarized image

c. Polarization enhanced image

d. Symmetry and polarization enhanced image

Figure 3.   Examples of projection images of two test objects.

θ with each view containing 128 frequency points.   Interpolation of the polar formatted data of a slice onto a rectangular grid followed by Fourier inversion yields in accordance to the projection slice theorem [22] a projection image of the scattering centers of the test object.   The

projection image represents the projection of the scattering centers of the target on a plane normal to the azimuthal axis of rotation. Fourier inversion of the frequency response for a given viewing angle yields the complex impulse response or complex range-profile of the target at that angle which resembles the echo or response of the target to impulsive plane wave illumination for the given viewing angle. The complex nature of the range-profile is caused by the fact that only positive spectral windows can be employed in practice. By plotting the modulus of the complex range-profiles in rectangular rather than polar format one obtains the sinogram representation of the object which is utilized and discussed further in section 3.

Figure 3 shows examples of projection images of two test objects and the process of their enhancement by polarization diversity and symmetrization. Circularly polarized plane wave illumination was used and both the co-polarized and cross-polarized components of the scattered field were measured and associated Fourier space slices were formed from which images were obtained. It is evident from the images formed, for the different polarization states, that these contain some complementary information. Therefore image enhancement can be anticipated by adding the intensities of the co-polarized and cross-polarized images as demonstrated by the images in (c) of Fig. 3. Because man-made objects of interest in imaging radars are invariably symmetrical and their plane or planes of symmetry can be inferred from their heading, symmetrization can be used to enhance the image further. Symmetrization of the polarization enhanced images in Fig. 3 (c) about the vertical line of symmetry running through the fuselage was performed digitally leading to the polarization and symmetry enhanced images shown in (d). Thus the image shown earlier in Fig. 2 (d) was polarization and symmetry enhanced in the fashion described. Also all images shown were actually magnified in the vertical direction by a factor $1/\cos \theta = 1.155$ in order to obtain a properly scaled projection image of the scattering centers as they would be seen in a top view of the test object shown for example in (a) of Fig. 2. It is seen that features of the test objects used are delineated clearly in correct geometrical relation and relative size to enable recognition and classification of the scatterer with the eye-brain system. The image resolution achieved is of the order of 2 cm employing a (6 - 17) GHz spectral window.

In the above measurement system, the target during data acquisition remained stationary. During that time the stepped frequency response is measured, then the target aspect is changed, and the next frequency response is measured and so on. To access the Fourier space of a moving target with the same accuracy as for stationary targets, the arrangement of Fig. 4 can be used. Here high-speed compensation the doppler shift caused by target motion is utilized. The arrangement exploits two capable frequency synthesis and high-speed frequency measurement. These capabilities are combined in the arrangement shown to realize rapid determination of the doppler shift F_D in the signal scattered from the moving scatterer (target) which can then be compensated for in the phase and amplitude measurement of the scattered field that is carried out by the coherent receiver (HP 8410B Network Analyzer) whose intermediate local oscillator is effectively made to equal $F_{IF} + F_D$ by using an externally synthesized signal at frequency $F_{IF} + F_D$ for the reference IF channel of the network analyzer. This eliminates the effect of doppler shift on phase measurement and provides data identical to that generated had the target been stationary and merely changed its aspect relative to the line of sight of the interrogating radar system. Note that what really matters in

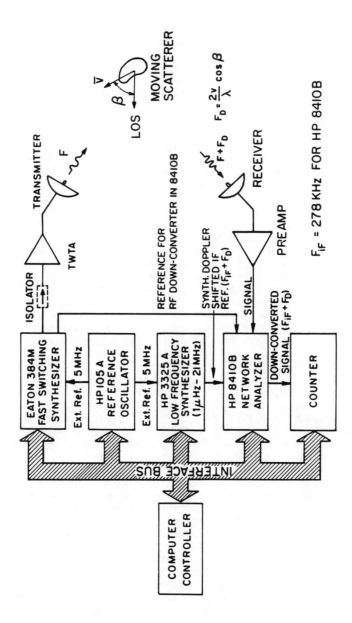

Figure 4. Doppler compensated arrangement for accessing the Fourier space of a moving scatterer.

the imaging of a moving target is its presenting different aspects to the interrogating system (station or stations) during motion. The same doppler compensation scheme can be utilized in microwave diversity imaging of rotating objects. Note the low frequency synthesizer source of the $F_{IF} + F_D$ signal, the HP 3325A synthesizer, is phase-locked to the RF source of illumination of the target. This synthesized source is carefully chosen because of its high frequency resolution which enables setting the doppler shift component of the IF reference frequency to precisely the measured value of F_D. This frequency resolution is necessary because any discrepancy between the intermediate frequency (IF) reference signal and the down converted IF version of the target echo signal in the network analyzer results in an undesirable time varying phase error. In this arrangement the complex frequency response of the moving scatterer can be measured rapidly in discrete frequency steps following the determination of the doppler shift and the immediate setting of the low frequency intermediate frequency reference synthesizer to $(F_{IF} + F_D)$. If we assume that 100 frequency stepped measurements per second, as limited by the bandwidth of the phase/amplitude measuring segment of the network analyzer, can be performed, then the Doppler shift during this interval is expected to change negligibly for most practical moving targets including aerospace targets. Therefore the Doppler measurement and doppler compensation operations need not be performed frequently by the system but only occasionally or at most once at the beginning of each digital frequency sweep.

2.3. <u>Attributes of microwave diversity images</u>. Microwave diversity images have several attributes that make them suitable for automated pattern recognition. First the images are naturally edge enhanced. This is a direct consequence of the nature of high-frequency electromagnetic scattering from complex shaped man made objects where flat parts of the target such as wings and tail sections etc. that are not oriented normal to the line of sight between the interrogating radar station and the object scatter the incident illumination specularly away from the receiving antennas. Only those parts of the target such as edges and protrusions that scatter the incident radiation widely will mostly contribute to the measured field and are therefore visible in the retrieved image. Second, because of TDR, the retrieved image is always centered within the image plane, and thirdly the images are tomographic in the sense that the object is viewed by the data acquisition system at a slant angle but the retrieved image appears as if we are looking at the object from below (or from the top depending on how the object is mounted on the positioner (see Fig. 2 (a)). The scattering mechanisms described earlier constitute a highly desirable process of reduction of dimensionality or compression of the data collected where unimportant detail about flat parts of the target that do not convey any useful information is discarded and only those parts that actually characterize the target such as edges, outlines and protrusions get retained in the image. This reduction of dimensionality taken together with the tomographic nature of the retrieved image, and the fact that it is always centered within the image plane, make the microwave diversity imaging system a smart sensor which performs a certain amount of preprocessing that helps automated pattern recognition. In fact given a grey level image such as that shown at the top left of Fig. 5, the first step in its automated pattern recognition is often computing its primal sketch [23], [24] or caricature (see top right of Fig. 5) which reduces the dimensionality of the image through straight line delineation of the

PRIMAL SKETCH

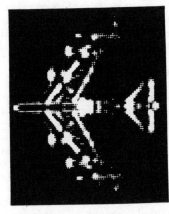

GRAY LEVEL IMAGE

Figure 5. Primal sketch nature of microwave diversity imagery. (Top part from ref. 23)

position and orientation of spatial changes of brightness in the original
image. A comparison of the optical and the microwave images of a B-52
test object, shown in the bottom row of Fig. 5, show clearly the primal
sketch nature of microwave diversity images.

The above attributes of microwave diversity imagery suggest that they
are well suited for automated pattern recognition by a machine. This may
be useful in certain circumstances. But when a human observer, the
ultimate in pattern recognition systems is available to analyze and
recognize the image, the benefit of automated recognition of the image
becomes questionable. Moreover, conventional pattern recognition works
best when a good image is available and may falter when the image is
incomplete or the amount of available information about the object or the
target is insufficient to form an image. Of course this is exactly the
challenge in practice, namely target recognition from partial or sketchy
information which taken by itself would not be adequate to form a
recognizable image. What we need therefore is an automated recognition
system that can identify objects or targets even when the amount of
available information is sketchy and is not sufficient to form a recognizable image.

Among its many fascinating capabilities such as robustness and fault
tolerance, the brain is also able to recognize objects from partial
information. We can recognize a partially obscured or shadowed face of an
acquaintance or a mutilated photograph of someone we know with little
difficulty and in reading text we easily are able to fill-in for misspelled
or mistyped words. The same is true with understanding spoken language.
The brain has a knack for supplementing missing information. Capitalizing
on this observation and on our knowledge of neural models and their
collective computational properties a study of "neural processing" for
recognizing objects from partial information was undertaken. Details and
results are given in the next section.

3. AUTOMATED TARGET RECOGNITION BASED ON MODELS OF NEURAL NETS
Neural net models and their analogs furnish a new approach to signal
processing that is collective, robust, and fault tolerant. Optical
implementations of neural nets (see for example [9], [10]) are attractive
because of the inherent parallelism and massive interconnection capabilities provided by optics and because of emergent optical technologies that
promise high resolution and high speed programmable spatial light modula-
tors (SLMs) and arrays of optical bistability devices (optical decision
making elements) that can facilitate the implementation and study of large
networks. Optical implementation of a one-dimensional network of 32
neurons exhibiting robust content-addressability and associative recall
has already been demonstrated to illustrate the above advantages [10]. By
one-dimensional we mean that in the architecture used the neurons are
deployed on a line. Extension to two-dimensional arrangements are of
interest because these are suitable for processing of 2-D image data or
2-D object representations directly and offer a way for optical implementation of denser networks.

In this section we will discuss content addressable memory (CAM)
architectures based on partitioning of the four dimensional memory or
interconnection matrix $T_{ijk\ell}$ encountered in the storage of 2-D
entities. A specific architecture and implementation based on the use of
partitioned unipolar binary (u.b.) memory matrix and the use of adaptive
thresholding in the feedback loop are described. The use of u.b. memory

masks greatly simplifies optical implementations and facilitates the realization of larger networks (10^3–10^4 neurons). Numerical simulations showing the use of such 2-D networks in the recognition of dilute point-like objects that arise in radar and other similar remote sensing imaging applications are described. Dilute objects pose a problem for CAM storage because of the small Hamming distance between them. The Hamming distance between two binary vectors or matrices of the same dimension is the number of bits in which they differ. We show that coding in the form of a sinogram representation or feature space of the dilute object can remove this limitation permitting recognition from partial versions of the stored entities. The advantage of this capability in super-resolved recognition of radar targets where the principles and methodologies of microwave diversity imaging are employed to form sinogram representations that are compatible with 2-D CAM storage and interrogation are discussed. Super-resolved automated recognition of scale models of three aerospace objects from partial information that can be as low as 10% of a learned entity is demonstrated employing hetero-associative storage and recall where the recognition outcome is a word label describing the recognized object. Capacity for error correction and generalization were also observed. The treatment here is similar to one we have given elsewhere [25].

3.1. <u>Two dimensional neural nets</u>. Storage and readout of 2-D entities in a content addressable or associative memory is described next. Given a set of M 2-D bipolar binary patterns or entities $v_{ij}^{(m)}$ m = 1,2,...M each of NxN elements i.e., NxN binary matrices, these can be stored in a manner that is a direct extension of the storage formula for 1-D entities [9], [10] as follows: For each element of a matrix a new NxN matrix is formed by multiplying the value of the element by all elements of the matrix including itself taking the self product as zero. The outcome is a new set of N^2 binary bipolar matrices each of NxN elements. A formal description of this operation is,

$$T_{ijk\ell}^{(m)} = \begin{cases} v_{ij}^{(m)} \, v_{k\ell}^{(m)} & i,j = 1,2,\ldots,N \\ \\ 0 & i=k, \ j=\ell \end{cases} \tag{1}$$

which is a four dimensional matrix. An overall or composite synaptic matrix or connectivity memory matrix is formed then by adding all 4-D matrices $T_{ijk\ell}^{(m)}$ i.e.,

$$T_{ikj\ell} = \sum_m T_{ijk\ell}^{(m)} \tag{2}$$

This symmetric 4-D matrix has elements that range in value between −M to M in steps of two and which assume values of +1 and −1 (and zeros for the self product elements) when the matrix is clipped or binarized as is usually preferable for optical implementations. Two dimensional unipolar

binary entities $b_{ij}^{(m)}$ are frequently of practical importance. These can be transformed into bipolar binary matrices by taking $v_{ij}^{(m)} = (2b_{ij}^{(m)} - 1)$ which are then used to form the 4-D connectivity matrix or memory matrix as described before. Also, as in the 1-D neural net case, the prompting or initializing entity can be unipolar binary $b_{ij}^{(m)}$, which would simplify further optical implementations in incoherent light.

Architectures for optical implementation of 2-D neural nets must contend with the task of realizing a 4-D memory or interconnectivity matrix. Here a scheme is presented that is based on partitioning the 4-D memory matrix into an array of NxN 2-D matrices each of which containing NxN elements. Thus a 2-D neural net of NxN = 32x32 neurons would contain N^4 interconnections i.e., over a million interconnections which shows why hardware implementations that use light rather than electronic interconnects are attractive. The opto-electronic architecture described below has the potential for realizing such massive interconnectivity in a relatively small volume. Provided that the number of entities stored is not excessive (see below) the 4-D interconnection matrix makes the stable states of the net the entities stored. This means that initiated from a given state, the network quickly converges, in a matter of few time constants of the "neuron", to the stored entity closest in the Hamming sense to the initiating vector or matrix.

The nearest neighbor search of the memory matrix for a given entity $b_{ij}^{(mo)}$ is done by forming the estimate,

$$\hat{b}_{ij}^{(mo)} = \sum_{k,\ell}^{N} T_{ijk\ell}\, b_{k\ell}^{(mo)} \quad \ldots \quad i,j,k,\ell = 1,2,\ldots N \qquad (3)$$

followed by thresholding to obtain a new unipolar binary matrix which is used to replace $b_{ij}^{(mo)}$ in eq. (3) and the procedure is repeated until the resulting matrix converges to the stored entity closest to the initializing matrix $\hat{b}_{ij}^{(mo)}$. The operation in eq. (3) can be interpreted as first partitioning of the 4-D $T_{ijk\ell}$ matrix into an array of 2-D submatrices each of NxN elements: $T_{11k\ell}$, $T_{12k\ell}$, $\ldots$, $T_{1Nk\ell}$; $T_{21k\ell}$, $T_{22k\ell}$, $\ldots$, $T_{2Nk\ell}$; $\ldots$, $T_{N1k\ell}$, $T_{N2k\ell}$, $\ldots$, $T_{NNk\ell}$ as depicted schematically in Fig. 6 (a) where the partition submatrices are arranged in a 2-D array. This first step is followed by multiplication of $\hat{b}_{k\ell}^{(mo)}$ by each of the partition submatrices, on an element by element basis, and summing the products for each submatrix to obtain the first estimate $b_{ij}^{(mo)}$. The multiplications and summation operations called for in eq. (3) can be carried out as shown in Fig. 6 (a) by placing a spatially integrating photodetector (PD) behind each submatrix of the partitioned memory mask which is assumed for the time being to be realized by pixel transmittance modulation in an ideal transparency capable of assuming negative transmittance values. The input entity $b_{ij}^{(mo)}$ is assumed to be displayed on a suitable LED array. The LED display of $b_{ij}^{(mo)}$ is multiplied by the ideal transmittance of each of the

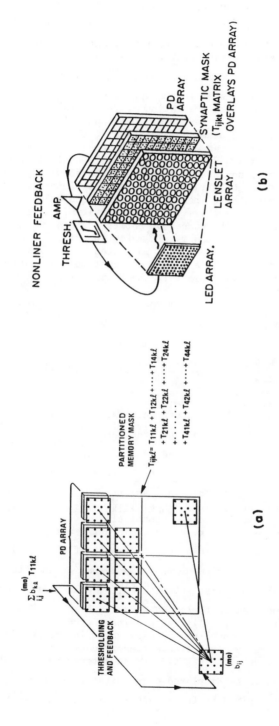

Figure 6. Optical analog of 2-D neural net. (a) Architecture based on partitioning of connectivity matrix, (b) Opto-electronic embodiment.

partition submatrices by imaging the display on each of these with exact registration of pixels by means of a lenslet array as depicted in Fig. 6 (b). The output of each PD, proportional to one of the components of eq. 3 is thresholded, amplified, and fed back to drive an associated LED. The (i,j)-th LED is paired with the (i,j)-th PD. This completes the interconnection of the 2-D array of NxN neurons in the above architecture where each neuron communicates its binary state (0 = LED off, 1 = LED on) to all other neurons through a prescribed four dimensional synaptic or memory matrix in which information about M 2-D binary entities (NxN matrices) had been stored distributively. The number of 2-D entities that can be stored in this fashion is $M \simeq N^2/8\ell nN$, which follows directly from the storage capacity formula for the 1-D neural net case [26] by replacing N by N^2.

The added complexity associated with having to realize a bipolar transmittance in the partitioned $T_{ijk\ell}$ memory mask of Fig. 6 can be avoided by using unipolar transmittance. This can lead however to some degradation in performance. A systematic numerical simulation study [29] of a neural net CAM in which statistical evaluation of the performance of the CAM for various types of memory masks (multivalued, clipped ternary, clipped u.b.) and thresholding schemes (zero threshold, adaptive threshold where energy of input vector is used as threshold, adaptive thresholding and relaxation) was carried out. The results indicate that a u.b. memory mask can be used with virtually no sacrifice in CAM performance when the adaptive thresholding and relaxation scheme is applied. The scheme assumes an adaptive threshold is used that is proportional to the energy (total light intensity) of the input entity displayed by the LED array at any time. In the scheme of Fig. 6 (b) this can be realized by projecting an image of the input pattern directly onto an additional PD element. The PD output being proportional to the total intensity of the input display is used as a variable or adaptive threshold in a comparator against which the outputs of the PD elements positioned behind the partitioned components of the $T_{ijk\ell}$ memory mask are compared. The outcomes, now bipolar, are attenuated and each is fed into a limiting amplifier with delayed feedback (relaxation). Each limiter/amplifier output is used to drive the LED that each photodetector is paired with. It was found [27] that this scheme yields performance equivalent to that of an ideal CAM with multivalued connectivity matrix and zero thresholding. Note that although the initializing 2-D entity $b_{ij}^{(mo)}$ is unipolar binary, the entities fed back after adaptive thresholding and limited amplification to drive the LED array would initially be analog resulting in multivalued iterates and intensity displays. However, after few iterations the outputs become binary assuming the extreme values of the limiter. The ability to use unipolar binary memory matrices in the fashion described means that simple black and white photographic transparencies or binary SLMs can be used respectively as stationary or programmable synaptic connectivity masks in accordance to Fig. 6.

3.2. Sinogram representation and hetero-associative storage.

Sinograms are object representations encountered in tomography [28], [29]. They are particularly useful when the object is point-like and sparse or dilute as is the case for microwave diversity imagery because these consist ordinarily of a countable number of isolated scattering centers (primal sketch nature). Given a set of 2-D dilute objects and their corresponding set of associated sinogram representations, the Hamming distances between the

sinogram representations will always be found to be greater than the Hamming distances between the objects themselves. This is assuming that objects and sinograms are quantized onto the same number and grid of binary pixels. The reason for this, as will be clarified below, is that each point of the object produces a distinct sinusoidal trace that spawns many points in the sinogram representation. Thus if two dilute objects differ in only two pixels, their sinogram representation will differ by two sinusoidal traces and hence in many pixels. The increased Hamming distance makes it easier for an associative memory to distinguish between the sinograms than to distinguish between the objects themselves [30]. This has the added attraction of making it less difficult to distinguish between similar objects, that is objects with small Hamming distance between them. Sinogram representations also have the advantage of being helpful in achieving distortion invariant (scale, rotation, and translation invariant) recognition.

The sinogram of a target is formed in our work by measuring the range-profile or differential range of the target as function of aspect angle and by arranging the measured range-profiles side-by-side as function of aspect angle (for example azimuthal angle ϕ in Fig. 7 (c)) as is illustrated in Fig. 7 for a planar object consisting of three points of unequal strength. This object is chosen to represent a highly simplified radar target. Every point (or scattering center) of the object generates a sinusoidal trace in the sinogram whose amplitude is determined by the radial distance of that point from the center of rotation, whose phase is determined by angular position of that point, and whose brightness or strength (represented in Fig. 7 (b) by line thickness) is proportional to the strength of the scattering center. Note that scatterer 3 whose position coincides with the center of rotation produced a zero amplitude line in the sinogram. A complete sinogram is produced by rotating the object 360°. It is important to note that the range-profile of an object is independent of its far field distance from the transmitter/receiver (T/R) in Fig. 7 (a). The range-profile depends however on object aspect and on the spectral window and polarization used in data acquisition (see section 2.2).

In laboratory work, the sinogram representation of a complex shaped test object is obtained as depicted in Fig. 7 (c) which is a highly simplified version of the measurement system of Fig. 1. The number of range-profiles N_ϕ needed to characterize the object and the number of samples N_R within each range-profile are determined by angular and spectral sampling considerations. Thus for an object of size L the number of angular samples is $N_\phi \simeq 4\pi L/\lambda_{min}$ and the number of samples within a range-profile is $N_R = N_f = 2\Delta f L/c$ where Δf is the width of the spectral window used, N_f is the number of frequency points, c is the velocity of light, and λ_{min} is the shortest wavelength used. The sinogram of an actual microwave target differs in appearance from the sinogram of the idealized object described above in that the intensity or brightness of its sinusoidal traces changes with aspect or viewing angle because of the anisotropic nature of the scattering centers on actual targets.

Figure 8 gives an example of the sinogram of a scale mode of a B-52 test object produced from a slice of its Fourier space (shown in Fig. 8 (a) obtained at an object inclination angle of $\theta = 30$ (see Fig. 7 (c)) employing the measurement facility of Fig. 1. Both intensity and 3-D perspective displays of the resulting sinogram are shown in (b) and (c) of Fig. 8 respectively. The sinogram shown demonstrates clearly how sinu-

SINOGRAM GENERATION — PRINCIPLE

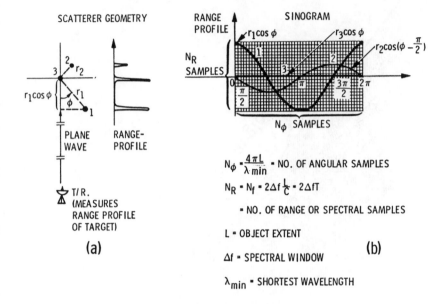

$$N_\phi = \frac{4\pi L}{\lambda\,min} = \text{NO. OF ANGULAR SAMPLES}$$

$$N_R = N_f = 2\Delta f \frac{L}{c} = 2\Delta f T$$

= NO. OF RANGE OR SPECTRAL SAMPLES

L = OBJECT EXTENT

Δf = SPECTRAL WINDOW

λ_{min} = SHORTEST WAVELENGTH

SINOGRAM GENERATION — EXPERIMENTAL

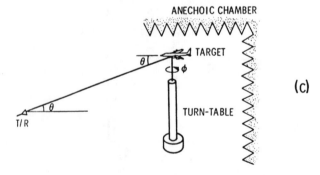

Figure 7. Sinogram representation. (a) Scattering geometry for an idealized planar object, (b) Sinogram, (c) Simplified arrangement for experimental generation of sinogram.

soidal traces of the different scattering centers fade in and out as function of target aspect and how point–like scattering centers such as the tips of engines and fuel tanks (see B-52 part of Fig. 3 (d)) produce more distinct traces than edge–like or extended scattering centers of the target.

382

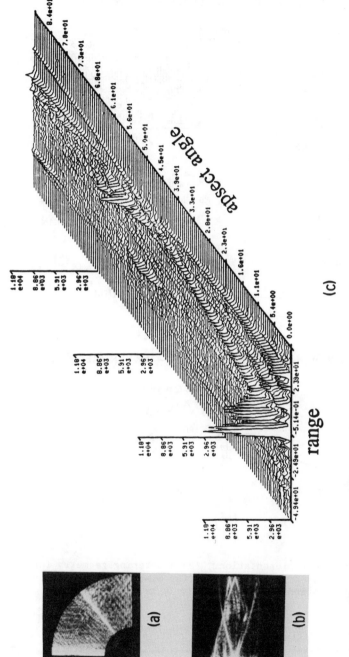

Figure 8. Sinogram of B-52 test object. (a) Fourier space slice from which the sinogram is generated. Intensity (b) and 3-D perspective (c) displays of the sinogram.

It is worth mentioning that the alignment of the range-profiles to produce distinct sinusoidal traces in the range-aspect displays (Fig. 8 (b) and (c)) is an essential requirement for image reconstruction by back-projection [20]. The alignment process defines also the center of rotation or phase center of the target in that had a point scatterer been located at the rotation center it would produce a constant range straight line in the sinogram. The alignment is also equivalent to the TDR procedure referred to in section 2.1 and described in more detail elsewhere [8], [21]. Thus formation of a distinct sinogram is not only needed for representing the target but is also an essential operation for removing the unknown range to the phase center of the target and the removal of undesirable effects associated with migration of its phase center with aspect. The crispness with which one or more sinusoidal traces appear in the sinogram can serve as a measure of how well the unknown range to a common reference point (center of rotation or phase center of the target) can be compensated for in the different aspect looks at the target. Quantization and thresholding of the sinogram pattern of Fig. 8 into a grid of NxN binary pixels yields the sinogram representation b_{ij} of the target that is suitable for the associative storage and recall process as described in section 3.1. In the top row of Fig. 9 are shown the sinogram representations of three aerospace test objects (B-52, AWAC, and Space Shuttle) digitized onto a grid of 32x32 binary pixels. These are treated as a learning set and stored hetero-associatively rather than autoassociatively by replacing $v_{k\ell}^{(m)}$ in eq. (1) by $r_{k\ell}^{(m)}$ $k,\ell=1,2,\ldots,32$; $m=1,2,3$ where $r_{k\ell}^{(m)}$ represents abbreviated word labels shown in the bottom row of Fig. 9 with which the three test objects are to be associated. In this fashion a 4-D hetero-associative memory or connectivity matrix $T_{ijk\ell}$ is formed to be used in numerical simulation which we describe next.

4. RESULTS

Numerical simulations of exercising the hetero-associatively formed memory matrix with complete and partial versions of the three entities (sinogram representations) stored in it following the procedure of section 3.1 were carried out. Complete and partial versions of the three stored sinogram representations were used to initialize the network (associative memory). The partial versions of the stored entities ranged down to a fraction $\eta = 10\%$ of the full representation. Reliable recognition was found to occur after one iteration for all entities stored down to $\eta = 0.2$. For $\eta = 0.1$ or less, successful recall of correct labels was found to depend on the angular location of the partial data the memory is presented with. In those cases the memory could not label the partial input correctly but converged instead after a few iterations onto a label that it did not learn before but appeared to be a generalization (mixture) of the three entities stored. These generalizations can be viewed as spurious stable states or attractors in the phase-space of the net on which the neural net converges when the initializing entity is far in Hamming distance from any of the entities stored. This is quite analogous to the generalization capability of the brain and can be useful in practice to reduce the number of sinogram representations needed to characterize the target for all relevant elevation angles. We have also observed that when insufficient partial information was presented, the generalization can be contrast reversed. This is expected as one can show that stable states of a memory with symmetric connectivity matrix are not only the entities stored but also their compliments.

384

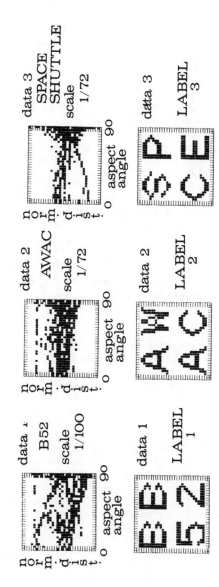

Figure 9. Hetero-associative storage. Sinogram representations (top) and associated word labels (bottom) of three aerospace test objects.

A representative example of the performance of the memory in hetero-
associative recall from complete and partial information is shown in
Fig. 10. Here correct recall is obtained when the memory matrix was
interrogated with a partial version of the B-52 sinogram with η = 10%.
   The example presented in Fig. 10 demonstrates the potential of neural
net processing in automated recognition of radar targets employing hetero-
associative storage and recall with the net performing the functions of
storage, processing, and recognition (labeling) simultaneously. The
processing is also robust and fault tolerant because of the distributed
storage and the collective, nonlinear, and iterative nature of processing
in such networks.

5. DISCUSSION

Methodologies of microwave diversity imaging extensively studied at the
Electro-Optics and Microwave Optics Laboratory of the University of Pennsylvania for more than a decade provide the basis for a new generation of
3-D formographic imaging radars that can furnish shape estimates of the
3-D distribution of scattering centers on remote aero-space targets with
near optical resolution. Resolution in such systems depends on the angular
and spectral windows, utilized for data acquisition and on polarization
diversity. Unprecedented centimeter resulation has been achieved and
reported employing GHz spectral windows, wide angular windows of ~ $\pi/2$

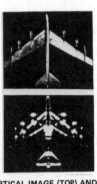 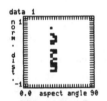

OPTICAL IMAGE (TOP) AND
RADAR IMAGE (BOTTOM) OF
SCALE MODEL OF B-52 AIR-
CRAFT PRODUCED AT THE
EXPERIMENTAL MICROWAVE
IMAGING FACILITY AT
THE UNIVERSITY OF
PENNSYLVANIA

(TOP) DIGITIZED SINOGRAM
REPRESENTATION OR SIGNA-
TURE OF B-52 TAUGHT WITH
OTHER SINOGRAMS (OF AN
AWAC AND A SPACE-SHUTTLE)
TO A HETERO-ASSOCIATIVE
MEMORY BASED ON MODELS
OF NEURAL NETS. (BOTTOM)
CORRECT LABEL PRODUCED
BY MEMORY WHEN IT WAS
PRESENTED WITH A COMPLETE
SINOGRAM OF THE B-52
SIGNIFYING THEREBY
RECOGNITION

SAME CPRRECT LABEL OF
B-52 (BOTTOM) WAS PRODUCED
BY MEMORY WHEN PRE-
SENTED WITH PARTIAL SINO-
GRAM INFORMATION (10% OF
THE SINOGRAM REPRESENTA-
TION (TOP).

Figure 10. Automated super-resolved recognition of a radar test object
from partial sinogram representation.

[Rad.] and image enhancement by polarization diversity and symmetrization. Image reconstruction algorithms based on Fourier inversion or by filtered backprojection are equally applicable and have been found to yield comparable results. The use of spectral, angular, and polarization degrees of freedom in such imaging systems has the advantage of increasing the information content of the oject scattered wavefields. This makes it possible for a broad-band, polarization selective array aperture to acquire more information about a scattering object than it could have monochromatically (at a single frequency) or at a single polarization. A useful trade-off between the spectral and angular degrees of freedom exists. It enables considerable thinning of the imaging array. Because angular degrees of freedom are associated with the number of elements or stations in the array, their replacement with less costly spectral degrees of freedom, associated with the number of frequency points used in data acquisition, can cut cost and lead to significant improvement in cost-effectiveness.

Despite these attractive attributes of microwave diversity imaging systems, there are circumstances when the base-line, whether physical or synthetic, required to realize the wide angular windows needed to achieve high resolution is not available or is not sufficient to form a recognizable image. One has to rely then on means of target identification other than image formation and analysis by the eye-brain system.

Microwave diversity imaging systems are shown here to have attributes that make them attractive for use in automated target recognition. Conventional pattern recognition techniques do not however alleviate the long base-line requirement since enough data to form an input image to the recognizer is required.

The "brain-like" processing approach to super-resolved, robust, and fault tolerant recognition described in the preceeding section is not only intelectually attractive but also has important practical significance. It can obviate the need for large expensive imaging array systems of the type needed in microwave diversity imaging systems and other more conventional approaches to radar target imaging and avoid the time spent for aperture synthesis by target motion in ISAR imaging. The implication of this for microwave (and other) automated object identification systems can be far reaching [31]. One can envision now for example a new generation of automated radar and sonar systems, that can identify remote targets from only a few looks and many findings of the work reported here carry over to the domain of machine vision and recognition for robotic applications. In such radar systems, suitable target representations, signatures, or feature spaces, such as the sinogram representation described above, would be generated cost-effectively from scale models of targets of interest in a controlled anechoic chamber environment employing measurement systems, of the type we have described. The representations would be "taught" to an associative memory or a neural network that can be used to recognize partial sinogram representations of actual targets collected by actual broad-band coherent radar systems. Realization of this scenario entails paying careful consideration to scaling issues and to the principle of "electromagnetic similitude" [32] in order to ensure that the scale-model sinogram representations resemble as closely as possible those of actual targets. This and other issues such as "fuzziness" of echos from actual airborne targets because of flexing, deformation, or wind-buffeting, the minimum number of looks (range-profiles) needed to represent an actual target i.e., characterize it for all practical

encounter aspects; the number of neural modules needed and their storage capacity, together with the use of sequential storage and recall, self-organization and learning are currently under investigation and will be reported elsewhere. The ultimate objective of this work is to achieve distortion tolerant recognition from one look.

6. ACKNOWLEDGMENT

Research work described in this paper was carried out under grant support from DARPA/NRL and with partial support from AFOSR, ARO, and the RCA (GE) corporation. Several of the authors' students, T. H. Chu, S. Miyahara, and Y. Shen, have contributed to the results used in this paper. Some of the work reported was carried out by the author when he was a summer distinguished visiting scientist at NASA–JPL.

REFERENCES

1. N. H. Farhat, "Principles of Broad-Band Coherent Imaging," J. Opt. Soc. Am., Vol. 67, Aug. 1977, pp. 1015–1020.

2. N. H. Farhat and C. K. Chan, "Three-Dimensional Imaging by Wave-Vector Diversity," Acoustical Imaging, Vol. 8 (Proceedings of the 1978 International Symposium on acoustical Imaging). A. Metherell (ed.), Plenum Press, New York, (1980), pp. 499–515.

3. C. K. Chan and N. H. Farhat, "Frequency Swept Imaging of Three Dimensional Perfectly Reflecting Objects," IEEE Trans. on Antennas and Propagation – Special Issue on Inverse Scattering, Vol. AP-29, March 1981, pp. 312–319.

4. N. H. Farhat, "Holography, Wavelength Diversity and Inverse Scattering," in Optics in Four Dimensions – 1980, M. A. Machado and L. M. Narducci (Eds.), American Inst. of Phys. publication, New York (1981), pp. 627–642.

5. N. H. Farhat, T. H. Chu and C. L. Werner, "Tomographic and Projective Reconstruction of 3-D Image Detail in Inverse Scattering," Proc. 10-th Int. Optical Computing Conference, IEEE Cat. No. 83CH7880-40, (1983), pp. 82–88.

6. N. H. Farhat, C. L. Werner and T. H. Chu, "Prospects for 3-D Tomographic Imaging Radar Networks," Proc. URSI Symp. on Electro-Magnetic Theory, Santiago De Compostela, Spain, Aug. 1983, pp. 279–301.

7. N. H. Farhat, "Projection Imaging of 3-D Microwave Scatterers with Near Optical Resolution," in Indirect Imaging, J. A. Roberts (Ed.), Cambridge University Press (1984).

8. N. H. Farhat, C. L. Werner, and T. H. Chu, "Prospects for Three-Dimensional Projective and Tomographic Imaging Radar Networks," Radio Science, Vol. 19, September–October, 1984, pp. 1347–1355.

9. D. Psaltis and N. Farhat, "Optical Information processing Based on an Associative-Memory Model of Neural Nets with Thresholding and Feedback," Opt. Lett., Vol. 10, 1985, pp. 98–100.

388

10. N. H. Farhat, et al., "Optical Implementation of the Hopfield Model," Applied Optics, Vol. 24, May 1985, pp. 1469-1475.

11. N. Bojarski, "Three-Dimensional Short Pulse Inverse Scattering," Syracuse University Research Report, Syracuse, N. Y., 1967.

12. N. Bojarski, "Inverse Scattering," Naval Air Command Final Report N000 19-73-C0312F (1974).

13. R. M. Lewis, "Physical Optics Inverse Diffraction," IEEE Trans. on Ant. and Prop., AP-17, May 1969, pp. 308-314.

14. S. R. Raz, "On Scatterer Reconstruction from Far Field Data," IEEE Trans. on Ant. and Prop., AP-24, Jan. 1976, pp. 66-70.

15. Y. Das and W. M. Boerner, "On Radar Shape Estimation Using Algorithms for Reconstruction From Projections," IEEE Trans. on Ant. and Prop., AP-26, March 1978, pp. 274-279.

16. W. M. Boerner, C-M. Ho, and B. Y. Foo, "Use of Radon's Projection Theory in Electromagnetic Inverse Scattering," IEEE Trans. on Ant. and Prop., AP-24, March 1981, pp. 360-367.

17. J. L. Walker, "Range-Doppler Imaging of Rotating Objects," IEEE Trans. on Aerospace and Electronic Systems, Vol. AES-16, Jan. 1980, pp. 23-25.

18. D. C. Munson, J. D. O'Brien, and W. K. Jenkins, "A Tomographic Formulation of Spotlight-Mode Synthetic Aperture Radar," Proc. IEEE, Vol. 71, Aug. 1983, pp. 917-925.

19. C. C. Cheng and H. C. Andrews, "Target Motion Induced Radar Imaging," IEEE Trans. on Aerospace and Electronic Systems, Vol. AES-16, Jan. 1980, pp. 2-14.

20. N. H. Farhat and T. H. Chu, "Tomography and Inverse Scattering," Proc. ICO-13, 13th Congress of the International Commission on Optics, Sapporo, Japan, 1984, pp. 62-63.

21. N. H. Farhat, C. K. Chan, and T. H. Chu, "A Target Derived Reference for Frequency Diversity Imaging," Poster paper presented at the North American Radio Science/IEEE Meeting, Quebec, Canada, 1980.

22. R. M. Mersereau and A. V. Oppenheim, "Digital Reconstruction of Multi-dimensional Signalf from Their Projections," Proc. IEEE, Vol. 62, Oct. 1978, pp. 1319-1339.

23. G. S. Stent, "Thinking About Seeing," Sciences, Vol. 20, May/June 1980, pp. 6-11.

24. D. Marr, Vision, W. H. Freeman, San Francisco, CA, 1982.

25. N. Farhat, et al., "Optical Analog of Two-Dimensional Neural Networks and Their Application in Recognition of Radar Targets," in Neural Networks for Computing, J. S. Denker (Ed.), Am. Inst. of Phys., New York (1986), pp. 146-152

26. R. J. McEliece, E. C. Posner, E. R. Rodermich and S. Venkatesh, "The Capacity of the Hopfield Associative Memory," IEEE Trans. on Information Theory (accepted for publication).

27. K. S. Lee and N. Farhat, "Content Addressable Memory with Smooth Transition and Adaptive Thresholding," OSA Annual Meeting, Wash. D.C., 1985.

28. G. R. Gindi and A. F. Gmitro, "Optical Feature Extraction Via the Radon Transform," Opt. Engrg., Vol. 23, Sept./Oct. 1984, pp. 499-506.

29. G. Herman, Image Reconstruction From Projections, Academic Press, N. Y. (1980), p. 11.

30. N. Farhat and S. Miyahara, "Super-Resolution and Signal Recovery Based on Models of Neural Networks," Technical Digest, Spring 86 OSA Topical Meeting on Signal Recovery and Synthesis II, April 1986, pp. 120-125.

31. Super-Resolution, Patent Disclosure filed by University Patents Inc., Westport, Ct, on behalf of the University of Pennsylvania, April 1987.

32. J. D. Stratton, Electromagnetic Theory, McGraw Hill, New York (1941), pp. 488-490.

THE DIRECT SCATTERING OF ELECTROMAGNETIC WAVES BY ONE- AND TWO-DIMENSIONAL OBJECTS AND ITS USE FOR INVERSE PROFILING

ANTON G. TIJHUIS

Laboratory of Electromagnetic Research
Department of Electrical Engineering
Delft University of Technology
P.O. Box 5031, 2600 GA Delft
The Netherlands

1. INTRODUCTION

The aim of **inverse scattering** is to determine physical properties of an unknown object or configuration that are not readily available from direct measurements. Typically, such direct measurements are impracticable either because of their costs or other consequences, our because the region of interest is simply inaccessible. Hence, we must probe that region indirectly by generating some wave phenomenon that can penetrate into it, and measuring the resulting scattered field. From the knowledge of that field, we then hope to recover the desired properties. In a general inverse-scattering problem, the unknown physical properties may involve geometrical aspects of the configuration, as well as the spatial distribution of certain material parameters. The probing wave may be of an electromagnetic, acoustic, or elastodynamic nature, and may be either pulsed or monochromatic.

Inverse-scattering problems can be subdivided into a number of classes, according to the kind of information that must be retrieved in the measurement. In the present paper, we will encounter two types of inverse-scattering problems. In the first place, we will consider the **identification problem**, which amounts to establishing whether the scattering configuration is or is not identical to some fully known reference configuration. In the second place, we have the **inverse-profiling problem**, where the geometry of the configuration is known, and where we wish to determine the spatial distribution of one or more material parameters. Such a distribution is commonly referred to as a **profile**.

From the definition of an inverse-scattering problem as given above, it will be clear that such a problem cannot be resolved without a detailed understanding of the mechanisms that govern the propagation and scattering effects experienced by the probing wave inside and outside the scattering domain. Hence, we will also have to investigate the **direct-scattering** problem of computing the response to the probing wave in a **known** configuration.

In the present paper, we will investigate electromagnetic scattering problems as indicated in Fig. 1. As far as the direct-scattering problem is concerned, we will concentrate on transient plane-wave scattering by one- and two-dimensional, inhomogeneous, penetrable obstacles. In the inverse case, we have limited the discussion to the class of one-dimensional problems. Thanks to these restrictions, we have been able to investigate the analytical as well as the computational aspects of a number of solution techniques in full detail. In formulating these solution techniques, on the other hand, we have attempted to preserve their capability of being generalized to more complicated geometries.

B. de Neumann (ed.), Electromagnetic Modelling and Measurements for Analysis and Synthesis Problems, 391–425.
© 1991 *Kluwer Academic Publishers. Printed in the Netherlands.*

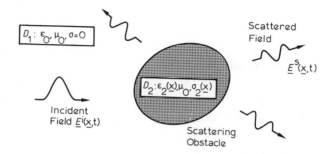

FIGURE 1. Transient plane-wave scattering by an inhomogeneous, lossy
dielectric obstacle embedded in vacuum.

   This paper is organized as follows. In Sections 2 and 3, we will consider
the direct-scattering problem. In Section 2, this problem is solved
indirectly via a Fourier or Laplace transformation to the real- or complex-
frequency domain. In Section 3, it is solved directly in the time domain.
In Sections 4 and 5, we will study the identification problem and the
inverse-profiling problem for a lossy, inhomogeneous dielectric slab
embedded in vacuum, respectively. In this context, we will base the
formulation of the method of solution on the frequency-domain analysis
described in Section 2, and we will use the time-domain results obtained
from the analysis of Section 3 as "known" data. In Section 5, we will
describe as well how the identification method outlined in Section 4 can be
used in inverse profiling. For reasons of clarity, we will state the
conclusions that we have arrived at in applying each method of solution
toward the end of the section dealing with that method. Finally, it should
be mentioned that, in view of space limitations, we have restricted the
presentation to summarizing the methods of solution, and to showing a few
representative results. For all details and derivations, and for more
results, the reader is referred to [1].

2. DIRECT SCATTERING: FREQUENCY-DOMAIN TECHNIQUES
2.1. Introduction
   In this section, we consider solving the direct-scattering problem of a
pulsed, polarized electromagnetic plane wave, normally incident on a one-
or two-dimensional obstacle with the aid of frequency-domain techniques.
These techniques have in common the feature that the transient response of
the obstacle is represented as a Laplace inversion integral of the type

$$\underline{E}(\underline{x},t) = (2\pi i)^{-1} \int_{\beta - i\infty}^{\beta + i\infty} \exp(st) \, \underline{E}(\underline{x},s) \, ds. \tag{1}$$

In (1) the contour of integration (the Bromwich contour) must be chosen in
the right half of the complex-frequency plane (s-plane), i.e. $\mathrm{Re}(\beta) \geq 0$.
The main advantage of such a representation is that in the equation(s)
governing its constituents (the Laplace-transformed field quantities), the
partial time differentiation from the corresponding time-domain equation(s)
is replaced by a scalar multiplication with the complex frequency s. As a
consequence, these equations decompose into a system of differential or

integral equations for the individual field constituents. In this section, we will concentrate in the latter kind. In operator form such equations can be written as

$$L(s) \; \underline{E}(\underline{x},s) = \underline{E}^i(\underline{x},s) \text{ for } \underline{x} \text{ in } \mathcal{D}, \qquad (2)$$

or similar equations for $\underline{H}(\underline{x},s)$ or $\{\underline{E}(\underline{x},s),\underline{H}(\underline{x},s)\}$. In (2), $\underline{E}^i(\underline{x},s)$ represents the Laplace-transformed incident field. For an incident field as specified above, this transformed field is an entire function in the complex-frequency plane. $L(s)$ denotes a linear operator of the form

$$L(s) \; \underline{f}(\underline{x}) = \int_{\mathcal{D}} K(x,x';s) \; \underline{f}(\underline{x}') \; dx'. \qquad (3)$$

The domain $\mathcal{D}$ denotes either the boundary of the scattering obstacle (for homogeneous or impenetrable scatterers), or its interior (for inhomogeneous, penetrable scatterers). The kernel $K(\underline{x},\underline{x}';s)$ is assumed to act linearly on $\underline{f}(\underline{x}')$, and may contain singularities and/or space differentiations. Formally, the solution of the system of equations can now be written as

$$\underline{E}(\underline{x},t) = (2\pi i)^{-1} \int_{\beta - i\infty}^{\beta + i\infty} \exp(st) \left(L^{-1}(s) \; \underline{E}^i(\underline{x},s) \right) ds. \qquad (4)$$

(4) directly shows the main disadvantage of frequency-domain techniques: the evaluation of a single time-domain response requires the solution of several frequency-domain problems.

Basically, there are two ways in which frequency-domain results of the type described above can be combined into a transient response, both characterized by the way in which the Laplace inversion integral is evaluated. In the next two subsections, we will formulate the two ways to evaluate this integral. Subsequently, we will describe two configurations to which the resulting solution techniques have been applied. For one of these, we will present and discuss some numerical results. We will end this section by stating our conclusions.

2.2. Contour deformation in the complex-frequency plane: singularity expansion method.

In the **singularity expansion method**, commonly referred to as SEM, the Bromwich integral in (4) is evaluated by contour deformation in the complex-frequency plane. The first formal treatment of a transient scattering problem with the aid of this method was given by Baum [2], who analyzed the plane-wave scattering by a perfectly conducting sphere. Since then, numerous papers dealing with its mathematical and physical aspects as well as with its application to specific problems have appeared. Insight into the development and the application of the SEM can, for instance, be obtained from [3-7,1].

The contour is closed in either the left or the right half of the complex-frequency plane, according to whether the integral along the closing contour in the relevant half-plane vanishes. When the contour can be closed to the right, the field represented by the inversion integral is zero, since the Laplace transform of a causal signal is regular in the right half of the complex-frequency plane. Upon closure to the left, we end

up with a field representation in terms of the contributions from the singularities in the complex s-plane, i.e. poles and, possibly, branch points. In the remainder of this subsection, we will consider the form of these contributions and the conditions under which this procedure can be followed.

2.2.1. Residual contributions from the poles. Let us first consider the residual contribution from the poles in the complex-frequency plane, also referred to as **natural frequencies**. These poles $\{s_\alpha\}$, with the subscript α being assigned according to some numbering system, occur either on the negative real s-axis or in complex conjugate pairs in the left half of the s-plane. Since the occurrence of poles of higher order can always be handled by considering limiting cases pertaining to simple poles, we restrict ourselves to the general case of simple poles only. In the vicinity of such a pole, we can write the Laplace-transformed field as

$$\underline{E}(\underline{x},s) = \underline{e}_\alpha(\underline{x})/(s - s_\alpha) + \underline{E}_\alpha^e(\underline{x},s), \tag{5}$$

where $\underline{e}_\alpha(\underline{x})$ denotes the residue of $\underline{E}(\underline{x},s)$ at $s = s_\alpha$, and where the components of $\underline{E}_\alpha^e(\underline{x},s)$ are functions that are analytic in some finite domain around $s = s_\alpha$. Substituting (5) in (2), and using the analytic properties of $L(s)$, $\underline{E}^i(\underline{x},s)$, and $\underline{E}_\alpha^e(\underline{x},s)$, we then find

$$L(s_\alpha) \, \underline{e}_\alpha(\underline{x}) = 0. \tag{6}$$

This result implies that, for $s = s_\alpha$, the operator equation (2) has a nontrivial solution even for vanishing incident field, and explains the connection between the poles and the natural modes of the scattering configuration. Since (6) is a homogeneous equation, we have to introduce an appropriate normalization. The normalized solution $\underline{E}_\alpha(\underline{x})$ is commonly designated as the **natural-mode field distribution**. The normalization should be chosen according to numerical convenience, and depends strongly on the configuration at hand.

Since we are only dealing with simple poles, the residual contribution $\underline{e}_\alpha(\underline{x})$ must then be a multiple of $\underline{E}_\alpha(\underline{x})$. Using the linearity of the problem, we can write

$$\underline{e}_\alpha(\underline{x}) = F(s_\alpha) \, D_\alpha \, \underline{E}_\alpha(\underline{x}). \tag{7}$$

In (7), $F(s)$ denotes the Laplace transform of the time distribution of the incident field. D_α is the so-called **coupling coefficient**, which determines the strength of the natural mode. Its special form, which is not important in the present discussion, can be obtained from the operator equation (2) by following a procedure that was first formulated for impenetrable scatterers by Marin [8], and later generalized to penetrable scatterers by Tijhuis and Blok [9,10]. Rather than deriving this form explicitly, we have, in Table 1, specified the dependence of the various modal quantities introduced above on the properties of the object, and on the incident field. From this table it is observed that the natural frequencies $\{s_\alpha\}$ in

TABLE 1. Dependence of modal quantities on the object properties and on the incident field.

Modal quantity	Depends on
s_α	Object properties only.
$\underline{E}_\alpha(\underline{x})$	Object properties only.
D_α	Object properties; space distribution of incident field.
$F(s)$	Time distribution of incident field.

particular depend on object properties only, and, hence, can be used for identifying the scattering object. We will come back to this possibility in Section 4.

Combining (5) and (7), we find that, in absence of branch-cut contributions, the Laplace-transformed field $\underline{E}(\underline{x},s)$ can be written in terms of the **Mittag-Leffler expansion**:

$$\underline{E}(\underline{x},s) = \sum_\alpha F(s_\alpha) \, D_\alpha \, \underline{E}_\alpha(\underline{x}) \, / \, (s - s_\alpha) + \underline{E}^e(\underline{x},s), \tag{8}$$

where the components of $\underline{E}^e(\underline{x},s)$ are entire functions.

2.2.2. <u>Contributions from branch cuts.</u> For some scattering geometries, we have to take into account contributions from branch cuts in the complex-frequency plane. By applying Cauchy's integral formula to a contour circumscribing the whole complex s-plane, excepting the singularities, we find directly that the branch-cut contributions can be incorporated into the Mittag-Leffler representation (8). To this end, we should allow the index α to run over continuous ranges as well, and we should identify the term $D_\alpha \underline{E}_\alpha(\underline{x})$ as being $(2\pi i)^{-1}$ times the jump discontinuity of $\underline{E}(\underline{x},s)$ across the branch cut at $s = s_\alpha$. This result indicates that the branch-cut contribution can be envisaged as a continuous spectrum of natural-mode contributions. This interpretation can also be reached in a different manner: the Laplace-transformed field $\underline{E}(\underline{x},s)$ on both sides of the branch cut constitutes a solution to the differential equations underlying the operator equation (2). The differential operators occurring in these equations, as well as the relevant multiplication factors and the Laplace-transformed incident field, do not change discontinuously across the branch cut. Hence, the jump discontinuity in $\underline{E}(\underline{x},s)$ constitutes a nontrivial solution to these differential equations for a vanishing incident field, and may be interpreted as a "continuous" natural mode.

It should be remarked that the combined pole/branch-cut spectrum is not unique: a different choice of the branch cut will generally result in a different set of discrete poles, with corresponding natural-mode field distributions. In conformity with the case of pole-contributions only, this ambiguity is usually resolved by requiring that the Schwartz reflection principle

$$\underline{E}(\underline{x},s^*) = \underline{E}(\underline{x},s)^*, \tag{9}$$

which holds in the right half of the complex s-plane for each $\underline{E}(\underline{x},s)$ corresponding to a causal signal, also hold for its analytical continuation into the left half-plane.

2.2.3. <u>Contributions from the singularities to the time-domain field.</u> From either of the inversion formulas (1) and (4), it now follows immediately that, in the evaluation of the Laplace inversion integral by contour deformation in the complex s-plane, the total contribution from the singularities is given by

$$\underline{E}_{\text{singularities}}(\underline{x},t) = \sum_\alpha D_\alpha \ F(s_\alpha) \ \underline{E}_\alpha(\underline{x}) \ \exp(s_\alpha t). \tag{10}$$

With (10), we have arrived at the **conventional SEM representation** for a transient electric field in a linearly reacting configuration.

2.2.4. <u>Contributions from the closing contours at infinity.</u> Whether this representation holds, depends on the possible presence of a nonvanishing contribution from the closing contours at infinity. These contributions are associated with the entire function $\underline{E}^e(\underline{x},s)$ that appears in the Mittag-Leffler expansion (8). In order to determine whether the contribution from the closing contour vanishes in the left or the right half of the complex-frequency domain, a high-frequency approximation to $\underline{E}(\underline{x},s)$ is needed, which holds over the entire range of arg(s). With the aid of such an asymptotic approximation and Jordan's lemma, it can then be established for each given space-time point $(\underline{x},t)$ whether the integrals along one or both of the closing contours vanish.

Until now, suitable asymptotic expression have only been obtained for special geometries. Examples can be found in [9-11]. Nevertheless, some general conclusions can be formulated on physical grounds. Let us first consider the total field in the object domain $\mathcal{D}$. Since the incident field is a pulsed plane wave, we can always choose the time variable t such that its front first reaches the scattering domain at t=0. This implies that, for each space point in $\mathcal{D}$, there will be some finite termination instant, before which the contribution from the closing contour vanishes upon closure to the right. This termination instant is exactly the instant where the influence of the scatterer first becomes noticeable. For an incident pulse of finite duration, there must also be an instant after which the incident field has completely passed the scattering domain. Hence, for each space point, there must also be an initial instant where the SEM representation starts to hold. Finally, it follows from the principle of causality that an incident pulse cannot excite a natural mode before it actually reaches the scattering obstacle. Consequently, there will be at least some nonvanishing subregion of $\mathcal{D}$ where there is a time interval in between these instants. In that time interval, we have a nonvanishing field that cannot be represented by modal contributions only. Its length will correspond to the period during which the incident pulse is reaching the scattering obstacle. For points outside $\mathcal{D}$, we must consider the scattered field. Realizing that the sources of this field correspond to the total field in $\mathcal{D}$, and taking into account the appropriate retardation leads to similar conclusions as stated above for the total field inside $\mathcal{D}$.

2.3 Direct evaluation of the contour integral.

Instead of evaluating the Laplace inversion integral by contour deformation, one can also compute it directly. In this context, it is convenient to factor out the spectrum of the incident pulse, i.e. F(s), and to write the complex frequency s as s = β + iω. This reduces (2) to

$$\underline{E}(\underline{x},t) = (2\pi)^{-1} \exp(\beta t) \int_{-\infty}^{\infty} F(\beta + i\omega) \; \underline{\hat{E}}(\underline{x},\beta + i\omega) \exp(i\omega t) \, d\omega, \tag{11}$$

where, for $\beta = 0$, $\underline{\hat{E}}(\underline{x},\beta + i\omega)$ denotes the response to a monochromatic plane wave with angular frequency ω. Now, the integral in (11) can be recognized as as Fourier inversion integral that can be evaluated with the aid of a standard FFT procedure. This way to evaluate the contour integral is known as **direct Fourier inversion, or Bromwich inversion.** One advantage is that the Fourier representation employed is also valid at instants where the SEM representation (10) does not hold. As such, its computation is certainly preferable to numerically evaluating the contribution from a closing contour in the complex s-plane. Generally, the parameter β is taken equal to zero; for perfectly conducting scatterers, however, it may be preferable to choose $\beta > 0$ in order to avoid the well-known non-uniqueness problems near the resonances of the corresponding interior problem (for a detailed analysis of such resonances, see [12]).

2.4. Results.

In order to obtain insight into the applicability of the SEM in practice, the general theory presented in Subsections 2.2 and 2.3 has been implemented for the case of an E-polarized, pulsed plane wave normally incident from the left on either a lossy, inhomogeneous dielectric slab or a lossy, radially inhomogeneous dielectric circular cylinder, both embedded in vacuum. Details of these applications can be found in [9-11,1]. Here, we restrict ourselves to a general indication. In both configurations specified above, the electric field can be written as

$$\underline{E}(\underline{x},t) = E(\underline{x},t) \, \underline{i}_z, \tag{12}$$

where the z-component $E(\underline{x},t)$ can be regarded as the fundamental field quantity, from which the remaining (magnetic) field components can be determined. The slab problem is one-dimensional, while the cylinder problem can be decomposed into one-dimensional subproblems by subjecting the time-domain fields to an angular Fourier transformation. This results in a Fourier representation that converges globally in the spatial domain, uniformly in time. For the one-dimensional (sub)problems, first-order WKB approximations can be found that allow the derivation of closing conditions as described above. In the interior of the scatterer, exact (numerical) solutions can be obtained by applying a Runge-Kutta-Verner integration to the relevant first-order differential equations. By matching these numerically obtained solutions to the available closed-form solutions in the exterior domain, a characteristic equation can be derived, whose zeros are the natural frequencies. To determine these zeros, we use a combination of a procedure based on argument-type integrals, and Muller's method. The natural-mode field distributions and the coupling coefficients are obtained automatically in the search process.

Some representative results are shown in Figs. 2-4. In view of space limitations, we only present results for the case of the cylinder. In Fig. 2, the conditions for vanishing contributions from the closing contours in the right and left half of the complex s-plane are illustrated graphically. Note that these conditions are in agreement with the general conclusions formulated in Subsection 2.2, and that they do not depend on the angle of observation. The latter result is caused by the decomposition into one-dimensional subproblems, and does not reflect the true causal nature of the

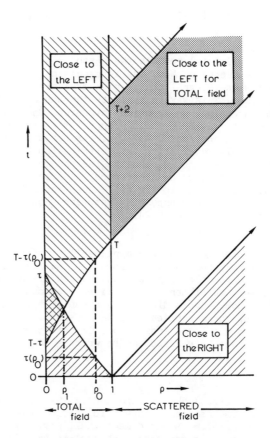

FIGURE 2. Space-time illustration of the conditions for vanishing contributions from the closing contours at infinity for a radially inhomogeneous, lossy dielectric circular cylinder embedded in vacuum. The normalized coordinates are $c_0 t/a$ and ρ/a, with c_0 being the electromagnetic wave speed in vacuo, and a being the radius of the cylinder. In the unshaded region, the Bromwich contour cannot be closed without a possible contribution along the supplementary arcs.

field. This is the price we have to pay for obtaining a globally converging representation, and, hence, for being able to analyze the role of the closing contours at all. In Fig. 3, some natural frequencies are shown for a cylinder with a parabolic permittivity profile and a constant conductivity. A special feature of this diagram is that there are apparently two complementary ways to order the natural frequencies and the corresponding natural modes. This effect was first observed for perfectly conducting scatterers, and has given rise to a very elegant physical interpretation of the scattering process [13,14]. For the present configuration, a similar interpretation has been given in [15,1]. Fig. 4 presents, for the same case as considered in Fig. 3, the decomposition of a representative time signal into contributions from the poles and from the branch cut.

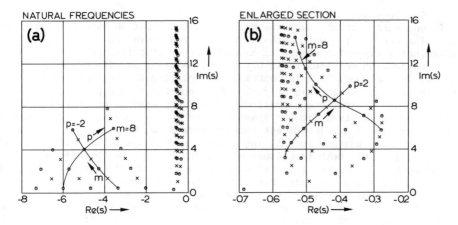

FIGURE 3. Normalized natural frequencies $s_{mp} a/c_0$ for a dielectric cylinder with a parabolic permittivity profile and constant conductivity. In Fig. 3b, the real s-axis was stretched to bring out the ordering in the poles with $m \geq 0$.

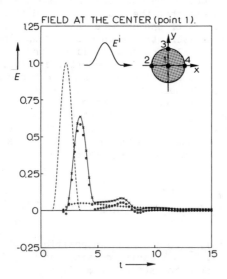

FIGURE 4. The electric field at the center (point 1) of the configuration previously considered in Fig.3 as caused by a sine-squared incident pulse. Normalized time as specified in Fig. 2.; dashed line: incident field; solid line: result of direct Fourier inversion; x: total SEM result; o: total pole contribution; +: branch-cut contribution; inset: the scattering configuration. Similar results were obtained for the fields at points 2-4.

2.5. Conclusions.

To end this section, we state the conclusions that we have reached after applying frequency-domain methods as described above. As far as the **singularity expansion method** is concerned, the principal advantage is that its application provides insight into the physical aspects of the scattering process. In this context, the ordering of the poles in the complex-frequency plane is of particular importance. The principal disadvantage is that the natural-mode field representation obtained is not valid for early times and for special pulse shapes such as a delta-function or a Gaussian pulse. The **direct Fourier inversion,** on the other hand, is valid at all times and for all integrable pulse shapes, but provides little physical insight. **Both techniques** have been observed to consume considerable amounts of computation time, most of which is spent in solving individual frequency-domain problems. In the SEM, such computations form part of the search process for the natural frequencies; in the FFT evaluation, the solutions occur as factors in the integrand in (11). As a consequence, the application of frequency-domain techniques will probably remain restricted to special (separable) geometries. For such geometries, however, the results obtained will be both stable and very accurate.

3. DIRECT SCATTERING: THE MARCHING-ON-IN-TIME METHOD
3.1. Formulation of the method.

In this section, we take leave of the idea of solving time-domain electromagnetic direct-scattering problems with the aid of frequency-domain techniques. Instead, we consider solving such problems with the aid of the **marching-on-in-time-method,** which utilizes the common property in the relevant integral equations that the scattered field in each space-time point is expressed in terms of one or more integrals of field values at previous instants. Around 1968, this method showed up almost simultaneously in the acoustic and in the electromagnetic literature [16-20]. Since then, it has been applied to a wide range of acoustic, elastodynamic, and electromagnetic transient scattering problems (for a review, see Section 3.1 of [1]). An abstract formulation of the technique was first proposed in [21]. In the present section, we will adhere to this formulation.

With our applications in mind, we restrict ourselves to a scalar wave field $\phi(\underline{x},t)$, which may be regarded as a fundamental electric or magnetic field component. However, the scalar nature of $\phi(\underline{x},t)$ is not essential. For a such a wave field, the time-domain equivalent of the system of equations (2)-(4) reads:

$$\phi(\underline{x},t) = \phi^i(\underline{x},t) + \int_D d\underline{x}' \int_0^{t-R/c} dt'\, K(\underline{x},\underline{x}';t-t')\, \phi(\underline{x}',t') \tag{13}$$

(see also [22]). In (13), $\underline{x}$ is a Cartesian position vector and $R = |\underline{x} - \underline{x}'|$. As in (3), D is a finite domain, and $K(\underline{x},\underline{x}';t-t')$ is a linear, time-invariant operator acting on $\phi(\underline{x}',t')$, which may contain singularities as well as space and time differentiations. $\phi^i(\underline{x},t)$ is a known incident field that for x in D vanishes when $t \leq 0$, and c is a wave speed parameter. When $\underline{x}$ in D, $0 \leq t < \infty$, (13) is an integral equation for $\phi(\underline{x},t)$.

In order to solve the integral equation (13) numerically, we must discretize in space and time. To this end, we construct a uniform spatial grid $\{\underline{x}_\alpha\}$ with mesh size h which covers D, and we take $t = n\Delta t$, where

$n = 0,1,2,\ldots,\infty$. Conventionally, this time step is chosen such that $\Delta t = \min(R_{\alpha\alpha'})/c$, where $R_{\alpha\alpha'} = |\underline{x}_\alpha - \underline{x}_{\alpha'}|$. Then we end up with discretized, algebraic equations of the type

$$\tilde{\phi}(\alpha,n) = \tilde{\phi}^i(\alpha,n) + \sum_{\alpha'} \sum_{n'=0}^{n} \tilde{K}(\alpha,\alpha';n-n')\,\tilde{\phi}(\alpha',n'),$$

$$n = 0,1,2,\ldots,\infty, \tag{14}$$

where $\tilde{\phi}^i(\alpha,n) = \phi^i(\underline{x}_\alpha,n\Delta t)$. (14) expresses, for each fixed instant n, the field values $\{\tilde{\phi}(\alpha,n)\}$ in terms of the given incident field at the instant $t = n\Delta t$ and the field values $\{\tilde{\phi}(\alpha,n')|n' < n\}$, i.e. the numerical solution at previous instants. The determination of the unknown $\{\tilde{\phi}(\alpha,n)\}$ requires at most the inversion of the so-called **weighting matrix**

$$W(\alpha,\alpha') = \delta_{\alpha\alpha'} - \tilde{K}(\alpha,\alpha';0), \tag{15}$$

which can be carried out prior to performing the first time-marching step. Furthermore, the upper limit of the time integration in (13) is always less than t, unless $R = 0$, which occurs only if $\alpha = \alpha'$. Hence, owing to the choice of Δt, the interpolations in the discretization can generally be organized such that $W(\alpha,\alpha') = 0$ if $\alpha \neq \alpha'$, which eliminates the necessity for a numerical inversion.

In terms of matrix calculus, the discretized equation (14) can be envisaged as a matrix equation of infinite dimension for the unknown vector $\{\tilde{\phi}(\alpha,n)\}$. In that perspective, the inversion of $W(\alpha,\alpha')$ can be understood as being part of the reduction of (14) to lower-triangular form.

3.2. Error accumulation and instabilities.

In the implementation of the numerical scheme outlined above, difficulties can be expected due to the discretization of the multiple integral in (13). Because of the error made in this discretization, $\tilde{\phi}(\alpha,n)$ will only be approximately equal to the actual field value $\phi(\underline{x}_\alpha,n\Delta t)$.

Since, in the numerical solution of (14), each approximate field value $\tilde{\phi}(\alpha,n)$ is computed from field values $\tilde{\phi}(\alpha',n')$ at previous instants, the computational errors caused in this manner may accumulate. As a consequence, the solution obtained may be unstable. In the majority of the papers dealing with the marching-on-in-time method, this difficulty is not addressed. Generally, the unwanted instabilities are eliminated by some ad hoc method that is applicable to the specific problem under consideration.

Only in recent years, some attention has been devoted to finding more fundamental approaches to solving this problem. Two categories can be distinguished. The first one attempts to control the unstable behavior by modifying the discretization. For example, in [21,23] it was demonstrated that, for some scattering problems, both the accuracy and the stability of a marching-on-in-time result can be controlled by reducing the mesh size in a systematic discretization. In [24,25], a marching-on-in-time result was stabilized by doubling the time step in the backward interpolation formula for a time differentiation occurring in the kernel $K(\underline{x},\underline{x}';t-t')$. The second category avoids the instabilities by modifying the method of solution. To

this end, $\{\tilde{\phi}(\alpha,n)\}$ is redefined as the set of approximate field values that vanish outside a finite interval $0 \leq n \leq N$, and that minimize a summed squared error in the equality sign of (14) over the semi-infinite range $0 \leq n < \infty$. This solution is then obtained by applying a gradient-type minimization method. Representative results can be found in [22] and in [26-28]. The condition that the field values $\{\tilde{\phi}(\alpha,n)\}$ vanish for $n > N$ is essential: the (unstable) marching-on-in-time result meets the equality sign in (14) exactly for all n.

Both categories have their specific disadvantages. Each iteration step in a gradient-type methods requires a numerical effort equivalent to two complete marching-on-in-time computations. Moreover, a large number of iteration steps may be needed to arrive at an acceptable result. Hence, the computation may be excessively time-consuming. Modifying the discretization may be more efficient, but is not always possible. In the present section, we summarize a procedure which avoids these disadvantages by combining the best of both approaches. More details can be found in Section 3.2 of [1].

3.2.1. Discretization error. In order to understand this procedure, we first have to gain some more insight into the process of error accumulation. To this end we consider two types of deviations. In the first place, we define the **discretization error** $D(\alpha,n)$ as the error resulting from approximating the integral in (13) for $\underline{x} = \underline{x}_\alpha$ and $t = n\Delta t$ by the discrete sum in (14) for the **exact** field $\phi(\underline{x},t)$, i.e.

$$D(\alpha,n) = \int_D d\underline{x}' \int_0^{n\Delta t - R/c} dt' \, K(\underline{x}_\alpha,\underline{x}';n\Delta t - t') \, \phi(\underline{x}',t')$$

$$- \sum_{\alpha'} \sum_{n'=0}^n \tilde{K}(\alpha,\alpha';n - n') \, \phi(\underline{x}_{\alpha'},n'\Delta t). \tag{16}$$

Although this error is not known in full detail, since $\phi(\underline{x},t)$ is unknown, its magnitude can be estimated from the continuity and/or differentiability properties of this field, which are known.

In the second place, we define the **marching-on-in-time error** $\Delta\phi(\alpha,n)$ as the deviation between the sampled actual field value and the corresponding marching-on-in-time result, i.e.

$$\Delta\phi(\alpha,n) = \phi(\underline{x}_\alpha,n\Delta t) - \tilde{\phi}(\alpha,n). \tag{17}$$

Finally, we define the so-called **error growth** as the difference between the marching-on-in-time errors at two consecutive instants at the same space point, i.e. as $\Delta\phi(\alpha,n) - \Delta\phi(\alpha,n - 1)$. By taking a suitable linear combination of (13), (14), and (16), we find that this error growth is governed by the equation

$$\Delta\phi(\alpha,n) - \Delta\phi(\alpha,n - 1) = D(\alpha,n) - D(\alpha,n - 1)$$

$$+ \sum_{\alpha'} \sum_{n'=0}^n \tilde{K}(\alpha,\alpha';n - n') \, [\Delta\phi(\alpha',n') - \Delta\phi(\alpha',n' - 1)], \tag{18}$$

which is of the same form as (14). In (18), the role of the incident field is played by the discretization-error difference $D(\alpha,n) - D(\alpha,n - 1)$.

3.2.2. Global error accumulation. Generally, the discretization error $D(\alpha,n)$ consists entirely of interpolation errors. Each of these errors is proportional to some power of the mesh size h and to some higher-order space or time derivative at a space-time point that lies within a specified space or time interval of length $O(h)$. For a smoothly varying $\phi(\underline{x},t)$ and a sufficiently small h, these errors will contain a systematic part corresponding to the average values of these derivatives, and a randomly varying part due to the arbitrariness of the exact location of the space-time points within the intervals. By considering the discretization-error difference, we have eliminated most of the systematic part. Hence, we may regard $D(\alpha,n) - D(\alpha,n-1)$ as an almost randomly varying variable. Since the systematic part of $D(\alpha,n)$ may be absent, we estimate its magnitude as

$$D(\alpha,n) - D(\alpha,n-1) = O\big(D(\alpha,n)\big). \tag{19}$$

In terms of the matrix interpretation given toward the end of Subsection 3.1, this observation implies that $D(\alpha,n) - D(\alpha,n-1)$ has components of equal order of magnitude along each of the basis vectors that span the space of possible marching-on-in-time solutions $\{\phi(\alpha,n)\}$. This provides us with the key to the error analysis for the discretized equation (14). Since (14) is a, presumably accurate, discretization of the well-posed integral equation (13), almost all eigenvalues of its system matrix will be of $O(1)$. Therefore, the components of the known vector $D(\alpha,n) - D(\alpha,n-1)$ along the corresponding eigenvectors will not be amplified in the solution vector $\Delta\phi(\alpha,n) - \Delta\phi(\alpha,n-1)$. In the subspace spanned by these "well-behaved" eigenvectors, we then have by induction:

$$\Delta\phi(\alpha,n) \leq O\big(n \max\{D(\alpha,n'|0 \leq n' \leq n\}\big). \tag{20}$$

Now, we should realize that, for scattering by passive obstacles, we are usually only interested in the field values in some finite time interval $0 < n\Delta t < T_{max}$. Moreover, the interpolation errors constituting $D(\alpha,n)$ will generally add up to an estimate of the type

$$D(\alpha,n) = O(h^m), \tag{21}$$

with m a positive integer. With these two observations in mind, we conclude directly that the error in the subspace spanned by the well-behaved eigenvectors of the system matrix of (14), (18) is at most proportional to

$$(T_{max}/h)\, h^m = T_{max}\, h^{m-1}. \tag{22}$$

Hence, it can be controlled by organizing the discretization such that $m > 1$, and by choosing h sufficiently small. Since this "well-behaved" subspace covers almost the entire space of possible solutions, we will designate the behavior found in (22) as **global** error accumulation.

3.2.3. Local error accumulation: instabilities. For some configurations, the system matrix of (14) and (18) may also have a few "problematic" eigenvectors with small eigenvalues. In the numerical solution of (14), the components of $D(\alpha,n) - D(\alpha,n-1)$ along these eigenvectors will be amplified, and the result obtained will be unstable. Since this effect will be localized within the small subspace spanned by these eigenvectors, an error of this type is called a **local** error. Note that this qualification

does not imply a local behavior of the error in the space-time domain. Rather, it refers to an error behavior of a special, unwanted character.

Greater comprehension of these local errors can only be obtained when the kernel $K(\underline{x},\underline{x}';t - t')$ is specified in more detail. In general, this kernel involves one or two time differentiations acting directly on $\phi(\underline{x}',t')$. In the discretization, the resulting derivatives need to be approximated by suitable backward difference formulas. Moreover, $K(\underline{x},\underline{x}';t - t')$ is singular at the point of observation for most multi-dimensional scattering problems. Restricting ourselves to kernels involving only a single time derivative, we can then come up with the following simplified model of the discretized equation (14). Let $f[n]$ and $g[n]$, with $n = 0,1,2,\ldots,\infty$, be real-valued time sequences that satisfy the difference equation

$$f[n] + w_0 \, C \, \{\tfrac{3}{2}f[n] - 2f[n - 1] + \tfrac{1}{2}f[n - 2]\}$$

$$+ w_1 \, C \, \{\tfrac{3}{2}f[n] - 2f[n - 1] + \tfrac{1}{2}f[n - 2]\} = g[n]. \tag{23}$$

In (23), w_0 and w_1 are positive weighting factors with $w_1 > w_0$, C is a positive contrast or curvature parameter, and the terms between braces originate from a three-term backward difference formula based on quadratic time interpolation. The interpretation of the terms in (23) is given in Table 2.

TABLE 2. Relation between the terms in the difference equation (23) and the terms in the discretized integral equation (14)

Term in (23)	Corresponds with
$f[n]$	Field at the point of observation.
Term with w_0	Contribution from "self term" in integral, or systematic contribution from space-time points with "even" delay.
Term with w_1	Contribution from "neighboring" space-time points, or systematic contribution from points with "odd" delay.
$g[n]$	Incident field plus contribution from remaining space-time points.

Obviously, the model proposed above is an oversimplification. In the first place, it presupposes a fixed relation between the field values occurring on the left-hand side. In the second place, assuming $g[n]$ to be a smoothly varying known signal restricts the instability analysis to the interaction between field values at a few selected space-time points. However, the only consequence of this simplification is that certain types of instabilities may be excluded. This need not keep us from regarding the instabilities predicted by our model as realistic.

For the simple difference equation (23), a closed-form stability analysis can be performed. Details can be found in Subsection 3.2.3 of [1]. It turns out that (23) has a stable solution for

$$0 \leq C \leq 1/(4w_1 - 4w_0). \tag{24}$$

For larger values of C, an instability of the form

$$\delta[n] = (z_1)^n, \quad n = 0,1,\ldots,\infty, \tag{25}$$

with z_1 being a negative real number with $|z_1| > 1$, will show up. The result given in (24) also explains the (partial) success of modifying the discretization described in [21,23-25]: each of the successful modifications reduces the weighting factors w_0 and w_1. From the above discussion, it is also clear that the occurrence of instabilities in marching-on-in-time solutions depends on the structure of the integral equation at hand, rather than on the accuracy of the discretization.

3.2.4. A modified marching-on-in-time method. (23) and (24) also provide us with the clue to the stabilization of the marching-on-in-time method: by increasing the coefficient of $\delta[n]$ in (23), we can increase the numerator in the quotient on on the right-hand side of (24). This idea forms the basis of the following procedure for the iterative solution of (14). Starting from a sufficiently smooth initial estimate $\bar{\phi}^{(0)}(\alpha,n)$, we carry out a number of iteration steps. In step number j, the approximate solution $\tilde{\phi}^{(j)}(\alpha,n)$ is the result of minimizing, for each fixed n, the squared error

$$\sum_\alpha \{\tilde{\phi}^{(j)}(\alpha,n) - \tilde{\phi}^i(\alpha,n) - \sum_{\alpha'} \sum_{n'=0}^{n} \tilde{K}(\alpha,\alpha';n - n') \, \tilde{\phi}^{(j)}(\alpha',n')\}^2$$

$$+ p^2 \sum_\alpha \{\tilde{\phi}^{(j)}(\alpha,n) - \bar{\phi}^{(j-1)}(\alpha,n)\}^2, \tag{26}$$

for known values of $\tilde{\phi}^i(\alpha,n)$, $\bar{\phi}^{(j-1)}(\alpha,n)$, $\tilde{\phi}^{(j)}(\alpha,n')$ with n'<n, and p. In (26), $\bar{\phi}^{(j-1)}(\alpha,n)$ denotes the result of subjecting the previous approximation $\tilde{\phi}^{(j-1)}(\alpha,n)$ to the **smoothing operation**

$$\bar{\phi}^{(j)}(\alpha,n) = \{\tilde{\phi}^{(j)}(\alpha,n-1) + 2\tilde{\phi}^{(j)}(\alpha,n) + \tilde{\phi}^{(j)}(\alpha,n+1)\}/4, \tag{27}$$

which is based on linear interpolation over time intervals of length $2\Delta t$.

The iterative procedure outlined above is designated as a **relaxation method,** and can be shown to converge strictly for each individual component of the solution $\{\phi(\alpha,n)\}$ along an eigenvector of the system matrix of (14) and (18). Resetting the component along some of the "problematic" eigenvectors that correspond to instabilities of the form (25) does not affect the convergence for the components along the remaining eigenvectors. Moreover, the unstable behavior of the type (25) can always be suppressed by increasing the parameter p in (26).

The essential difference between the relaxation method and a gradient-type method is that the relaxation method produces, in each iteration step, a **partial** improvement to the components along **all** the well-behaved eigenvectors. A gradient-type method, on the other hand, produces a **total** improvement in a **single** direction. In view of the vast dimension of the space of possible solutions, the latter method is bound to be less efficient, especially for fine discretizations. The common element of both

methods is that they will produce similar global errors, which can only be
controlled by reducing the discretization step.

3.3. Results.

Until now, the marching-on-in-time method as described above has been
implemented for three types of configurations. First, we investigated
plane-wave scattering by a lossy, inhomogeneous dielectric slab, embedded
either in vacuum or in between two homogeneous, lossless dielectric half-
spaces [23,1]. Second, we reconsidered the problem originally solved by
Bennett, i.e. that of plane-wave scattering by an impenetrable cylinder
embedded in a homogeneous, lossless dielectric [21,1]. Third, we considered
plane-wave scattering by an inhomogeneous, lossy dielectric cylinder
embedded in vacuum [21,1]. All of these problems have in common the feature
that the relevant integral equations pertain to a single field component.
With some ingenuity, the integrals occurring in these scalar equations can
be discretized such that the error estimate (21) holds with m = 2. The
first two problems can be formulated such that refining the discretization
eliminates the unstable behavior. For the third one, however, such a
formulation could not be attained. This is explained by the fact that, for
the weighting factors in (23), we have the estimate

$$w_i = O[\ln(h)], \quad \text{for } i = 0,1 \text{ as } h \downarrow 0. \tag{28}$$

From this estimate and (24), it is observed that, for this configuration,
the stability range will shorten as h ↓ 0. This effect is illustrated in
Fig. 5, which demonstrates that refining the discretization may turn a
stable result into an unstable one. The effectiveness of the relaxation
method is demonstrated in Fig.6, which gives results obtained for the
discretized equation considered in Fig. 5b, but for a considerably larger
susceptibility contrast.

3.4. Conclusions.

Based on the description given above, Refs. [16-21,23-25], and the
author's personal experience, it can be stated that the marching-on-in-time
method is a numerically efficient, "brute-force" method for the computation
of transient electromagnetic fields. As such, it provides little insight
into the physical aspects of the scattering process. Generally, the results
obtained are less accurate than those obtained by frequency-domain methods;
the global accuracy, however, can be controlled by reducing the
discretization step. The marching-on-in-method seems to be applicable to a
wide range of geometries. For a long time, the fundamental problem in its
application has been the unstable behavior of the results. With the aid of
the relaxation method discussed in [1] and summarized in Subsection 3.2,
this problem can now be resolved at the cost of a reduction in numerical
efficiency. In this context, we should keep in mind that, for example, for
the third type of scattering problem discussed in Subsection 3.3, a
reduction of the mesh size h by a factor of 2 increases the computational
effort required by a factor of 32. Hence we can carry out 32 iteration
steps in the relaxation method (or 16 iteration steps in a gradient-type
method) with the same effort.

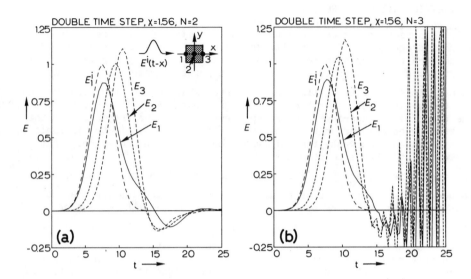

FIGURE 5. Results of applying the explicit marching-on-in-time method to a homogeneous, lossless square cylinder with susceptibility $\chi = 1.56$ and with edge-length $2a$, illuminated by an E-polarized Gaussian pulse incident from the left. The time coordinate is $c_0 t/a$, and the mesh size is $h = a/N$, with N as indicated in the figure headings.

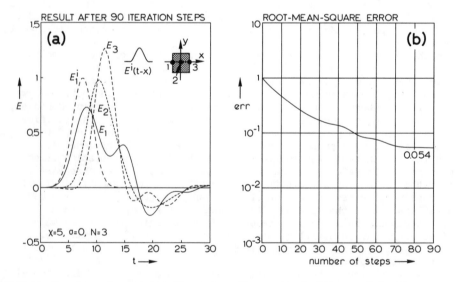

FIGURE 6. Results of applying the relaxation method with $p = 4$ to the discretized integral equation considered in Fig. 5b, but with $\chi = 5$. (a): Result after 90 iteration steps. (b): Root-mean-square error in the equality sign of the discretized equation as a function of the number of iteration steps.

4. IDENTIFICATION: A PRONY-TYPE METHOD

4.1. Introduction : the role of a Prony-type method in identification.

In this section, we consider the identification of a scatterer from its natural frequencies. To this aim, we try to determine the relevant modal parameters from the transient field excited by an incident pulse of finite duration. As we saw in Section 2, at each point of observation $\underline{x} = \underline{x}_\beta$, there exists a finite initial instant after which either the total or the scattered field can be written in the form given in (10). The dependence of the quantities occurring in this representation on the properties of the scattering object and on the incident field has been indicated in Table 1. This dependence suggests the following identification procedure. First the field $\underline{E}(\underline{x}_\beta, t)$ should be decomposed into exponentials with frequencies $\{s_\alpha\}$ and amplitudes $\{F(s_\alpha) \ D_\alpha \ \underline{E}(\underline{x}_\beta, s_\alpha)\}$ by the application of a **Prony-type method**. For a known incident field, the time distribution can then be factored out by dividing by $F(s_\alpha)$. This provides us with the set of configuration-dependent parameters $\{s_\alpha, \ D_\alpha \ \underline{E}(\underline{x}_\beta, s_\alpha)\}$, which may be regarded as the "signature" of the scattering obstacle. Next, the scatterer should be identified from that signature, and, possibly, a priori information.

The use of natural-mode parameters in target identification was first proposed by Moffatt and Mains [29]. The application of Prony's algorithm for the decomposition of a transient signal into exponentials, which dates back from 1795 [30], was suggested by Van Blaricum and Mittra [31]. Since then, a variety of papers dealing with Prony's method have appeared. The practical application of the technique to a realistic signal was treated in [32-35]. The connection with current methods from system theory and from speech processing was studied in [36-40]. Possible alternatives have been presented in [41,42]. Comparatively, only a few successful applications to actual signals, either measured or synthesized by some time-domain code, have been reported. Measured data were used in [43], and synthetic ones in [44].

In the present section, we consider the application of Prony's method to the reflected and transmitted fields caused by a sine-squared incident pulse, incident on a lossy, inhomogeneous dielectric slab embedded in vacuum. With the marching-on-in-time procedure outlined in Section 3, we are able to determine such signals directly in the time domain. Moreover, by choosing the space-time step small enough, we can attain a specified accuracy. On the other hand, the analysis summarized in Section 2 provides us with an accurate procedure to determine the natural frequencies, and with initial instants for the validity of the SEM representation at each space point. Therefore, we are in a unique position to test Prony's algorithm for an actual, albeit simple scattering configuration. We will not consider the reconstruction of the slab configuration from the modal parameters. It is the opinion of this author, for one, that such a reconstruction should be undertaken by directly using the time-domain reflected field. This will be discussed in more detail in Section 5. In that context, we will also describe how the Prony results can be used for identification purposes.

The presentation in this section is organized as follows. First, we will formulate Prony's algorithm, and we will indicate some of the problems associated with its application. Next, we will apply the algorithm to a slab signal. Some illustrative results will be presented and discussed. Finally, we will state our conclusions.

4.2. Prony's algorithm.

Let us start off by formulating Prony's algorithm. To this end, we consider a real-valued signal $\delta(t)$ consisting of M linearly independent exponentials. Such a signal can be written in the form

$$\delta(t) = \sum_{m=1}^{M} A_m \exp(s_m t), \tag{29}$$

where A_m denotes a **complex amplitude** and s_m a **complex frequency**. The parameters $\{A_m\}$ and $\{s_m\}$ completely determine the signal and, hence, can be designated as the characteristic parameters. The problem we want to resolve is the determination of these parameters from N + 1 **samples** $\delta[n]$ of $\delta(t)$ taken with **sample period** Δt at the instants $t = n\Delta t$, $n = 0,1,\ldots,N$. At these instants, (29) reduces to

$$\delta[n] = \delta(n\Delta t) = \sum_{m=1}^{M} A_m (z_m)^n, \qquad n = 0,1,2,\ldots,N, \tag{30}$$

with

$$z_m = \exp(s_m \Delta t), \qquad m = 1,2,\ldots,M. \tag{31}$$

The relation (30) constitutes a system of N + 1 nonlinear equations for the 2M unknown variables $\{A_m\}$ and $\{z_m\}$. The essence of Prony's method [30] is that it provides a conversion of this system into two systems of linear equations for M unknowns. This conversion is achieved by introducing the **characteristic polynomial**

$$P(z) = \prod_{m=1}^{M} (z - z_m) = - \sum_{\ell=0}^{M} p_\ell z^{M-\ell}, \text{ with } p_0 = -1, \tag{32}$$

whose zeros are $z = z_m$, $m = 1,2,\ldots,M$. From (30) and (32), it follows that

$$\sum_{\ell=0}^{M} \delta[n - \ell] p_\ell = \sum_{m=1}^{M} A_m (z_m)^{n-M} P(z_m) = 0, \tag{33}$$

for $M \leq n \leq N$. With $p_0 = -1$, (33) can be rewritten as

$$\sum_{\ell=1}^{M} \delta[n - \ell] p_\ell = \delta[n], \qquad M \leq n \leq N, \tag{34}$$

which constitutes a system of N + 1 - M linear equation for the **polynomial coefficients** $\{p_\ell\}$.

With (34), we have now available the basis of the procedure to determine the characteristic parameters $\{A_m\}$ and $\{s_m\}$. The total procedure should run as follows. First, the polynomial coefficients $\{p_\ell\}$ should be determined from the system (34). This allows the determination of the zeros $\{z_m\}$ from the characteristic equation $P(z) = 0$. The complex frequencies $\{s_m\}$ are then directly available from (31). Finally, once the zeros $\{z_m\}$ are known, (30) can be regarded as a system of N + 1 linear equations for the unknown

amplitudes $\{A_m\}$. This reconstruction procedure is known as **Prony's algorithm.** In this procedure, it is implicitly assumed that the number of exponentials M is known. In Subsection 4.3, we shall see that this is indeed what happens in a practical situation.

The principal advantage of Prony's algorithm is that it can be implemented entirely using available standard software (see Subsection 4.2.2 of [1]). The principal problems associated with its application are the choice of the sample period Δt, and the influence of noise. The selection of the sample period should be made on the basis of the relation (31) between z_m and s_m. On the one hand, ambiguity problems will arise in the determination of the complex frequencies $\{s_m\}$ unless the sampling is organized such that

$$-\pi < \text{Im}(s_m)\ \Delta t < \pi, \tag{35}$$

for m=1,2,...,M. In system theory, this condition is known as the **Nyquist criterion.** The condition (35) implies that, for too large a value of Δt, one or more of the complex frequencies $\{s_m\}$ may be recovered with a shift of $2\pi i/\Delta t$ (or multiples). This is the so-called **aliasing effect.** On the other hand, for too small a value of the sample period Δt, we may have $z_m \approx 1$ for some or all of the $\{z_m\}$. This will lead to difficulties in the determination of the polynomial coefficients $\{p_\ell\}$. Hence, we can only hope to apply Prony's algorithm with some success when the signal is sampled such that the $\{z_m\}$ are more or less evenly dispersed in the entire unit disk $|z| < 1$ (see also [35]).

The second problem associated with the application of Prony's algorithm is that even a small amount of noise on the data may deteriorate the accuracy of the parameter reconstruction considerably. One way of understanding this phenomenon is that, in decomposing the nonlinear system of equations (30) into two systems of linear equations, we have forfeited the possibility to compute $\{A_m\}$ and $\{z_m\}$ such that they "best fit" the sampled data with respect to some error criterion. Alternatively, this phenomenon can be understood from the occurrence of values of $\delta[n]$ in the system matrix of (34), the elements of which are inaccurate when the elements of the sequence $\delta[n]$ are. As demonstrated in [36], such an inaccuracy can produce large errors in the zeros $\{z_m\}$ obtained.

4.3. Results: application to a slab signal.

As mentioned above, Prony's method has been applied to determine natural-mode parameters from transient fields reflected from and transmitted by a lossy, inhomogeneous dielectric slab embedded in vacuum. The main obstacle to a successful application of the algorithm to such a slab signal seems to be the total residual contribution from the higher-order poles. Whether this contribution is regarded as a high-frequency term subject to the aliasing effect, or as multiplicative noise, the considerations given above show that it may considerably disturb the reconstruction.

For the configuration at hand, this problem can be circumvented as follows. We start from the response of the slab to a sine-squared pulse of given duration T. This response is synthesized by computing it with the aid of the marching-on-in-time-method. Next, the spectrum of the time signal thus obtained is determined with the aid of an FFT algorithm. This spectrum

has zeros at the known values $\omega_m = 2\pi m/T$ with $m = \pm 2, \pm 3, \ldots, \pm\infty$. Using this feature, we compute a new synthetic signal by setting all components with $|\omega| > \omega_{2,3}$ equal to zero, and carrying out the inverse Fourier transformation. This synthetic signal can then, from some initial instant $t = t_0(z)$, with z being the one-dimensional space coordinate, be represented by a sum of M exponentials as defined in (29), with the amplitudes $\{A_m\}$ and the complex frequencies $\{s_m\}$ being good approximations to the unknown modal parameters. Hence, these parameters can be determined by sampling the synthetic signal as outlined in Subsection 4.2, and subjecting the resulting time sequence to Prony's algorithm. Obviously, the number of exponentials M can be adjusted by varying the incident-pulse duration T.

As an illustration, the course of a typical Prony-type computation is illustrated in Fig. 7. Fig. 7a shows a reflected field computed with the marching-on-in-time method. In Fig. 7b, the corresponding frequency spectrum has been plotted. The absorption peaks in this spectrum correspond to the natural frequencies, and the dashed lines mark the cut-off at

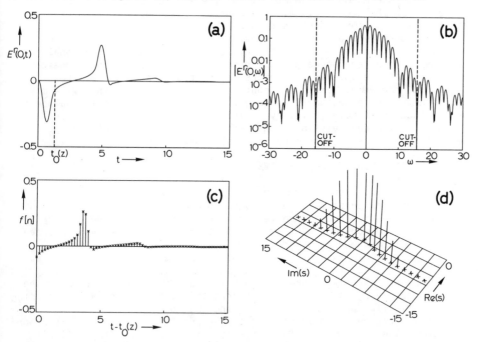

FIGURE 7. Results of reconstructing the SEM parameters of a homogeneous, lossless slab of thickness d with $\varepsilon_r = 6.25 - 16(z/d - 0.5)^2$ from the reflected field caused by a sine-squared incident pulse. (a): Marching-on-in-time result; the time coordinate is $c_0 t/d$. (b) Frequency spectrum and cut-off; the frequency coordinate is $\omega d/c_0$. (c) Vertical lines: sampled data; x: reconstruction. (d): Magnitude of the amplitudes $\{A_m\}$, plotted in the complex-frequency plane; the dimensionless frequency is sd/c_0.

$\omega = \pm \omega_3$. With these observations in mind, we notice that the filtered signal consists of 21 modal contributions, i.e. M = 21. The sampled data values that are put into Prony's algorithm are indicated by the vertical lines in Fig. 7c. Fig. 7d presents a three-dimensional picture of the modal parameters obtained. Finally, Fig. 7c also shows the result of evaluating the right-hand side of (30) for the complex amplitudes and frequencies shown in Fig. 7d.

Computations as presented in Fig. 7 were also carried out for noisy signals. Both multiplicative and additive noise were superimposed on the marching-on-in-time results. As could be expected, the presence of noise renders the result of a single computation unreliable. However, repeated computations do produce results that tend to cluster around the proper values (see also [34]). Possibly, this effect can be utilized to determine natural-mode parameters from noisy data.

4.4. Conclusions.

Two important conclusions seem to protrude from our analysis of and our numerical experiments with Prony's method. In the first place, it is essential to use an optimum sampling rate, even when exact data are available. In the second place, the transient field must consist of no more than the residual contributions from a specified number of poles. For the one-dimensional problem of scattering by a slab, a preprocessing procedure was devised which yields such a field. For two- and three-dimensional problems, such a preprocessing will probably involve combining transient fields at multiple observation points, as proposed in [45]. For two-dimensional problems, branch-cut contributions as observed in Fig. 4 may even then prevent a successful recovery. Future research should include a better handling of noise, e.g. by applying alternative methods [41,42]. In addition, a cluster analysis of the results obtained from repeated scattering experiments producing noisy data seems imminent.

5. INVERSE PROFILING: BORN-TYPE ITERATIVE APPROACH.
5.1. Introduction.

In this section, finally, we consider solving the one-dimensional inverse-profiling problem for a lossy, inhomogeneous dielectric slab embedded in vacuum. In this problem, the reflected field caused by one or more plane, electromagnetic waves, normally incident on one or both sides of the slab is assumed to be known. The aim of the computation is to reconstruct one or both of the constitutive parameters of the slab, i.e. the dielectric susceptibility $\chi(z)$ and the conductivity $\sigma(z)$. In particular, we investigate whether this inverse-scattering problem can be solved by utilizing the knowledge gathered in solving the direct-scattering and identification problems studied in the previous sections.

In the literature, this inverse-profiling problem has received attention for a long time. The problem for a lossless slab was first theoretically treated by Kay [46]. Through a Liouville transformation, he reduced it to an equivalent, uniquely solvable, quantum-mechanical potential-scattering problem. For the solution of the latter problem, a number of differential- and integral-equation approaches have been developed, reviews of which have been given in [47-50]. The main difficulty in applying one of these schemes appears to be the Liouville transformation, which involves a double space differentiation of the unknown refractive index. When this refractive index is discontinuous, numerical difficulties can be expected in the reconstruction of the equivalent quantum-mechanical scattering potential near the discontinuities. Nevertheless, Kay's analysis is, to the best of

the author's knowledge, one of the few uniqueness proofs in inverse scattering. The generalization of the quantum-mechanical approach to the case of a lossy slab was first dealt with in [51,52]. Since then, a series of papers has been devoted to this approach. Recent examples are [53-55] and references cited therein. As in the lossless case, the main disadvantage lies in the necessary Liouville transformation.

In the electromagnetic literature we can, in addition to the quantum-mechanical approach, distinguish two trends in developing solution methods for the one-dimensional inverse-profiling problem. In the first one, the integral relations pertaining to the corresponding direct-scattering problem are being used as additional relations between the known reflected-field data and the unknown fields and constitutive parameters inside the slab. In conjunction with the corresponding direct-scattering integral or differential equations, these relations are then used to resolve the inverse-scattering problem. One way to achieve this objective is to accurately trace the wavefront in the time-domain case [56-58]. This method has the disadvantage that the field at the wavefront is computed with a large relative error, especially for nonvanishing susceptibility contrast. An alternative is to follow the **Born-type** approach, which amounts to substituting some educated guess for the unknown total field inside the slab into the integral relations pertaining to the known reflection data. This yields one or two approximate integral equations of the first kind, whose solution results in an approximation to the unknown susceptibility and/or conductivity profiles. The Born-type inversion can either be performed in a **single step**, in combination with a suitable correction procedure [59,1], or within the framework of an **iterative scheme** that leads to successive approximations of the unknown total field as well as of the unknown constitutive parameter(s) [23,60-62,1].

The second trend has been to follow the so-called **optimization approach**, where one applies a conventional minimization procedure to find the smallest value of a so-called **cost function** [63-67,1]. This cost function consists of the total integrated squared deviation between the actual field(s) and the reflected field(s) resulting from the incident pulse(s) in a known reference configuration.

5.2. Integral relations for the frequency-domain field.

In this section, we will restrict ourselves to outlining the approach that we consider the most promising. It is a modified version of the Born-type iterative approaches proposed in [23,61]. For a more detailed analysis of the method and for more results, the reader is referred to Chapters 5 and 6 of [1]. Starting point of the analysis is the second-order differential equation for the electric-field strength. In terms of the dimensionless quantities introduced in [9,10,1], this equation reads

$$\{\partial_z^2 - s^2[\varepsilon_r(z) + \sigma(z)/s]\}\, \hat{E}_\pm(z,s) = 0, \tag{36}$$

where z is the one-dimensional space coordinate, and where $\hat{E}_\pm(z,s)$ denote the normalized solutions excited by the unit-amplitude plane waves

$$\hat{E}_+^i(z,s) = \exp(-sz), \quad \hat{E}_-^i(z,s) = \exp(-s(1-z)). \tag{37}$$

In these definitions, the subscripts + and - refer to the direction of propagation. Now, we introduce a **reference medium** with constitutive

coefficients $\{\bar{\varepsilon}_r(z), \bar{\sigma}(z)\}$, and with $\bar{E}_\pm(z,s)$ being the normalized solutions that would be present in this reference medium. Furthermore, we introduce the **Green's function** $G(z,z';s)$ of the reference medium as the solution of the inhomogeneous second-order differential equation

$$\{\partial_z^2 - s^2[\bar{\varepsilon}_r(z) + \bar{\sigma}(z)/s]\}\ G(z,z';,s) = -\ \delta(z-z').\tag{38}$$

Combining (36) and (38), and using the superposition principle, we then find the integral relation

$$\hat{E}_\pm(z,s) = \bar{E}_\pm(z,s) - s^2 \int_0^1 C(z',s)\ G(z,z';s)\ \hat{E}_\pm(z',s)\ dz',\tag{39}$$

where $C(z,s) = \varepsilon_r(z) - \bar{\varepsilon}(z) + [\sigma(z) - \bar{\sigma}(z)]/s$ denotes the so-called **contrast function.** When we locate the point of observation on the slab interface illuminated by the incident field, the relation (39) assumes a special form. We then arrive at

$$r^\pm(s) - \bar{r}^\pm(s) = \hat{E}_\pm(z_\pm,s) - \bar{E}_\pm(z_\pm,s)$$

$$= -\ \frac{s}{2} \int_0^1 C(z,s)\ \bar{E}_\pm(z,s)\ \hat{E}_\pm(z,s)\ dz,\tag{40}$$

with $z_+ = 0$ and $z_- = 1$, and with $r_\pm(s)$ and $\bar{r}_\pm(s)$ being the **plane-wave reflection coefficients** for incidence from the left and right in the actual and in the reference medium, respectively. With (39) and (40), we now have available the integral relations on which the Born-type iterative scheme is based.

5.3. The Born-type approximation.

In order to formulate the Born-type iterative scheme, we should first state the inverse-profiling problem more clearly. We assume one or both of the reflection coefficients $r^\pm(s)$ to be known for all $s = i\omega$, with $-\infty < \omega < \infty$. These known reflection data may have been obtained from repeated frequency-domain measurements or from a spectral analysis of a measured time-domain response. The aim of the computation is to reconstruct the spatial distribution of one or both of the constitutive parameters that describe the material properties of the slab, i.e. $\chi(z)$ and $\sigma(z)$.

If the total field inside the slab $\hat{E}_\pm(z,s)$ were known, the integral relation (40) would, for each frequency $s = i\omega$, constitute an integral equation of the first kind for the contrast function $C(z,s)$. However, this field can only be determined once $C(z,s)$ is known. Hence, we have to content ourselves with deriving an approximate integral equation of the same type. To this end, we substitute (39) in (40). This yields

$$r^{\pm}(s) - \bar{r}^{\pm}(s) = -\frac{s}{2} \int_0^1 C(z,s) \, \bar{E}_{\pm}(z,s)^2 \, dz + \Delta r_B^{\pm}(s),$$ (41)

with

$$\Delta r_B^{\pm}(s) = \frac{s^3}{2} \int_0^1 C(z,s) \, \bar{E}_{\pm}(z,s) \int_0^1 C(z',s) \, G(z,z';s) \, \hat{E}_{\pm}(z',s) \, dz' \, dz.$$ (42)

Now the **Born-type** approximation amounts to neglecting the Born error $\Delta r_B^{\pm}(s)$ in (41). By neglecting this term, we indeed obtain approximate linear equations of the first kind for $C(z,s)$. Consequently, we can obtain an approximation to this unknown contrast function by minimizing an integrated squared error of the form

$$\sum_{\alpha} \int_{-\infty}^{\infty} d\omega \, w^2(\omega) \, \left| \, r^{\alpha}(i\omega) + \frac{i\omega}{2} \int_0^1 C(z,i\omega) \, \bar{E}_{\alpha}(z,i\omega)^2 \, dz \, \right|^2,$$ (43)

where α runs over one or both of the indices $+$ and $-$, and where $w(\omega)$ is a real-valued weighting function. This weighting function can, for example, be identified as the magnitude of the Laplace-transformed incident-pulse shape $F(i\omega)$. In addition, it can be used to account for possible band limitations in the measured reflected-field data. The approximate profiles $\chi_B(z)$ and $\sigma_B(z)$ found from minimizing (43) are usually referred to as **distorted Born approximations** [68].

As remarked above, we can either carry out the Born-type procedure once, or incorporate it in an iterative scheme. In the latter case, we start from some initial guess, and we take, in each iteration step, the reference medium identical to the approximation obtained previously. The essential difference between this scheme and the one analyzed in [23,61] is that, in those references, the procedure was formulated for a contrast-source integral representation of the type (40) with a vacuum background medium.

5.4. Implicit assumptions in the Born-type approximation.

Before we discuss the implementation of the Born-type scheme outlined above, we should justify two assumptions that were made implicitly in its formulation. In the first place, we have assumed that one or both of the equations that are obtained by neglecting $\Delta r_B^{\pm}(s)$ in (41) suffices for a unique determination of some approximation to $\chi(z)$ and/or $\sigma(z)$. In the second place, we have assumed that, for $\bar{\chi}(z)$ and $\bar{\sigma}(z)$ sufficiently close to their actual counterparts, the error made by neglecting $\Delta r_B^{\pm}(s)$ is small enough for $\chi_B(z)$ and $\sigma_B(z)$ to be better approximations.

In order to investigate both assumptions, we need closed-form expressions for the fields $\hat{E}_{\pm}(z,s)$, and the Green's function $G(z,z';s)$ in a slab of arbitrary inhomogeneity. However, such expressions are not available. Hence, we have to be content with using approximate expressions. In [1], the first-order WKB expressions derived in [10] were used for this purpose. These expressions remain accurate down to surprisingly low frequencies, and, consequently, will at least provide us with a good qualitative

description. Moreover, for a homogeneous, lossless slab, the WKB expressions are identical to the actual solutions. Hence, there exists at least one class of reference configurations for which any conclusion following from the approximate theory holds rigorously. In particular, this applies to possible ambiguities in the determination of $\chi_B(z)$ and $\sigma_B(z)$.

Since the asymptotic analysis is fairly complicated, we will restrict ourselves to summarizing the results. The differences $r^\pm(s) - \bar{r}^\pm(s)$ are each dominated by a term that can be identified with a linear combination of a distorted version of $d_z[\chi(z) - \bar{\chi}(z)]$ and a distorted version of $[\sigma(z) - \bar{\sigma}(z)]$. This result may be considered as a generalization of the result obtained by Tabbara [59] for the case of a lossless slab with a small $\chi(z)$. In the corresponding time signal, the dominating term shows up for $0 < t < 2\bar{\tau}$, with $\bar{\tau}$ being the one-way travel time in the reference medium. Moreover, in that interval, only the dominant term contributes. For the Born error $\Delta r_B^\pm(s)$, we have the estimate

$$\Delta r_B^\pm(s) = \Delta r_1^\pm(s) + \Delta r_2^\pm(s), \tag{44}$$

with

$$\Delta r_1^\pm(s) = s \, O[(\chi - \bar{\chi})^2], \text{ and } \Delta r_2^\pm(s) = s \, O(\chi - \bar{\chi}) \, O(\chi), \tag{45}$$

for $s = i\omega$, with $|\omega| \to \infty$. In the time signal corresponding to the Born error, the first term shows up for $t > 0$, and the second term for $t > 2\bar{\tau}$.

These results lead to the following conclusions for the feasibility of the the the Born-type approach:

- The error $\Delta r_2^\pm(s)$ is of the same order of magnitude as the improvements $[\chi(z) - \bar{\chi}(z)]$ and $[\sigma(z) - \bar{\sigma}(z)]$ that one would like to obtain. Moreover, as $|\omega| \to \infty$, this error dominates over $r^\pm(s) - \bar{r}^\pm(s)$, which is of $O[(\chi - \bar{\chi})]$. Hence, a computation neglecting this error may suffer from an **inherent band limitation**, which shows up in the reconstructed profiles. The first term in (45) is of $O[(\chi - \bar{\chi})^2]$, and, hence, will vanish in an iterative reconstruction procedure as outlined in Subsection 5.3, as the computation progresses.

- In order to avoid the possible band limitation due to $\Delta r_2^\pm(s)$, therefore, we must consider the **time-domain signal** corresponding to $r^\pm(s) - \bar{r}^\pm(s)$ in the finite time interval $0 < t < 2\bar{\tau}$ only. If we were to use **frequency-domain data**, a band limitation as described above could not be avoided. Owing to this restriction, we can at best reconstruct a **single** unknown profile from **one-sided** reflection data. In that case, the other constitutive parameter must be known. For a unique reconstruction of **both** profiles, at least **two-sided** reflection data must be available.

- From the correspondence between $r^\pm(s) - \bar{r}^\pm(s)$ and the configuration

    parameters $\chi(z) - \bar{\chi}(z)$ and $\sigma(z) - \bar{\sigma}(z)$ given above, it follows as well
    that a band limitation in the incident field will give rise to a band
    limitation in the reconstructed profiles. This band limitation should not
    be confused with the one discussed above, which is inherent in the Born-
    type approximation. In particular, when high-frequency information is
    missing from the reflected-field data, the short-range variation of the
    unknown profile(s) will be unresolved. This means that this short-range
    variation must be derived from a **priori** information.
In [1], similar conclusions were obtained for the Born-type iterative
procedures proposed in [23,61] as well. The main difference is that, for
these procedures, neglecting the Born error will **always** give rise to a band
limitation in the results. This band limitation is inherent in the method
of solution, and shows up even when wide-band reflected-field information
is available.

5.5. Numerical implementation.
   With the conclusions stated above, we are now in a position to discuss
the numerical implementation of the Born-type iterative procedure. In order
to make the simulation as realistic as possible, we performed our numerical
experiments for the case where a band-limited time-domain response is known
for one or two directions of incidence. This response was synthesized in
the same manner as in Section 4, i.e. by carrying out a marching-on-in-time
computation, and filtering out high-frequency components.
   The ambiguities in the short-range variation due to the band limitation
in this response and due to the inherent Born error were handled as
follows. In the first place, the unknown profiles were approximated by a
piecewise-linear expansion of the form

$$\chi(z) = \sum_{m=0}^{M} \chi_m \, \phi_m(z), \qquad \sigma(z) = \sum_{m=0}^{M} \sigma_m(z) \, \phi_m(z), \tag{46}$$

where the $\{\phi_m(z)\}$ are **triangular expansion functions**, and where the
expansion coefficients $\{\chi_m\}$ and $\{\sigma_m\}$ can be interpreted as being
approximations to values of the corresponding profiles at $z = m/M$,
$m = 0,1,2,\ldots,M$. Substituting these expansions in (41), enforcing the Born
approximation, multiplying the result with the band-limited Laplace-
transformed pulse shape $F(s)$, and carrying out the inverse Fourier
transformation result in the following approximate equations for these
coefficients :

$$E_\pm^r(z_\pm,t) - \bar{E}_\pm^r(z_\pm,t) = -\frac{1}{2} \sum_{m=0}^{M} \Delta\chi_m \, \partial_t E_m^\pm(t) - \frac{1}{2} \sum_{m=0}^{M} \Delta\sigma_m \, E_m^\pm(t), \tag{47}$$

where $0 < t < \infty$, $\Delta\chi_m = \chi_m - \bar{\chi}_m$, and $\Delta\sigma_m = \sigma_m - \bar{\sigma}_m$. In (47), the left-hand
side represents the difference between the actual time-domain response and
its counterpart in the reference medium, while the **matrix element** $E_m^\pm(t)$ is
defined as the time signal corresponding to

$$E_m^\pm(s) = F(s) \int_0^1 \phi_m(z) \, \bar{E}_\pm(z,s)^2 \, dz. \tag{48}$$

Note that (47) is in agreement with the conclusions stated in Subsection 5.4: when one of the unknown constitutive parameters is known, the corresponding sum on the right-hand side vanishes, and we are left with an approximate linear equation for the expansion coefficients for the unknown profile. Since (47) only holds approximately, the unknown expansion coefficients $\{\chi_m\}$ and $\{\sigma_m\}$ should be determined by minimizing an integrated squared error in its equality sign. In drawing up this error, we should take into account the conclusions found in Subsection 5.4 as well as the fact that we are dealing with an incident pulse of duration T. For the case where only $\chi(z)$ is unknown and where reflected-field information is available for incidence from the left, this leads to the **cost function**

$$\int_{T/2}^{2\bar{\tau}+T/2} \{E_+^r(0,t) - E_+^r(0,t) + \frac{1}{2} \sum_{m=0}^{M} \Delta\chi_m \, \partial_t \, E_m(t)\}^2 \, dt$$

$$+ \, \delta \, \sum_{m=1}^{M-1} \{\Delta\chi_{m-1} - 2\Delta\chi_m + \Delta\chi_{m+1}\}^2. \tag{49}$$

In (49), the second term is a small **regularization term**, which serves to relieve any ambiguities that may still be present in the piecewise-linear expansion (46), with δ being a small, nonnegative **regularization parameter**. It is based on the same linear interpolation that was used in the smoothing operation (27), and suppresses a Gibbs-type phenomenon in the successive approximate profiles obtained. Similar cost functions can be drawn up when only $\sigma(z)$ or both $\chi(z)$ and $\sigma(z)$ are unknown. As mentioned above, two-sided reflection data is required in the latter case.

The selection of the parameters M and δ runs as follows. In the absence of time-distortion effects in the correspondence between the time-domain reflected fields and the unknown profiles, the number of cells M could be chosen according to **Shannon's sampling theorem**. For filtering at $|\omega| = \omega_2 = 4\pi/T$, this would result in $M \approx 8\tau/T$. However, we do have time-distortion effects, which vary nonlinearly with the local value of $\bar{\chi}(z)$. Therefore, we need to choose M somewhat larger. Obviously, this will introduce some ambiguities in the short-range variation of the unknown profiles, which have to be eliminated by the influence of the regularization term. In that term, the parameter δ should be chosen by numerical experimentation. Generally, an appropriate value can be found by varying it in the first iteration step. Note that this does not involve a repeated computation of the matrix coefficients $E_m^{\pm}(t)$, which is the time-consuming part of the numerical procedure.

Finally, it should be remarked that the matrix elements $E_m^{\pm}(t)$ are computed by direct Fourier inversion of the corresponding frequency-domain result, as outlined in Section 2. In particular, the normalized field distributions $\bar{E}_{\pm}(z,s)$ occurring in (48) are determined by Runge-Kutta-Verner integration of the frequency-domain first-order differential equations. In this manner, we avoid the possibility of the solution of the inverse-profiling problem being biased by numerical errors made in solving the corresponding direct-scattering problem.

5.6. Results.

With the numerical procedure mentioned above, a wide range of slab configurations has been reconstructed. Both the case of one and of two unknown profiles was investigated. Within the band limitations imposed by the spectrum of the incident field, accurate reconstructions have been obtained for all the configurations considered. Compared with the iterative procedures proposed in [23,61], a considerable improvement was observed. In fact, the results were comparable to those obtained by applying a nonlinear optimization approach as described in [66,1] to a cost function of the type (49) pertaining to the time interval $0 < t < \infty$, and to the unfiltered reflected field. As an illustration, we present in Figs. 8 and 9 reconstructions obtained for two different discontinuous susceptibility profiles. In these figures, we have also listed the root-mean-square deviations between the actual and the reconstructed profiles and between the corresponding reflected fields. Comparing these errors, we observe that the error in $\chi(z)$ is comparable for the results of applying the two approaches. The difference in field error can be explained from the fact that the reconstructions shown in Fig. 8 were obtained from low-pass filtered data and those shown in Fig. 9 from unfiltered data. This difference in field error provides an additional confirmation of the conclusions stated in Subsection 5.4. As shown in Fig. 9, its origin can be observed from comparing the local behavior of transient fields reflected by the actual and the reconstructed slab. Especially in the fields directly reflected from the discontinuities inside the slab, which cannot be represented accurately by the piecewise-linear expansion (46), a deviation is observed.

5.7. Using the natural frequencies.

This paper would not be complete without reconsidering the natural frequencies that were analyzed in Sections 2 and 4. Now that we have solved

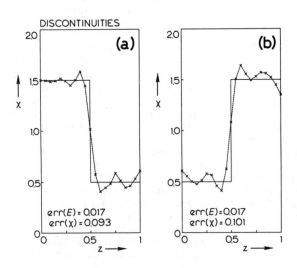

FIGURE 8. Actual discontinuous susceptibility profiles (solid lines) compared with the reconstructions obtained after ten iteration steps (dashed lines) of the Born-type iterative procedure proposed in this section for M = 20, T=1 and δ = 0.001.

420

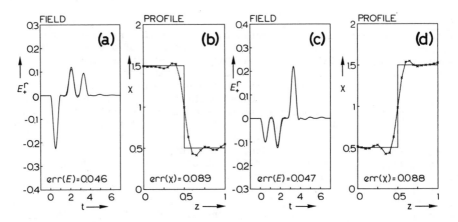

FIGURE 9. Results of reconstructing the same susceptibility profiles as considered in Fig. 8 by employing the conjugate-gradient method to minimize a squared error in the unfiltered reflected field, integrated over the interval $0 < t < \infty$. The dashed lines pertain to the reconstructed configuration and the solid lines to the actual one.

the inverse-profiling problem, we have two ways of finding the natural frequencies of a slab configuration from time-domain reflected-field data. In the first place, we can apply the Prony-type method outlined in Section 4. In the second place, we can determine the natural frequencies indirectly by applying the Born-type reconstruction procedure summarized in the present section, and, subsequently, computing the natural frequencies of the reconstructed configuration as indicated in Section 2.

By comparing the two sets of results obtained with the actual natural frequencies and with each other, we can resolve three issues at once. First, we can verify the conclusion drawn above that reconstructed configurations as shown in Fig. 8 have the same reflection coefficient as the actual configuration over the frequency-range excited. If this conclusion is correct, we should observe good agreement between the natural frequencies obtained. Second, we can decide whether the piecewise-linear expansions employed are flexible enough. If not, the difference between the natural frequencies obtained directly and indirectly will increase considerably with their increasing imaginary parts. Third, we can find out which of the natural-mode contributions are relevant to the profile reconstructions obtained.

An example of a comparison as proposed above is presented in Table 3. This table contains results for a lossless, piecewise-homogeneous three-layer medium that was previously considered in [64,61,1]. With the obvious exception of the real part of the natural frequency nearest to the cut-off of the low-pass filtering, the Prony results turn out to be slightly better than the ones obtained indirectly via profile reconstruction. This suggests that a slight improvement of that reconstruction should, in principle, be possible. This would be in agreement with the root-mean-square field error of magnitude 0.047 remaining in the final iteration step.

5.8. Conclusions.

In this section, we have analyzed a new version of the Born-type iterative scheme proposed in [23,61]. The essential difference with the

TABLE 3. Comparison of the (dimensionless) natural frequencies of a lossless three layer slab with susceptibility $\chi(z) = 0.5 + 1.5\text{rect}(3z - 1.5;1)$ compared with the ones recovered from the time-domain reflected field excited by a sine-squared incident pulse of duration $T = 0.7$.

Actual natural frequency	Approximation from profile reconstruction	Approximation from Prony-type method
-1.3636	-1.3820	-1.3639
-1.7288+i2.6023	-1.7492+i2.5926	-1.7299+i2.6045
-1.8719+i4.3320	-1.8885+i4.3154	-1.8753+i4.3269
-1.4461+i6.4973	-1.4588+i6.4814	-1.4438+i6.4902
-1.5208+i9.3500	-1.5435+i9.3289	-1.5026+i9.3373
-1.8974+i11.3444	-1.9425+i11.3084	-1.8556+i11.3863
-1.6441+i13.1536	-1.7558+i13.0785	-1.6726+i13.1590
-1.3782+i15.8968	-1.3529+i15.4588	-1.1806+i15.8341

versions applied in those papers is that we have now used a contrast-source representation for the reflection coefficient in which the approximation obtained previously is treated as the reference medium instead of a vacuum. From the analysis of this scheme, we observed that it only makes sense to apply this procedure in the time domain. By considering the proper time interval, we may then hope to eventually avoid the band limitation in the reconstructed profiles which is inherent in the Born-type approximation employed. In this respect, the present version is better than the one using the vacuum reference medium. A disadvantage of the procedure is that, in its present implementation, the method is relatively time-consuming, mainly because of the repeated frequency-domain computations. However, faster methods for evaluating the field(s) in the reference medium are available. The quality of the reconstructions obtained seems comparable to that attained by an optimization approach. Finally, the contention that the band limitation inherent in the Born-type methods vanishes as the iterative procedure progresses was verified by comparing the natural frequencies of the actual and reconstructed slab configurations with those obtained by applying Prony's algorithm. In this context, it was observed as well that the agreement between the latter results and the natural frequencies of a reconstructed slab configuration can be used to assess the accuracy of the reconstruction.

ACKNOWLEDGMENT.
    The author would like to thank Professor H. Blok of the Delft University of Technology for many helpful discussions and for his permanent interest during the progress of this work. In addition, he would like to thank Professor A.T. de Hoop of the Delft University of Technology and Professor J. Boersma of the Eindhoven University of Technology for their suggestions and remarks.

REFERENCES

1. Tijhuis, A.G.: Electromagnetic inverse profiling; theory and numerical implementation. Utrecht, the Netherlands: VNU Science Press, 1987.

2. Baum, C.E.: On the singularity expansion method for the solution of electromagnetic interaction problems. Interaction Note 88, Air Force Weapons Laboratory, Albuquerque, New Mexico, 1971.

3. Baum, C.E.: The singularity expansion method. In: Transient electromagnetic fields, Felsen, L.B. (Ed.), Berlin: Springer Verlag, 1976, Chap. 3.

4. Baum, C.E.: Emerging technology for transient and broadband analysis and synthesis of antennas and scatterers. Proc IEEE 64, 1976, 1598-1616.

5. Baum, C.E.: Toward an engineering theory of scattering: the singularity and eigenmode expansion methods. In: Electromagnetic scattering, Uslenghi, P.L.E. (Ed.), New York: Academic Press, 1978, Chap. 15.

6. Pearson, L.W., and Marin, L. (Eds.): Special issue on the singularity expansion method. Electromagnetics 1, 1981.

7. Langenberg, K.J. (Ed.): Special issue on transient fields. Wave Motion 5, 1983.

8. Marin, L.: Natural-mode representation of transient fields. IEEE Trans. Antennas Propagat. 21, 1973, 809-818.

9. Tijhuis, A.G., and Blok, H.: SEM approach to the transient scattering by an inhomogeneous, lossy dielectric slab; Part I: the homogeneous case. Wave Motion 6, 1984, 61-78.

10. Tijhuis, A.G., and Blok, H.: SEM approach to the transient scattering by an inhomogeneous, lossy dielectric slab; Part II: the inhomogeneous case. Wave Motion 6, 1984, 167-182.

11. Tijhuis, A.G., and Van der Weiden, R.M.: SEM approach to the transient scattering by a lossy, radially inhomogeneous cylinder. Wave Motion 8, 1986, 43-63.

12. Yaghjian, A.D.: Augmented electric- and magnetic-field integral equations. Radio Science 16, 1981, 987-1001.

13. Überall, H., and Gaunaurd, G.C.: The physical content of the singularity expansion method. Appl. Phys. Lett. 39, 1981, 362-364.

14. Heyman, E., and Felsen, L.B.: Creeping waves and resonances in transient scattering by smooth, convex objects. IEEE Trans. Antennas Propagat. 31, 1983, 426-437.

15. Tijhuis, A.G.: Angularly propagating waves in a radially inhomogeneous, lossy dielectric cylinder and their connection with the natural modes. IEEE Trans. Antennas Propagat. 34, 1986, 813-824.

16. Mitzner, K.M.: Numerical solution for transient scattering from a hard surface of arbitrary shape - Retarded potential technique. J. Acoust. Soc. Am. 42, 1967, 391-397.

17. Shaw, R.P.: Diffraction of acoustic pulses by obstacles of arbitrary shape with a Robin boundary condition. J. Acoust. Soc. Am. 41, 1967, 855-859.

18. Shaw, R.P.: Diffraction of plane acoustic pulses by obstacles of arbitrary cross section with an impedance boundary condition. J. Acoust. Soc. Am. 44, 1968, 1062-1068.

19. Bennett, C.L.: A technique for computing approximate impulse response of conducting bodies. Ph. D. Thesis, Lafayette, Indiana: Purdue University, 1968.

20. Bennett, C.L., and Weeks, W.L.: Transient scattering from conducting cylinders. IEEE Trans. Antennas Propagat. 18, 1970, 627-633.

21. Tijhuis, A.G.: Toward a stable marching-on-in-time method for two-dimensional transient electromagnetic scattering problems. Radio Science 19, 1984, 1311-1317.

22. Herman, G.C., and Van den Berg, P.M.: A least-square iterative technique for solving time-domain scattering problems. J. Acoust. Soc. Am. 72, 1982, 1947-1953.
23. Tijhuis, A.G.: Iterative determination of permittivity and conductivity profiles of a dielectric slab in the time domain. IEEE Trans. Antennas Propagat. 29, 239-245.
24. Rynne, B.P.: Stability and convergence of time marching methods in scattering problems. IMA J. Appl. Math. 35, 1985, 297-310.
25. Rynne, B.P.: Instabilities in time marching methods for scattering problems. Electromagnetics 6, 1986, 129-144.
26. Sarkar, T.K.: The application of the conjugate gradient method to the solution of operator equations arising in the electromagnetic scattering from wire antennas. Radio Science 19, 1984, 1156-1172.
27. Sarkar, T.K., and Rao, S.M.: The application of the conjugate gradient method for the solution of electromagnetic scattering from arbitrarily oriented wire antennas. IEEE Trans. Antennas Propagat. 32, 1984, 398-403.
28. Van den Berg, P.M.: Iterative schemes based on the minimization of the error in field problems. Electromagnetics 5, 1985, 237-262.
29. Moffatt D.L., and Mains R.K.: Detection and discrimination of radar targets. IEEE Trans. Antennas Propagat. 23, 1975, 358-367.
30. Prony, R.: Essai expérimental et analytique sur les lois de la dilatabilité des fluides élastiques et sur celles de la force expansive de la vapeur de l'eau et da la vapeur de l'alkool, à différentes températures. Journal de l'École Polytechnique 1, Cahier 2, 1795, 24-76.
31. Van Blaricum, M.L., and Mittra, R.: A technique for extracting the poles and residues of a system directly from its transient response. IEEE Trans. Antennas Propagat. 23, 1975, 777-781.
32. Poggio, A.J., Van Blaricum, M.L., Miller, E.K., and Mittra, R.: Evaluation of a processing technique for transient data. IEEE Trans. Antennas Propagat. 26, 1978, 165-173.
33. Van Blaricum, M.L., and Mittra, R.: Problems and solutions associated with Prony's method for processing transient data. IEEE Trans. Antennas Propagat. 26, 1978, 174-182.
34. Miller, E.K.: Natural mode methods in frequency and time domain analysis. In: Theoretical methods for determining the interaction of electromagnetic waves with structures, Skwyrzinski, J.K. (Ed.), Alphen aan den Rijn, The Netherlands: Sijthoff and Noordhoff, 1981, pp. 173-212.
35. Kulp, R.W.: An optimum sampling procedure for use with the Prony method. IEEE Trans. Electromagn. Compat. 23, 1981, 67-71.
36. Dudley, D.G.: Parametric modeling of transient electromagnetic systems. Radio Science 14, 1979, 387-395.
37. Sarkar, T.K., Weiner, D.D., Nebat, J., and Jain, V.K.: A discussion of various approaches to the identification/approximation problem. IEEE Trans. Antennas Propagat. 29, 1982, 373-379.
38. Dudley, D.G.: Parametric identification of transient electromagnetic systems. Wave Motion 5, 1983, 369-384.
39. Sarkar, T.K., Dianat, S.A., and Weiner, D.D.: A discussion of various approaches to the linear system identification problem. IEEE Trans. Acoustics, Speech, and Signal Processing 32, 1984, 654-656.
40. Knockaert, L.: Parametric modeling of electromagnetic waveforms. Electromagnetics 4, 1984, 415-430.
41. Jain, V.K., Sarkar, T.K., and Weiner, D.D.: Rational modeling by the pencil-of-functions method. IEEE Trans. Acoustics, Speech and Signal Processing 31, 1983, 564-573.

42. Drachman, B., and Rothwell, E.: A continuation method for identification of the natural frequencies of an object using a measured response. IEEE Trans. Antennas Propagat. 33, 1985, 445-450.
43. Pearson, L.W., and Roberson, D.R.: The extraction of the singularity expansion description of a scatterer from transient surface current response. IEEE Trans. Antennas Propagat. 28, 1980, 182-190.
44. Lin, C.A., and Cordaro, J.T.: Determination of the SEM parameters for an aircraft model from transient surface current. Electromagnetics 3, 1983, 65-75.
45. Dudley, D.G.: A state-space formulation of transient electromagnetic scattering. IEEE Trans. Antennas Propagat. 33, 1985, 1127-1130.
46. Kay, I: The inverse scattering problem. Research Report EM-74, New York: New York University, 1955.
47. Chadan, K., and Sabatier, P.C.: Inverse problems in quantum scattering theory. New York: Springer Verlag, 1977, Chap. 17.
48. Burridge, R.: The Gelfand-Levitan, the Marchenko, and the Gopinath-Sondhi integral equation of inverse scattering theory, regarded in the context of impulse-response problems. Wave Motion 2, 1980, 305-323.
49. Newton, R.G.: Inverse scattering. I. One dimension. J. Math. Phys. 21, 1980, 493-505.
50. Sabatier, P.C.: Theoretical considerations for inverse scattering. Radio Science 18, 1983, 1-18.
51. Weston, V.H.: On the inverse problem for a hyperbolic dispersive partial differential equation. J. Math. Phys. 13, 1972, 1952-1956.
52. Weston, V.H., and Krueger, R.J.: On the inverse problem for a hyperbolic partial differential equation. II. J. Math. Phys. 15, 1974, 209-213.
53. Corones, J., and Krueger, R.J.: Obtaining scattering kernels using invariant imbedding. J. Math. Anal. Appl. 95, 1983, 393-415.
54. Kristensson, G., and Krueger, R.J.: Direct and inverse scattering in the time domain for a dissipative wave equation. Part I: scattering operators. J. Math. Phys. 27, 1667-1682.
55. Kristensson, G., and Krueger, R.J.: Direct and inverse scattering in the time domain for a dissipative wave equation. Part II: simultaneous reconstruction of dissipation and phase velocity profiles. J. Math. Phys. 27, 1683-1693.
56. Bolomey, J.-Ch., Durix, Ch., and Lesselier, D.: Determination of conductivity profiles by time-domain reflectometry. IEEE Trans. Antennas Propagat. 27, 1979, 244-248.
57. Lesselier, D.: Determination of index profiles by time-domain reflectometry. J. Optics 9, 1978, 349-358.
58. Bojarski, N.N.: One-dimensional direct and inverse scattering in causal space. Wave Motion 2, 1980, 115-124.
59. Tabbara, W.: Reconstruction of permittivity profiles from a spectral analysis of the reflection coefficient. IEEE Trans. Antennas Propagat. 27, 1979, 241-244.
60. Roger, A., Maystre, D., and Cadilhac, M.: On a problem of inverse scattering in optics: the dielectric inhomogeneous medium. J. Optics 9, 1978, 83-90.
61. Tijhuis, A.G., and Van der Worm, C.: Iterative approach to the frequency-domain solution of the inverse-scattering problem for an inhomogeneous, lossless dielectric slab. IEEE Trans. Antennas Propagat. 32, 1984, 711-716.
62. Uno, T., and Adachi, S.: Electromagnetic inverse scattering method for one-dimensional inhomogeneous layered media. Proceedings of the International Symposium on Antennas and Propagation, Kyoto, August

20-22, 1985, pp. 887-890.

63. Mostafavi, M., and Mittra, R.: Remote probing of inhomogeneous media using parameter optimization techniques. Radio Science 7, 1972, 1105-1111.

64. Coen, Sh., Mei, K.K., and Angelakos, D.J.: Inverse scattering technique applied to remote sensing of layered media. IEEE Trans. Antennas Propagat. 29, 1981, 298-306.

65. Roger, A.: Problemès inverses de diffraction en électromagnétisme. Théorie, traitement numérique, et applications à l'optique. Ph. D. Thesis, Marseille: l'Université d'Aix-Marseille III, 1981.

66. Lesselier, D.: Optimization techniques and inverse problems: reconstruction of conductivity profiles in the time domain. IEEE Trans. Antennas Propagat. 30, 1982, 59-65.

67. Lesselier, D.: Diagnostic optimal de la lame inhomogène en régime temporel. Applications à l'électromagnétisme et à l'acoustique. Ph. D. Thesis, Paris: l'Université Pierre et Marie Curie, 1982.

68. Devaney, A.J., and Oristaglio, M.L.: Inversion procedure for inverse scattering within the distorted-wave Born approximation. Phys. Rev. Lett. 51, 1983, 237-240.

Signature Creation in Pattern Recognition for Infra-Red and Possibly for Microwave Systems

Dr. Jozef K. Skwirzynski, MBE, FIEE
14 Galleywood Road
Great Baddow, Essex CM2 8DH, U.K.

We shall consider several methods for creating signatures of various objects, i.e. tanks, aircraft, etc. Eventually I will describe my own method and here I will give you a short introduction. Consider a tank; its silhuette is shown in Fig. 1. Then take all points where the boundary angles change. Record then both the radius (which we shall discuss shortly), and the angle. How to do this? First take all the points, where the boundary changes the direction, take the x,y coordinates of this change, add them together, and produce the "central gravity" by dividing both coordinates by the number of them. Then you have the "centre of gravity" of your silhuette. Next find the distance from your gravity centre to each point of the silhuette, where boundaries are broken, and also find the angle of this point, related to some zero angle line. Next you should record the radius and angle of each such point, and place these on a single segment, normalised to unity. Typical example of this segment is shown in Fig.2, for the silhuette shown in Fig.1. There are two such segments, one for the radius of breaking points from the chosen centre of gravity, the other of angle changes. This is the signature of this particular tank. The question now arises what to do if you see this tank at an angle, not entirely placed at right angle to you. Then you do the same thing as before, and again you normalise this to unit segments of both radius and phase. Then there will be few differences in your signature, but not significant. So if you store the silhuettes of all tanks in your computer, you can easily recognise which tank you are seeing. The same thing can be done with aircraft, provided you see it at a proper angle.

First we shall consider the aircraft identification by moment invariants, produced by S.A. Dudani, K.J. Breeding and R.B. McGhee in the IEEE Transactions on Computers, vol. C.26, No.1, Jan. 1977. Here a white plastic model airplane is placed in a fixture in front of television camera and is viewed against a black background. The resulting video output from the camera is digitised into 180 binary resolution elements per horizontal scan line by a simple treshold circuit. The binary image formed in this way consists of 256 lines from one-half frame of the interlaced scan. The one-cells of the binary image correspond to video signals above the brightness threshold; the zero-cells, to those below the threshold. The resulting 256 X 180 binary image is next preprocessed to clean up small areas of noise if present. Finally, the boundary of the resulting aircraft silhouette is extracted using techniques similar to those described by Dr. A. Rosenfeld: Picture Processing by Computer, New York Academic Publishers, 1969. Some typical examples of the aircraft boundaries obtained by this processing are shown in Fig. 3.

B. de Neumann (ed.), Electromagnetic Modelling and Measurements for Analysis and Synthesis Problems, 427–447.
© 1991 Kluwer Academic Publishers. Printed in the Netherlands.

428

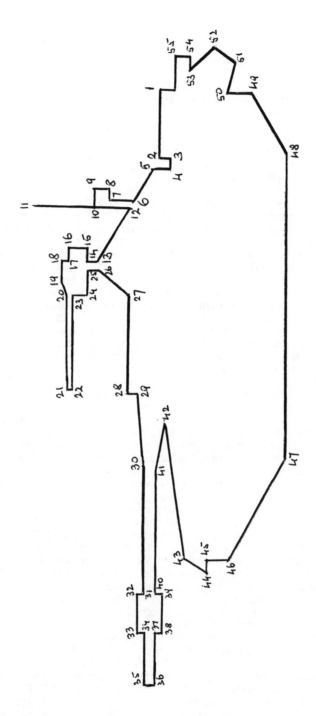

Fig. 1. The silhuette of the tank considered here for production of its signature.

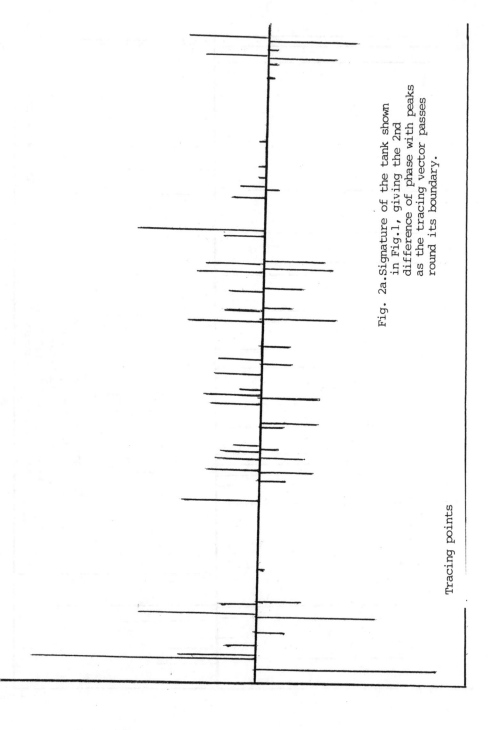

Fig. 2a.Signature of the tank shown
in Fig.1, giving the 2nd
difference of phase with peaks
as the tracing vector passes
round its boundary.

Tracing points

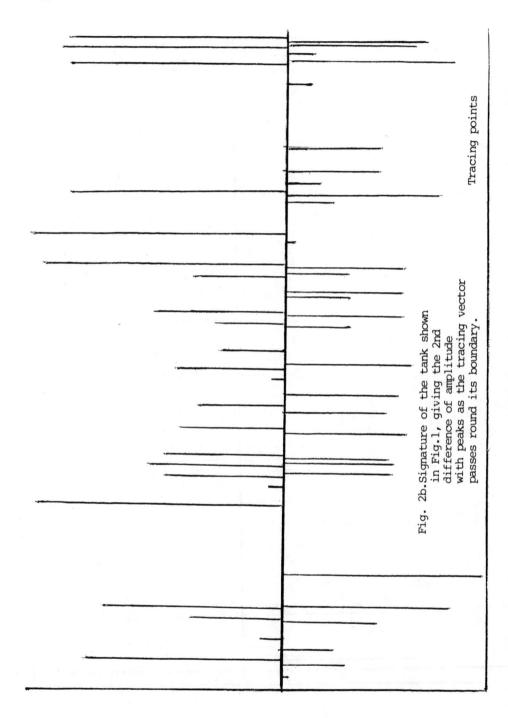

Fig. 2b. Signature of the tank shown in Fig.1, giving the 2nd difference of amplitude with peaks as the tracing vector passes round its boundary.

Tracing points

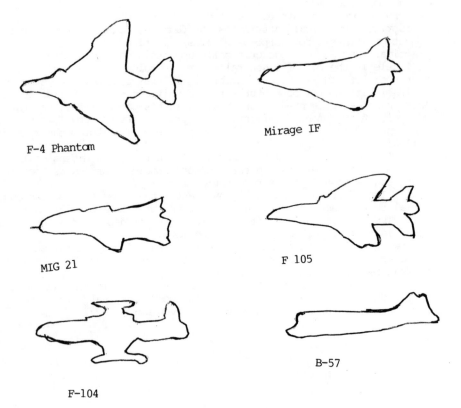

F-4 Phantom

Mirage IF

MIG 21

F 105

F-104

B-57

Fig. 3. Typical images obtained
with the experimental
image acquisition system
for moment method usage.

One of the more difficult problems in the design of a pictorial pattern recognition system relates to the selection of a set of appropriate numerical attributes or features to be extracted from the object of interest for purposes of classification. The success of any practical system depends critically upon this decision. Although there is little in the way of a general theory to guide in the selection of features for an arbitrary problem, it is possible to state some desirable attributes of features for identification of solid objects. This has been attempted by Dr. M.K. Hu in his paper on ""Visual Pattern Recognition by Moment Invariants" (IRE Trans. Information Theory, Vol. IT8, pp.179-187, Feb., 1962). Among these attributes there are the following:

1) The features should be informative. That is, the dimensionality of a vector of measurements (feature vector) should be as low as possible, consistent with acceptable recognition accuracy.

2) The features should be invariant with translation of the object normal to the camera optical axis and with rotation about this axis.

3) The features should either be invariant or depend in a known way upon distance of the object from the camera.

Now it will be shown that certain functions of the central moments of a binary image satisfy the above criteria, and in addition are a class of features which can be rather readily computed.

Consider a solid silhuette or its boundary which is represented by a matrix of ones and zeros in discretised two-dimensional image plane. For such an image, the central moments are given by:

$$\mu_{pq} = \frac{1}{N} \sum_{i=1}^{N} (u_i - u)^p (v_i - v)^q$$

where u and v are the mean values of the image coordinates u and v, respectively, and the summation is over all image points. Hu (mentioned abve) derived a set of moment functions which have the desired property of invariance under image translation and rotation. Similar invariants have been proposed for use in automatic ship recognition by J.M. Wozencraft and I.M. Jacobs in their book on 'Principles of Communication Engineering', New York, Wiley, 1965, pp.148-153. A set of moment invariant functions based only on the second- and third-order moments and which appear to be suitable for the present problem are given below, following Hu:

$$M_1 = (\mu_{20} - \mu_{02})$$

$$M_2 = (\mu_{20} - \mu_{02})^2 + 4\mu_{11}^2$$

$$M_3 = (\mu_{30} - 3\mu_{12})^2 + (3\mu_{21} - \mu_{03})^2$$

$$M_4 = (\mu_{30} + \mu_{12})^2 + (\mu_{21} + \mu_{03})^2$$

$$M_5 = (\mu_{30} - 3\mu_{12})(\mu_{30} + \mu_{12})$$

$$\cdot \left[(\mu_{30} + \mu_{12})^2 - 3(\mu_{21} + \mu_{03})^2\right]$$

$$+ (3\mu_{21} - \mu_{03})(\mu_{21} + \mu_{03})$$

$$\cdot \left[3(\mu_{30} + \mu_{12})^2 - (\mu_{21} + \mu_{03})^2\right]$$

$$M_6 = (\mu_{20} - \mu_{02})\left[(\mu_{30} + \mu_{12})^2 - (\mu_{21} + \mu_{03})^2\right]$$

$$+ 4\mu_{11}(\mu_{30} + \mu_{12})(\mu_{21} + \mu_{03})$$

$$M_7 = (3\mu_{21} - \mu_{03})(\mu_{30} + \mu_{12})$$

$$\cdot \left[(\mu_{30} + \mu_{12})^2 - 3(\mu_{21} + \mu_{03})^2\right]$$

$$- (\mu_{30} - 3\mu_{12})(\mu_{21} + \mu_{03})$$

$$\cdot \left[3(\mu_{30} + \mu_{12})^2 - (\mu_{21} + \mu_{03})^2\right]$$

The functions M_1 through M_6 are invariant under rotation, reflection or a combination of rotation and reflection. However, the function M_7 is invariant only in its absolute magnitude under a reflection as it changes its sign for a given reflection.

Although the concept of Euler angles is used in many fields, there is no general agreement on its definition. There are in fact a large number of choices for the three angles required to define an Euler set of angles, as explained in the paper by R.I. Pio on 'Euler angle transformations' (IEEE Trans. Automatic Control, Vol. AC 11, pp.707-715, Oct., 1966). For the purpose of this research, a particular set of angles, called camera orientation angles, was selected with the aim of obtaining moment invariance with one of the angles of the set. Let xyz be a fixed coordinate system attached to a particular aircraft model with the origin at the centre of gravity, with positive x-axis towards the nose, positive y-axis along the right wing, and positive z-axis towards the bottom surface; let XYZ be another reference system such that XZ is the image plane and the Y axis is the same as the optical

axis of the camera. Now starting with the xyz and the XYZ coordinate systems initially aligned, an aircraft or other solid object is oriented by first rotating it about the camera Y-axis (elevation), then rotating it about the rotated aircraft z-axis (azimuth), and finally rotating it about the aircraft x-axis (roll).

When the elevation angle of a certain object is changed, the locus of the projection of each point of the object on the image plane is a circle with its centre at the origin. Consequently, it follows that the result on the image of varying the elevation angle of a certain object is to produce merely a rotation in the image plane, with no change in size and shape. The moment functions M_1 through M_7 mentioned above are therefore invariant with the elevation angle .

It is clear that as an object is moved along the optical axis of the camera, the first-order effect on the image is just a change in size. The second-order effect is that a few small portions of the image may appear or disappear when the object is moved in this direction. This second-order effect diminishes as the distance of the object from the camera increases.

The radius of gyration r of a planar pattern is defined as follows:

$$r = (\mu_{20} + \mu_{02})^{1/2}$$

The radius of gyration is directly proportional to the size of the image or inversely proportional to the distance of the object along the optical axis. Thus, the product of the radius of gyration of the image and the distance B along the optical axis of the object is a constant:

$$(\mu_{20} + \mu_{02})^{1/2} \cdot B = \text{constant.}$$

Therefor, the radius of gyration r should be used to normalise the moment functions M_2 through M_7 to be in the desired size invariance.

Thus, the following moment functions are invariant with respect to the elevation angle and the distance B along the optical axis:

$$M_1 = (\mu_{20} + \mu_{02})^{1/2} \cdot B = r.B$$

$$M_2 = M_2/r^4$$

$$M_3 = M_3/r^6$$

$$M_4 = M_4/r^6$$

$$M_5 = M_5/r^{12}$$

$$M_6 = M_6/r^8$$

$$M_7 = M_7/r^{12}$$

It has been shown that the moment functions M_1 through M_7 are invariant with elevation angle and distance of the object from the camera. Therefore, these moment features can be used to represent or characterise, the image for a given object viewing aspect (see J. Sklansky and G.A. Davidson, 'Recognising three-dimensional objects by their silhuettes', J. Soc. Photo-Opt. Instrum. Eng., Vol. 10, pp.10-17, Oct., 1971), which is defined uniquely by a pair of values for the azimuth angle and roll angle .

The central moments of an image defined earlier (R.L. Cosgriff, 'Identification of shape', Ohio State University Res. Foundation, Columbus, OH, Rep. 820-11, ASTIA AD-254 792, Dec., 1960) can be computed either from the image boundary or the solid silhuette. Each of these two sets of moments - the set computed from the image boudary and the set from silhuette - contains some information not carried by the other set. The moments derived from the boundary contain more information about the high-frequency portions of the optical image than those derived from the silhuette. Therefore, the minute details of the optical image, such as the shape of the nose or the tail of an aircraft, are better characterised by the moments computed from the boundary. On the other hand, the gross structural features of the aircraft, such as the fuselage and the wings, can be better charactrised by those moments derived from the silhuette; also, these moments are less susceptible to noise. Hence, in order to take advantage of the information content of both the boundary and the silhuette, two sets of seven moment invariant functions $(M_1', M_2', M_3', M_4', M_5', M_6', M_7')$, one set derived from the boundary and the other from the silhuette, are computed. Therefore, the feature vector $r = (r_1, r_2, r_3, r_4, \ldots, r_{14})$ can therefore be defined as follows:

$$r_i = M_i' \quad i=1,7$$

(these are computed using the boundary of the optical image), and

$$r_{i+7} = M_i \quad i=1,7$$

(these are computed using the silhuette of the optical image).

It should be noted that the components r_1 and r_8 of the feature vector r require information about the distance B of the object along the optical axis. For situations where this distance B is not available,

these two components should not be used in constructing the feature vector. Hence, the feature vector in those cases will have only twelve components.

Nearing the conclusion of this method, we shall now consider certain moment features which were introduced to provide parametric characterisation in a fourteen-dimensional vector space for the image of an object for any given values of azimuth and roll angles. In order to make use of this feature vector for recognition purposes, it is necessary to construct a training sample set using the perspective projections (optical images) of the given object for various values of azimuth ψ and roll ϕ angles selected from the significant range. The significant range of values of azimuth and roll is the one which at least covers all distinct views of the object. Two views of an object are called distinct if and only if they produce images which do not merely differ from each other by a rotation, reflection, or a combination of rotation and reflection. It is shown by S.A. Dudani ('An experimental study of moment methods for automatic identification of three-dimensional objects from television images', Ph.D. Dissertation, Ohio State Univ., Columbus, OH, Aug. 1971) that for three-dimensional objects which possess symmetry about a plane, such as an aircraft, the significant range for azimuth and roll angles is as follows:

$$-\pi/2 \;<\; \psi \;<\; \pi/2$$

$$0 \;\;<\; \phi \;<\; \pi/2$$

In practice, the significant range of values of azimuth and roll angles is descretised; then, for all the possible combinations of these discretised values of azimuth and roll angles, the images for the given object are obtained, and from each of these images a set of fourteen features is computed. The collection of these feature vectors, computed from the distinct views of the object, forms the training sample set for that object. Such a training sample must be generated for every object in the given class of objects to be recognised.

For the system under consideration, a lengthy experiment was carried out to construct the training sample set. Six different aircraft types were chosen to be included in the recognition class; namely, an F-4 Phantom, a Mirage IIIC, a MIG-21, an F 105 and a B-57. The significant ranges of azimuth and roll angles were discretised at intervals of 5 degrees each. This interval was selected so that as much information as possible would be included in the training set. The intervals smaller than 5 degrees could not be measured accurately with the equipment in our hands. Next, for each of the different combinations of these discretised values of the two angles, live images were obtained by holding the aircraft model in the proper orientation with the help of a fixture which has provisions for independent adjustment of each of these orientation angles. For each of the images obtained the corresponding vector r was computed. The complete trainig sample set, thus constructed, was based on over 3000 live images for the six different types of aircraft. To compensate for scaling effects, all of the teaining samples were subjected to a linear transformation to obtain zero mean and unit variance for each component when averaged over the entire set.

The complete tables for the normalised trainig sets of the six aircraft can be found in the following paper: S.A. Dudani: 'User's Manual and Tables of Moment Invariance for an On-Line Automatic Aircraft Identification System', Commun. Contr.System Laboratory, Ohio State University, Columbus, OH, Techn. Note 15, Jan., 1974.

For each of the images in the training set, 14 different feature components were computed and stored. In order to cut down the amount of core storage for the training set, a feature ordering technique was employed. A transformation is made from the original feature space to a new space defined by the set of eigenvectors corrsponding to the training sample covariance matrix. The new set of feature components obtained by the above transformation can be ordered in terms of their information content; for this see the paper by S. Watanabe: 'A Method of Self-Feature Information Compression in Pattern Recognition', Proc. 1966 Bionics Meeting, 1968, pp. 697-707. For this problem, a subset of only the five most informative feature componenets were used to form the training set and later was used for classification purposes. The aggregate information content of these five new feature components amounted to 95% of the toal information available from the original training set.

Now we are going to describe still another method for producing object signatures. This was produced by Charles W. Richard, Jr. and Hooshang Hemani in IEEE Trans. on Systems, Man and Cybernetics, in July 1974, where they have used the identification of three-dimensional objects using Fourier descriptors of the boundary curve. They take a closed boundary curve of the image of an aircraft and place it on the complex plane S. The closed contour Z is a continuous mapping from the unit interval onto the set of complex numbers with parametric representation given by

$$Z(t) = (x(t), y(t)), \ t \in [0,1], \ Z(0) = Z(1)$$

where x and y are continuous real valued functions on the interval 0,1 . The derivatives x', y' are piecewise continuous on (0,1). The parameter t will be chosen to be proportional to arc length, so that the speed dZ/dt is a constant. The orientation of the curve is taken to be counterclockwise; i.e. the interior of the region enclosed by the contour will be to the left as the curve is transversed for increasing values of the parameter t. The complex valued function Z for t $(-\infty, \infty)$ shall be considered to be periodic extension of Z for t in the basic interval $[0,1]$.

Now, suppose we have two contours of the same object, slightly differing from each other. Then addition of two contours in S and multiplication by a complex scalar are defined in the ususal way. The innre product of two contours Z_1 , Z_2 S is introduced as

$$(Z_1, Z_2) = \int_0^1 Z_1(t) \overline{Z_2(t)} \, dt$$

Since every $Z \in S$ is continuous, this Rieman integral always exisits, and it is clear that S is a subspace of the complex Hilbert space $L_2(0,1)$.

This inner product induces a norm and metric on S:

$$\| Z \| = \sqrt{(Z_1, Z_2)}$$

$$\rho(Z_1, Z_2) = \| Z_1 - Z_2 \|$$

The metric $\rho(Z_1, Z_2)$ is the root-mean-square (RMS) value of $\Delta Z(t)$, where $\Delta Z(t) = Z_1(t) - Z_2(t)$.

In order to characterise closed contours with the same geometric shape but which may differ in position, orientation, and size, the shape preserving operations are listed in terms of the parametric representations.

a) Translation:

$$Z'(t) = Z(t) + B, \text{ B complex.}$$

b) Rotation about the origin by a real angle α :

$$Z'(t) = \exp(i\alpha) Z(t)$$

c) Change of scale:

$$Z'(t) = K Z(t), \quad K > 0$$

d) Reflection about the x-axis:

$$Z'(t) = \overline{Z(1 - t)}$$

e) Shift in initial point:

$$Z'(t) = Z(t + t_0)$$

The minus sign appears in the reflection to preserve the counterclockwise orientation of the contour.

The closed contour Z' will be said to be similar to the closed contour Z in case the image of Z can be mapped into the image of Z' by a sequence of reflection, rotation, translation, and change of scale transformations. From these parametric representations we have the result that Z' is similar to Z in case there are real positive parameters t_0, α, K and a complex parameter B such that for all $t \in [0,1]$ either

$$Z'(t) = K \exp(i\alpha) Z(t + t_0) + B$$

or

$$Z'(t) = K \exp(i\alpha) \overline{Z(-t - t_0)} + B$$

Since $K = 0$, the general linear transformation is invertible; that is, if Z' is similar to Z, then Z is similar to Z'.

This relation is a valid equivalence relation (reflexive, symmetric, and transitive) and may be used to partition S into equivalence classes. Each class would be represented by its prototype, and the distance between any two classes would be measured as the distance between the prototypes. The prototype is characterised by zero first moment (centroid at origin), and second moment of unity (unit norm). The angular orientation and starting point on the contour will be arbitrary but fixed for each prototype. If $S_1 \subset S$ is an equivalence class of similar contours, and $Z_1 \in S_1$ is chosen as a reference, then Z_1 would be normalised to form Z_1^* as follows:

$$Z_1^*(t) = \frac{Z_1(t) - \mu_1}{\| Z_1 - \mu_1 \|}$$

where

$$\mu_1 = \int_0^1 Z_1(t)\, dt$$

is the centroid of the contour and

$$\| Z_1 - \mu_1 \|^2 = \int_0^1 | Z_1(t) - \mu_1 |^2 \, dt$$

for the purpose of two-dimensional pattern recognition, where scale, rotation, or translation are not important, distance between the contours Z_1 and Z_2 is defined such that the effects of scale, rotation, and translation are eliminated. To measure the distance between two contours Z_1 and Z_2, the contours are first normalised, and then one is reflected, rotated, and the initial point shifted for a best fit in the mean-square sense. The distance is defined to be:

$$d(Z_1, Z_2) = \min(d_a, d_b)$$

where

$$d_a = \min_{\beta, \tau} \| z_1^*(t) - \exp(i\beta)\, z_2^*(t + \tau) \|$$

$$d_b = \min_{\beta, \tau} \| z_1^*(t) - \exp(i\beta)\, \overline{z_2^*(-t - \tau)} \|$$

The minimisations are taken over the ranges $0 \leq \beta < 2\pi$, $0 \leq \tau < 1$. From the triangle inequality for a general norm it follows that $0 \leq d(Z_1, Z_2) \leq 2$. Based on these definitions two contours are similar if and only if $d(Z_1, Z_2) = 0$.

If $d_a = 0$, Z_2 fits Z_1. If $d_b = 0$, Z_2 fits the reflection of Z_1. For three-dimensional identification, translation, rotation, and relative scale of one contour with respect to another (reference) become important. These parameters are recoverable from the following relations:

$$\alpha = \beta^*$$

$$K = \frac{\|Z_1 - \mu_1\|}{\|Z_2 - \mu_2\|}$$

$$B = \mu_1 - K \exp(i\alpha)\mu_2$$

where β^* is the value of β in the equations above that defines d in the equation of contour distance. Alternate expressions for d_a and d_b that eliminate the minimisation with respect to the rotation angle may be derived.

$$d_a^2 = 2\left[1 - \max_{\tau}\left|\int_0^1 Z_1^*(t)\, Z_2^*(t + \tau)\, dt\right|\right]$$

$$d_b^2 = 2\left[1 - \max_{\tau}\left|\int_0^1 Z_1^*(t)\, Z_2^*(-t - \tau)\, dt\right|\right]$$

where

$$\beta^* = \arg \int_0^1 Z_1^*(t)\, \overline{Z_2^*(t + \tau)}\, dt$$

or

$$\beta^* = \arg \int_0^1 Z_1^*(t)\, Z_2^*(-t - \tau)\, dt$$

where arg is the principal value of the angle of the complex number. The equivalent formulae above, in terms of squares of distances d, express the fact our distance measure is just the Euclidean distance at the point of maximum correlation.

We now come to the next problem, namely the role of Fourier transform in this method of pattern recognition. For any closed contour in S, the periodic function $Z(t)$ can be represented by its Fourier series

$$Z(t) = \sum_{k=-\infty}^{\infty} c_k \exp(i2\pi kt)$$

where

$$c_k = \int_0^1 Z(t) \exp(-i2\pi kt)\, dt$$

Writing the complex coefficient $c_k = a_k \exp(i\alpha_k)$, the real numbers a_k and α_k are the kth harmonic "amplitude" and "phase angle", respectively

The set $\{a_k\}$ is called the amplitude spectrum and $\{\alpha_k\}$ the phase spectrum

The piecewise smoothness of $Z(t)$ is sufficient to insure that the Fourier series converges absolutely at any point t and converges uniformly on any closed interval. The amplitude spectrum is invariant with respect to translations, rotations, reflections, or shift in starting point of the contour. Moreover, for continuous piecewise smooth functions, the a_k decrease at a rate proportional to $1/k^2$ so that most of the shape information is contained in the low-order harmonics.

Some salient properties of the Fourier descriptions are summarised in the following. Let $C = \{c_k\}$ and $C' = \{c'_k\}$ be the Fourier coefficients for the contours Z and Z', respectively.

a) The inner product is preserved:

$$(Z', Z) = (C', C)$$

where

$$(C', C) = \sum_{k=-\infty}^{\infty} c'_k \overline{c_k}\,.$$

b) The norm is preserved:

$$\|Z\|^2 = (Z, Z) = (C, C) = \|C\|^2$$

c) The average value or centroid of Z is given by the zeroth harmonic

$$\mu = \int_0^1 Z(t)\, dt = c_0$$

d) Let C^* be the Fourier descriptions for the normalised contour Z^*. From b) and c) above we have:

$$\| c^* \| = 1, \quad c_0^* = 0$$

and

$$c_k^* = c_k \left[\left(\sum_{l=-\infty}^{+\infty} |c_l|^2 - |c_0^2| \right) \right]^{-1/2}, \quad \text{for } k \neq 0$$

e) Let Z' be similar to Z. If there is no reflection

$$c_k' = K \exp\left[i(\alpha + 2\pi k t_0) \right] c_k + \delta_{k,0} B$$

If there is reflection

$$c_k' = K \exp\left[i(\alpha + 2\pi k t_0) \right] \overline{c}_k + \delta_{k,0} B$$

where $\delta_{k,0} = 1$, if $k = 0$, and $\delta_{k,o} = 0$, if $k \neq 0$.

f) Amplitude spectra for normalised similar contours are the same.

g) Distance measure in the Fourier space: the distance between two contours Z_1 and Z_2 can be measured in the transform space, Using the two distance d relations quoted above, we obtain:

$$d_a^2 = 2 \left[1 - \max \left| \sum_{k=-\infty}^{\infty} c_k^{*'} \overline{c_k^*} \exp(-i2\pi k\tau) \right| \right]$$

$$d_b^2 = 2 \left[1 - \max \left| \sum_{k=-\infty}^{\infty} c_k^{*'} c_k^* \exp(-i2\pi k\tau) \right| \right]$$

The two series above are just Fourier expansions of the corresponding cross correlation functions quoted above. The first relation in e) above, which states how the Fourier coefficients transform under

translation, rotation, shift of initial point, and change of scale, was first introduced by G.H. Granlund ('Fourier Preprocessing for Hand Print Character Recognition', IEEE Trans. Comp., Vol. C21, pp. 193-201, Feb., 1972).

We shall now consider the classification of contours and polygonal closed curves. In applications of pattern recognition it is assumed that there is a finite set of known equivalent classes of contours, and the objective is not only to classify an unknown contour in one of the classes but also to recover the orientation and size of the unknown contour. The identification is performed in the Fourier domain using the minimum distance criterion. Let Z_i^* be the normalised prototype for class S_i, $i = 1,2,\ldots,P$, with some arbitrary but fixed initial point and angular orientation. Let $C_i^* = \left\{ c_{kt}^* \right\}$ be the set of Fourier coefficients for Z_i^*. The finite set of Fourier descriptions $\left\{ c^*, k = 0.+1,+2,..,+m \right\}$ is stored for each prototype Z_i^*, $i = 1,2,\ldots,P$. The unknown contour Z with Fourier coefficients c_k is classified in class S_j, in case $d(Z,Z_j^*) < d(Z,Z_i^*)$, $i = 1,2,\ldots,P$, $i \neq j$. The equations above, for Fourier transforms of both d, are used for distance calculation with finite k. The size, rotation angle about the centroid, reflection, and centroid of unknown contour Z, with respaect to the closed prototype Z_j^*, are estimated as before.

Now we consider the subset $S_N \subset S$ of closed contours, which is used in practical applications. Let Z in S_N consists of N line segments continuously connected to form a simple closed curve. A polygonal contour Z in S_N may arise as a first approximation to a contour Z(t) formed by selecting N equidistant points on the unit interval and connecting the points $z_j = Z(j/N)$ on the contour with line segments. Polygonal curves also arise naturally by connecting the finite set of boundary points of a simply connected digital picture with line segments. It is assumed that each contour in S_N onsists exactly of N line segments, where N is relatively large but fixed. In practice, N is chosen to be a power of two for efficiency in using the Cooley-Tukey algorithm (J. Cooley and J. Tukey, 'An Algorithm for the Machine Calculation of Complex Fourier Series', Math. Comp., Vol. 19, pp.297-304, Apr., 1965) to compute the Fourier descriptions of objects. In the case of digital pictures, points may be added or deleted from the boundary set to give N approximately distant points.

A polynimial contour in S can be represented as an ordered N-tuple over the field od complex numbers $Z = (z_1, z_2, \ldots, z_N)$, where the boldface is used to denote a vector. The point z_j is the initial point of the jth line segment and the terminal point of the (j-1)th line segment. Now S_N is an N-dimensional vector space over the complex field, and the inner product is analogous to the continuous case. Entirely analogous results hold for S_N as were developped for the continuous space S with integrals and infinite sums replaced by finite sums. The Fourier coefficients are now given by

$$c_k = \frac{1}{N} \sum_{j=1}^{N} z_j \exp(-i2\pi kj/N)$$

The only approximation involved here is in representing a nonpolygonal contour by a polygonal contour. The distance between two contours now is

$$d(Z', Z) = \min(d_a, d_b)$$

where

$$d_a^2 = 2\left[1 - \max_j \left|\sum_{k=-n_1}^{n_2} c_k^{*'} c_k^{*} \exp(-i2\pi kj/N)\right|\right]$$

$$d_b^2 = 2\left[1 - \max_j \left|\sum_{k=-n_1}^{n_2} c_k^{*'} c_k^{*} \exp(-i2\pi kj/N)\right|\right]$$

with $n_1 + n_2 + 1 = N$. Here the summation in the first eaquation above for d_a and in the second equation for d_b are just discrete Fourier transforms of U and V respectively, where

$$u_k = c_k^{*'} \overline{c_k^{*}} \ , \quad v_k = c_k^{*'} c_k^{*}$$

The maximisation is performed by taking the fast Fourier transform of U and V and finding the jth component with the largest magnitude.

Now we consider the problem of aircraft identification. The method was applied to the problem of identifying an aircraft and estimating its position and orientation in space from its projection on an optical image plane. Aircraft were selected because they represent complex and nonconvex solid objects. The body fixed coordinate system is the usual one for aircraft with the origin at the CG, the positive x axis towards nose, positive y axis along the starboard wing, and positive z axis towards the bottom surface. The image plane is the XY plane of the reference system and is considered to be the ground plane, X north, Y east, and positive Z down. The Euler angles for the aircraft are the roll angle about the x axis, pitch angle of the xy plane with respect to the ground plane, and azimuth or yaw angle of the projection of the x axis in the ground plane. Four model aircraft, an F-4 Phantom, a Mirage HIC, a MIG 21, and an F-105 were considered. The solid objects – in this case aircraft - were represented by a wire frame model constructed by selecting nodes on the surface of a 1:72 scale model with line segments and connecting the nodes with line segments. For a given position and orientation in space the wire frame model constructed onto the image plane, and the boundary points are traced using a numerical algorithm (J.G. Advani, 'Computer Recognition of Three-Dimensional Objects from Optical Images', Ph.D. Dissertation, Ohio State University, Aug., 1971). A typical projection is shown in Fig. 4 with the boundary shown in Fig. 5. Points are added to the outline to give N = 512 boundary points. The Fourier coefficients are calculated, and 39 low-order normalised components (m = 19) are stored for each aircraft at 5 degree increments in roll and pitch angles. The range of pitch angle is from -90 to 90 degrees, and the range in roll angle is from 0 to 90 degrees. Fourier descriptions for negative roll angles need not be stored since the outlines are the reflections of the corresponding positive angles. Because of the bilateral symmetry of an aircraft, a silhouette of a pitch angle $\phi +$ 180° is identical to a silhouette for ϕ. This quantisation in roll and pitch gives 666 reference curves for each aircraft or a total of 2664 classes.

Although 39 Fourier coefficients were stored for each class, the simulation results indicate that most of the significant shape information is contained in less than 21 low-order components. The relatively large number of points (N = 512) was chosen to approximate a continuous boundary curve in the image plane in order to reduce the quantisation error and to increase phase sensitivity. The error in aligning the starting points resulting from two quantisations of the same contour (or phase shift error in the Fourier domain) is proportional to 1/N. This classification problem, with 2664 classes using smaller values of N (N = 256, 128), gave variation in the distance between neighbouring classes in the reference set.

In order to test this approach in identification of type and estimation of parameters of the aircraft, two experiments were conducted. In the first experiment, a total of 300 silhouette boundaries were generated as input data with known true parameters of position and attitude of the aircraft. Noise was added to the boundary and the noisy boundary was compared with the reference data to identify the aircraft type and to estimate its parameters. For comparison, mean absolute errors in each of the parameters of position (X, Y, height) and attitude

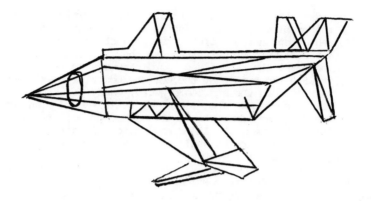

Fig. 4. Projection of wire frame
of aircraft model F-4.

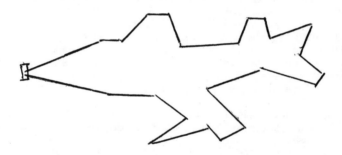

Fig. 5. Boundary curve derived
from the model in Fig. 4.

(yaw, pitch, and roll) were computed. In the second experiment, one of the four aircraft was deleted from the reference set. The deleted aircraft was then used to generate input data. The identification routine was repeated for 100 samples of the unknown aircraft. The following describes the two experiments in more detail.

The silhuette of an unknown aircraft was simulated by choosing one of the four models at random and selecting random values for the six parameters uniformly distributed over ranges

CG: -25 < X, Y < 2.5 ft,   2500 < height < 10 000 ft

0 < yaw < 360°,   -90° < pitch < 90°,   -90° < roll < 90°

Small variations were chosen for X and Y in comparison with the variations in height (or range) since it is assumed that the unknown silhuette is visible though not centered with a relatively narrow viewing window.

For the aircraft in Fig. 4 we have the following input parameters:

Type	X	Y	Ht.	Yaw	Pitch	Roll
F-4	0.00	0.00	6900.0	190.0°	30.0	-50.0°

Points on the boundary of Fig. 4 were selected and random noise (zero mean and standard deviation of 0.1) was added to the coordinates of each point to simulate measurement noise. The result is a set of scattered points in Fig. 5. The latter set of points was treated as input to the system where a best match was made to the stored reference set. The solid contour in Fig. 5 is the closest reference with the following estimated parameters:

Type	X	Y	Ht.	Yaw	Pitch	Roll
F-4	1.93	0.30	6696.0	189.7°	30.0	-50.0°

The results of this experiment are tabulated in the following table:

	0 Percent Noise	10 Percent Noise	20 Percent Noise
ID errors	0	1	8
X error (ft)	1.9	1.9	4.1
Y error	1.3	1.3	2.0
Ht error	75	204	700
Yaw error (deg.)	1.7	2.1	3.7
Pitch error	1.7	4.8	10.9
Roll error	2.5	3.3	9.6
Average minimum distance d	0.031	0.086	0.175

Number of Identification Errors, Average Parameter Error, and Minimum Distance for 100 Samples

Acknowledgement

The editor wishes to express his deep appreciation to the
Scientific Affairs Division of NATO, and the following
organisations and people, for their support and cooperation,
without which the Advanced Study Institute, and the
completion of this account, would not have been possible.

Craig Sinclair
Barbara Kester
Henny AMP Hoogervorst
Tjaddie Ammerdorffer
Adrian Wright
Barry Stuart
Bev Littlewood
John Williams

City University, London, UK.
Centre for Software Reliability, UK.
GEC Research Laboratories, UK.
Wilder Mann Enterprises Ltd, UK.

The Institute was also supported by:
National Science Foundation - USA
European Research Office of the US Army - UK
European Office of Aerospace Research & Development, US Air
      Force - UK